T0183340

Lecture Notes in Computer Science **12065**

More information about this series at http://www.springer.com/series/7407

Alessandra Di Pierro · Pasquale Malacaria ·
Rajagopal Nagarajan (Eds.)

From Lambda Calculus to Cybersecurity Through Program Analysis

Essays Dedicated to Chris Hankin on the Occasion of His Retirement

Springer

Editors
Alessandra Di Pierro
University of Verona
Verona, Italy

Rajagopal Nagarajan
Department of Computer Science
Middlesex University
London, UK

Pasquale Malacaria
School of Electronic Engineering
and Computer Science
Queen Mary University of London
London, UK

ISSN 0302-9743 ISSN 1611-3349 (electronic)
Lecture Notes in Computer Science
ISBN 978-3-030-41102-2 ISBN 978-3-030-41103-9 (eBook)
https://doi.org/10.1007/978-3-030-41103-9

LNCS Sublibrary: SL1 – Theoretical Computer Science and General Issues

Cover illustration: Chris Hankin

This Springer imprint is published by the registered company Springer Nature Switzerland AG
The registered company address is: Gewerbestrasse 11, 6330 Cham, Switzerland

Chris Hankin

Preface

This Festschrift is a collection of scientific contributions related to the topics that have marked the research career of Professor Chris Hankin. The contributions have been written to honor Chris' career and on the occasion of his retirement. As a celebration, a scientific workshop was organized on September 19, 2019, at Imperial College in London and these papers were presented at the event.

Chris' scientific contributions over the years can be divided into three themes. In his early work he focused on the theory of λ-calculus, on which he published various introductory books, e.g. [16, 18], and semantics-based program analysis, using lattice, domain theory, abstract interpretation, and game semantics to build mathematical models of programs for the static analysis of programs properties [1, 3, 17, 20, 21, 25, 30]. This work was motivated by the need to perform semantically correct code optimization which can be used, for example, by compilers. His contributions in semantics-based code optimization and program analysis are numerous [4, 22–24, 28, 29] and he co-authored the authoritative textbook in the area, *Principles of Program Analysis* [31], which is still today the reference for theoreticians and practitioners in the area.

In the late 90s Chris' interests moved towards quantitative analysis of programs and systems, where again he was involved in several influential works [5–7, 9–11] aiming to combine the classical abstract interpretation framework with probability theory and statistics, using linear algebra and operator theory, and finding insightful connections between the fundamental abstract interpretation concept of Galois connection and its quantitative correspondent in the form of Moore-Penrose pseudo-inverse. The re-formulation of various program analysis techniques in the setting of probabilistic abstract interpretation was then used to address the problem of the 'speculative analysis' of security, where Chris' most important achievement has been the analysis of approximate noninterference [8].

More recently Chris' research has focused on cybersecurity and data analytics [13–15, 26, 32, 33]. In this field he has worked on game theoretical analysis for cybersecurity investments and decision support tools. Visual analytics here support cyber-situational awareness, helping security professionals to take appropriate actions to defend their systems. Finally, in the most recent years Chris' research has focused on industrial control systems [2, 12, 19, 27]. He has led the Research Institute in Trustworthy Industrial Control Systems (RITICS) and the Institute for Security Science and Technology at Imperial College. In these roles he has been involved in several government initiatives. He chaired the Lead Expert Group for the Government Office of Science's Policy Futures Foresight report on Future Identities.

Beyond his scientific contributions, Chris has taken on several positions of high responsibilities within Imperial College where he was Deputy Principal of the Faculty of Engineering from 2006 until 2008, and prior Pro Rector (Research) from 2004 until 2006. He was Dean of City and Guilds College from 2000 until 2003. He has also been involved with major international scientific organizations like Inria, serving as President of the Inria Scientific Council, and ACM for which he served as Editor in Chief of *ACM Computing Surveys* and Chair of the ACM Europe Council.

Having known Chris for several years and having appreciated his scientific and human qualities we are very happy to have been involved in this project. We would like to express our gratitude to all authors and reviewers who were involved in producing this book. We thank EasyChair and Springer for their help in publishing this volume and Imperial College for providing support for the workshop.

December 2019

Alessandra Di Pierro
Pasquale Malacaria
Rajagopal Nagarajan

References

1. Abramsky, S., Hankin, C. (eds.): Abstract Interpretation of Declarative Languages. Ellis Horwood (1987)
2. Barrère, M., Hankin, C., Barboni, A., Zizzo, G., Boem, F., Maffeis, S., Parisini, T.: CPS-MT: A real-time cyber-physical system monitoring tool for security research. In: RTCSA. pp. 240–241. IEEE Computer Society (2018)
3. Burn, G.L., Hankin, C., Abramsky, S.: The theory of strictness analysis for higher order functions. In: Programs as Data Objects. Lecture Notes in Computer Science, vol. 217, pp. 42–62. Springer (1985)
4. Clark, D., Errington, L., Hankin, C.: Static analysis of value-passing process calculi. In: Theory and Formal Methods. pp. 307–320. Imperial College Press (1994)
5. Di Pierro, A., Hankin, C., Wiklicky, H.: Probabilistic confinement in a declarative framework. Electr. Notes Theor. Comput. Sci. **48**, 108–130 (2001)
6. Di Pierro, A., Hankin, C., Wiklicky, H.: Approximate non-interference. In: CSFW. pp. 3–17. IEEE Computer Society (2002)
7. Di Pierro, A., Hankin, C., Wiklicky, H.: Quantitative relations and approximate process equivalences. In: CONCUR. Lecture Notes in Computer Science, vol. 2761, pp. 498–512. Springer (2003)
8. Di Pierro, A., Hankin, C., Wiklicky, H.: Approximate non-interference. Journal of Computer Security **12**(1), 37–82 (2004)
9. Di Pierro, A., Hankin, C., Wiklicky, H.: Measuring the confinement of probabilistic systems. Theor. Comput. Sci. **340**(1), 3–56 (2005)
10. Di Pierro, A., Hankin, C., Wiklicky, H.: Quantitative static analysis of distributed systems. J. Funct. Program. **15**(5), 703–749 (2005)
11. Di Pierro, A., Hankin, C., Wiklicky, H.: Reversible combinatory logic. Mathematical Structures in Computer Science **16**(4), 621–637 (2006)
12. Fatourou, P., Hankin, C.: Welcome to the europe region special section. Commun. ACM **62** (4), 28 (2019)
13. Fielder, A., Li, T., Hankin, C.: Defense-in-depth vs. critical component defense for industrial control systems. In: ICS-CSR. Workshops in Computing, BCS (2016)

14. Fielder, A., Panaousis, E.A., Malacaria, P., Hankin, C., Smeraldi, F.: Decision support approaches for cyber security investment. Decision Support Systems **86**, 13–23 (2016)
15. Garasi, C.J., Drake, R.R., Collins, J., Picco, R., Hankin, B.E.: The MEDEA experiment - can you accelerate simulation-based learning by combining information visualization and interaction design principles? In: VISIGRAPP (3: IVAPP). pp. 299–304. SciTePress (2017)
16. Hankin, C.: An Introduction to Lambda Calculi for Computer Scientists. Texts in computing, Kings College (2004), https://books.google.co.uk/books?id=kzdmQgAACAAJ
17. Hankin, C.: Abstract interpretation of term graph rewriting systems. In: Functional Programming. pp. 54–65. Workshops in Computing, Springer (1990)
18. Hankin, C.: Lambda Calculi: A Guide, Handbook of Philosophical Logic. Handbook of Philosophical Logic, vol. 15, pp. 1–66. Springer, Dordrecht (2011)
19. Hankin, C.: Game theory and industrial control systems. In: Semantics, Logics, and Calculi. Lecture Notes in Computer Science, vol. 9560, pp. 178–190. Springer (2016)
20. Hankin, C., Burn, G.L., Peyton Jones, S.L.: A safe approach to parallel combinator reduction (extended abstract). In: ESOP. Lecture Notes in Computer Science, vol. 213, pp. 99–110. Springer (1986)
21. Hankin, C., Burn, G.L., Peyton Jones, S.L.: A safe approach to parallel combinator reduction. Theor. Comput. Sci. **56**, 17–36 (1988)
22. Hankin, C., Métayer, D.L.: Deriving algorithms from type inference systems: Application to strictness analysis. In: POPL. pp. 202–212. ACM Press (1994)
23. Hankin, C., Métayer, D.L.: Lazy type inference for the strictness analysis of lists. In: ESOP. Lecture Notes in Computer Science, vol. 788, pp. 257–271. Springer (1994)
24. Hankin, C., Métayer, D.L.: A type-based framework for program analysis. In: SAS. Lecture Notes in Computer Science, vol. 864, pp. 380–394. Springer (1994)
25. Hunt, S., Hankin, C.: Fixed points and frontiers: A new perspective. J. Funct. Program. **1**(1), 91–120 (1991)
26. Khouzani, M.H.R., Malacaria, P., Hankin, C., Fielder, A., Smeraldi, F.: Efficient numerical frameworks for multi-objective cyber security planning. In: ESORICS (2). Lecture Notes in Computer Science, vol. 9879, pp. 179–197. Springer (2016)
27. Li, T., Hankin, C.: Effective defence against zero-day exploits using bayesian networks. In: CRITIS. Lecture Notes in Computer Science, vol. 10242, pp. 123–136. Springer (2016)
28. Malacaria, P., Hankin, C.: Generalised flowcharts and games. In: ICALP. Lecture Notes in Computer Science, vol. 1443, pp. 363–374. Springer (1998)
29. Malacaria, P., Hankin, C.: A new approach to control flow analysis. In: CC. Lecture Notes in Computer Science, vol. 1383, pp. 95–108. Springer (1998)
30. Martin, C., Hankin, C.: Finding fixed points in finite lattices. In: FPCA. Lecture Notes in Computer Science, vol. 274, pp. 426–445. Springer (1987)
31. Nielson, F., Nielson, H.R., Hankin, C.: Principles of Program Analysis. Springer-Verlag, Berlin, Heidelberg (1999)
32. Simmie, D.S., Vigliotti, M.G., Hankin, C.: Ranking twitter influence by combining network centrality and influence observables in an evolutionary model. J. Complex Networks **2**(4), 495–517 (2014)
33. Thapen, N.A., Simmie, D.S., Hankin, C.: The early bird catches the term: combining twitter and news data for event detection and situational awareness. J. Biomedical Semantics **7**, 61 (2016)

Organization

Additional Reviewers

Martin Berger	University of Sussex, UK
Chiara Bodei	Università di Pisa, Italy
Marco Comini	Università di Udine, Italy
Andrew Fielder	Imperial College, UK
Simon Gay	University of Glasgow, UK
Sebastian Hunt	City University, UK
Martin Lester	University of Reading, UK
Andrea Masini	Università di Verona, Italy
Flemming Nielson	Technical University of Denmark, Denmark
Federica Paci	Univresità di Verona, Italy
Helmut Seidl	Technische Universität München, Germany
David Sands	Chalmers University of Technology, Sweden
Nikos Tzevelekos	Queen Mary University of London, UK
Herbert Wiklicki	Imperial College, UK
Enca Zaffanella	Università di Parma, Italy

Contents

Logic

Program Analysis

Security

Logic

Cables, Trains and Types

Simon J. Gay(✉)

School of Computing Science, University of Glasgow, Glasgow, UK
Simon.Gay@glasgow.ac.uk

Abstract. Many concepts of computing science can be illustrated in ways that do not require programming. *CS Unplugged* is a well-known resource for that purpose. However, the examples in *CS Unplugged* and elsewhere focus on topics such as algorithmics, cryptography, logic and data representation, to the neglect of topics in programming language foundations, such as semantics and type theory.

This paper begins to redress the balance by illustrating the principles of static type systems in two non-programming scenarios where there are physical constraints on forming connections between components. The first scenario involves serial cables and the ways in which they can be connected. The second example involves model railway layouts and the ways in which they can be constructed from individual pieces of track. In both cases, the physical constraints can be viewed as a type system, such that typable systems satisfy desirable semantic properties.

1 Introduction

There is increasing interest in introducing key concepts of computing science in a way that does not require writing programs. A good example is *CS Unplugged* [2], which provides resources for paper-based classroom activities that illustrate topics such as algorithmics, cryptography, digital logic and data representation. However, most initiatives of this kind focus on "Theoretical Computer Science Track A" [4] topics (algorithms and complexity), rather than "Track B" topics (logic, semantics and theory of programming). To the extend that logic is covered, the focus is on gates and circuits rather than deduction and proof.

In the present paper, we tackle Track B by describing two non-programming scenarios illustrating the principles of static type systems. The first scenario involves serial cables, and defines a type system in which the type of a cable corresponds to the nature of its connectors. The physical design of the connectors enforces the type system, and this guarantees that the semantics (electrical connectivity) of a composite cable is determined by its type.

The second scenario is based on model railway layouts, where there is a desirable runtime safety property that if trains start running in the same direction, there can never be a head-on collision. Again, the physical design of the pieces

Supported by the UK EPSRC grant EP/K034413/1, "From Data Types to Session Types: A Basis for Concurrency and Distribution (ABCD)".

A. Di Pierro et al. (Eds.): Festschrift Hankin, LNCS 12065, pp. 3–16, 2020.
https://doi.org/10.1007/978-3-030-41103-9_1

of track enforces a type system that guarantees runtime safety. The situation here is more complicated than for serial cables, and we can also discuss the way in which typability is only an approximation of runtime safety.

We partially formalise the cables example, in order to define a denotational semantics of cables and prove a theorem about the correspondence between types and semantics. A fully formal treatment would require more machinery, of the kind that is familiar from the literature on semantics and type systems, but including it all here would distract from the key ideas. We treat the railway example even less formally; again, it would be possible to develop a more formal account.

I am only aware of one other non-technical illustration of concepts from programming language foundations, which is Victor's *Alligator Eggs* [12] presentation of untyped λ-calculus. When I have presented the cables and trains material in seminars, audiences have found it novel and enjoyable. I hope that these examples might encourage other such scenarios to be observed—and there may be a possibility of developing them into activities along the lines of *CS Unplugged*.

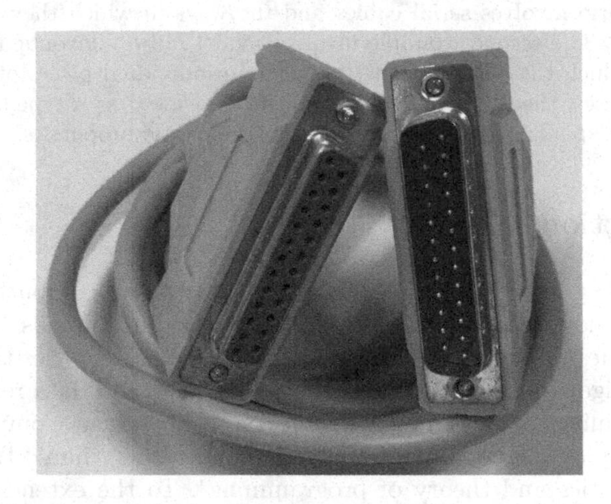

Fig. 1. A serial cable with 25-pin female (left) and male (right) connectors.

2 Cables and Types

The first example involves serial cables. These were widely used to connect computers to peripherals or other computers, typically using the RS-232 protocol, until the emergence of the USB standard in the late 1990s. Figure 1 shows a serial cable with 25-pin connectors, and illustrates the key point that there are two polarities of connector, conventionally called *male* and *female*. Figure 2 shows a serial cable with 9-pin connectors, both female. The physical design is such that two connectors can be plugged together if and only if they are of different male/female polarity and have the same number of pins. From now on we will ignore the distinction between 9-pin and 25-pin connectors, and assume that we are working with a particular choice of size of connector.

For our purposes, the interesting aspect of a serial cable is that it contains two wires for data transmission. These run between the *send* (SND) and *receive* (RCV) pins of the connectors. There are other wires for various power and control signals, but we will ignore them.

There are two ways of connecting the send/receive wires. If SND is connected to SND and RCV is connected to RCV, then the cable is called a *straight through* cable (Fig. 3). This is just an extension cable. Alternatively, if SND is connected to RCV and RCV is connected to SND, then the cable enables two devices to communicate because the SND of one is connected to the RCV of the other. This is called a *null modem* cable (Fig. 4).

Fig. 2. A serial cable with 9-pin female connectors.

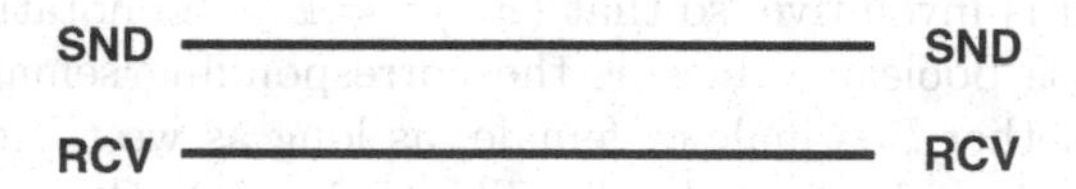

Fig. 3. A straight through cable.

Fig. 4. A null modem cable.

With two ways of wiring SND/RCV, and three possible pairs of polarities for the connectors, there are six possible structures for a serial cable. They have different properties in terms of their electrical connectivity and their physical pluggability. When choosing a cable with which to connect two devices, clearly it is important to have the correct connectors and the correct wiring. Because the wiring of a cable is invisible, there is a conventional correspondence between the choice of connectors and the choice of wiring.

- A straight through cable has different connectors at its two ends: one male, one female.
- A null modem cable has the same connectors at its two ends: both male, or both female.

It is easy to convince oneself that this convention is preserved when cables are plugged together to form longer cables. By thinking of the electrical connectivity of a cable as its semantics, and the nature of its connectors as its type, we can see the wiring convention as an example of a type system that guarantees a semantic property. In the rest of this section, we will sketch a formalisation of this observation.

Figure 5 gives the definitions that we need. Syntactically, a *Cable* is either one of the fundamental cables or is formed by plugging two cables together via the $\cdot$ operator. The fundamental cables are the straight through cable, $\mathsf{straight}$, and two forms of null modem cable, null_1 and null_2. Recalling that a null modem cable has the same type of connector at both ends, the forms null_1 and null_2 represent cables with two male connectors and two female connectors. It doesn't matter which cable is male-male and which one is female-female.

To define the type system, we use the notation of classical linear logic [8]. Specifically, we use *linear negation* $(-)^\perp$ to represent complementarity of connectors, and we use *par* ($\parr$) as the connective that combines the types of connectors into a type for a cable. This is a special case of a more general approach to using classical linear logic to specify typed connections between components [6]. We use $\mathbb{B}$ to represent one type of connector, and then $\mathbb{B}^\perp$ represents the other type. As usual, negation is involutive, so that $(\mathbb{B}^\perp)^\perp = \mathbb{B}$. The notation $\mathbb{B}$ is natural because we will use boolean values as the corresponding semantic domain. It doesn't matter whether $\mathbb{B}$ is male or female, as long as we treat it consistently with our interpretation of null_1 and null_2. The typing rule PLUG, which is a special case of the *cut* rule from classical linear logic, specifies that cables can be plugged together on complementary connectors. In this rule, A, B and C can each be either $\mathbb{B}$ or $\mathbb{B}^\perp$.

Example 1. The cable $\mathsf{straight} \cdot \mathsf{straight}$ represents two straight through cables connected together. It is typable by

$$\frac{\mathsf{straight} : \mathbb{B} \parr \mathbb{B}^\perp \qquad \mathsf{straight} : \mathbb{B} \parr \mathbb{B}^\perp}{\mathsf{straight} \cdot \mathsf{straight} : \mathbb{B} \parr \mathbb{B}^\perp}\ \textsc{Plug}$$

This composite cable has the same type as a single straight through cable, and we will see that it also has the same semantics.

Syntax

$$Cable ::= \mathsf{straight} \mid \mathsf{null}_1 \mid \mathsf{null}_2 \mid Cable \cdot Cable \qquad \text{cables}$$

$$A, B, C ::= \mathbb{B} \mid \mathbb{B}^{\perp} \qquad \text{types}$$

Type equivalence

$$(\mathbb{B}^{\perp})^{\perp} = \mathbb{B}$$

Typing rules

$$\mathsf{straight} : \mathbb{B} \parr \mathbb{B}^{\perp} \qquad \mathsf{null}_1 : \mathbb{B} \parr \mathbb{B} \qquad \mathsf{null}_2 : \mathbb{B}^{\perp} \parr \mathbb{B}^{\perp}$$

$$\frac{c : A \parr B \qquad d : B^{\perp} \parr C}{c \cdot d : A \parr C} \ \textsc{Plug}$$

Semantics

Writing $\mathbb{B}^{[\perp]}$ to represent either $\mathbb{B}$ or $\mathbb{B}^{\perp}$, the denotational semantics of $c : \mathbb{B}^{[\perp]} \parr \mathbb{B}^{[\perp]}$ is

$$[\![c]\!] \subseteq \{\mathtt{true}, \mathtt{false}\} \times \{\mathtt{true}, \mathtt{false}\}$$

defined inductively on the syntactic construction of c by:

$$[\![\mathsf{straight}]\!] = \{(\mathtt{false}, \mathtt{false}), (\mathtt{true}, \mathtt{true})\} \qquad \text{identity, id}$$

$$[\![\mathsf{null}_1]\!] = \{(\mathtt{false}, \mathtt{true}), (\mathtt{true}, \mathtt{false})\} \qquad \text{inversion, inv}$$

$$[\![\mathsf{null}_2]\!] = \{(\mathtt{false}, \mathtt{true}), (\mathtt{true}, \mathtt{false})\} \qquad \text{inversion, inv}$$

$$[\![c \cdot d]\!] = [\![c]\!] \circ [\![d]\!] \qquad \text{relational composition}$$

Fig. 5. Formalisation of cables.

Example 2. The cable $\mathsf{null}_1 \cdot \mathsf{null}_2$ is two null modem cables connected together, which will also be semantically equivalent to a straight through cable. It is typable by

$$\frac{\mathsf{null}_1 : \mathbb{B} \parr \mathbb{B} \qquad \mathsf{null}_2 : \mathbb{B}^{\perp} \parr \mathbb{B}^{\perp}}{\mathsf{null}_1 \cdot \mathsf{null}_2 : \mathbb{B} \parr \mathbb{B}^{\perp}} \ \textsc{Plug}$$

Example 3. The cable $\mathsf{straight} \cdot \mathsf{null}_1$ is a null modem cable extended by plugging it into a straight through cable. Semantically it is still a null modem cable. It is typable by

$$\frac{\mathsf{straight} : \mathbb{B} \parr \mathbb{B}^{\perp} \qquad \mathsf{null}_1 : \mathbb{B} \parr \mathbb{B}}{\mathsf{straight} \cdot \mathsf{null}_1 : \mathbb{B} \parr \mathbb{B}} \ \textsc{Plug}$$

To complete the formalisation of the syntax and type system, we would need some additional assumptions, at least including commutativity of $\parr$ so that we can flip a straight through cable end-to-end to give $\mathsf{straight} : \mathbb{B}^\perp \parr \mathbb{B}$. However, the present level of detail is enough for our current purposes.

We define a denotational semantics of cables, to capture the electrical connectivity. We interpret both $\mathbb{B}$ and $\mathbb{B}^\perp$ as $\{\texttt{true}, \texttt{false}\}$ so that we can interpret a straight through cable as the identity function and a null modem cable as logical inversion. Following the framework of classical linear logic, we work with relations rather than functions. Plugging cables corresponds to relational composition.

Example 4. Calculating the semantics of the cables in Examples 1–3 (for clarity, including the type within $[\![-]\!]$) gives

$$[\![\mathsf{straight} \cdot \mathsf{straight} : \mathbb{B} \parr \mathbb{B}^\perp]\!] = \mathsf{id} \circ \mathsf{id} \quad = \mathsf{id} \quad = [\![\mathsf{straight}]\!]$$

$$[\![\mathsf{null}_1 \cdot \mathsf{null}_2 : \mathbb{B} \parr \mathbb{B}^\perp]\!] = \mathsf{inv} \circ \mathsf{inv} = \mathsf{id} \quad = [\![\mathsf{straight}]\!]$$

$$[\![\mathsf{straight} \cdot \mathsf{null}_1 : \mathbb{B} \parr \mathbb{B}]\!] = \mathsf{id} \circ \mathsf{inv} \quad = \mathsf{inv} = [\![\mathsf{null}_1]\!]$$

This illustrates the correspondence between the type of a cable and its semantics.

The following result is straightforward to prove.

Theorem 1. *Let A be either $\mathbb{B}$ or $\mathbb{B}^\perp$ and let c be a cable.*

1. *If $c : A \parr A$ then $[\![c]\!] = \mathsf{inv}$.*
2. *If $c : A \parr A^\perp$ then $[\![c]\!] = \mathsf{id}$.*

Proof. By induction on the typing derivation, using the fact that $\mathsf{inv} \circ \mathsf{inv} = \mathsf{id}$. □

This analysis of cables and their connectors has several features of the use of static type systems in programming languages. The semantics of a cable is its electrical connectivity, which determines how it behaves when used to connect devices. The type of a cable is a combination of the polarities of its connectors. There are some basic cables, which are assigned types in a way that establishes a relationship between typing and semantics. The physical properties of connectors enforce a simple local rule for plugging cables together. The result of obeying this rule is a global correctness property: for *every* cable, the semantics is characterised by the type.

It is possible, physically, to construct a cable that doesn't obey the typing rules, by removing a connector and soldering on a complementary one. For example, connecting $\mathsf{straight} : \mathbb{B} \parr \mathbb{B}^\perp$ and $\mathsf{straight} : \mathbb{B}^\perp \parr \mathbb{B}$, by illegally joining $\mathbb{B}^\perp$ to $\mathbb{B}^\perp$, gives a straight through cable with connectors $\mathbb{B} \parr \mathbb{B}$. Such cables are available as manufactured components, called *gender changers*. Usually they are very short straight through cables, essentially two connectors directly connected back to back, with male-male or female-female connections. They are like type casts: sometimes useful, but dangerous in general. If we have a cable that has been constructed from fundamental cables and gender changers, and if we can't see exactly which components have been used, then the only way to verify that its connectors match its semantics is to do an electrical connectivity test—i.e. a runtime type check.

Typically, a programming language type system gives a *safe approximation* to correctness. Every typable program should be safe, but usually the converse is not true: there are safe but untypable programs. Cable gender changers are not typable, so the following typing derivation is not valid.

$$\dfrac{\dfrac{\textit{untypable}}{\mathsf{changer}_1 : \mathbb{B} \parr \mathbb{B}} \quad \dfrac{\textit{untypable}}{\mathsf{changer}_2 : \mathbb{B}^\perp \parr \mathbb{B}^\perp}}{\mathsf{changer}_1 \cdot \mathsf{changer}_2 : \mathbb{B} \parr \mathbb{B}^\perp}\ \textsc{Plug}$$

However, the semantics is defined independently of typing, and

$$\begin{aligned} [\![\mathsf{changer}_1 \cdot \mathsf{changer}_2]\!] &= [\![\mathsf{changer}_1]\!] \circ [\![\mathsf{changer}_2]\!] \\ &= \mathsf{id} \circ \mathsf{id} \\ &= \mathsf{id} \end{aligned}$$

so that the typing $\mathsf{changer}_1 \cdot \mathsf{changer}_2 : \mathbb{B} \parr \mathbb{B}^\perp$ is consistent with Theorem 1.

3 Trains and Types

The second example of a static type system is based on model railway layouts. Specifically, the simple kind that are aimed at young children [1,5], rather than the elaborate kind for railway enthusiasts [3]. The examples in this paper were constructed using a "Thomas the Tank Engine" [7] set.

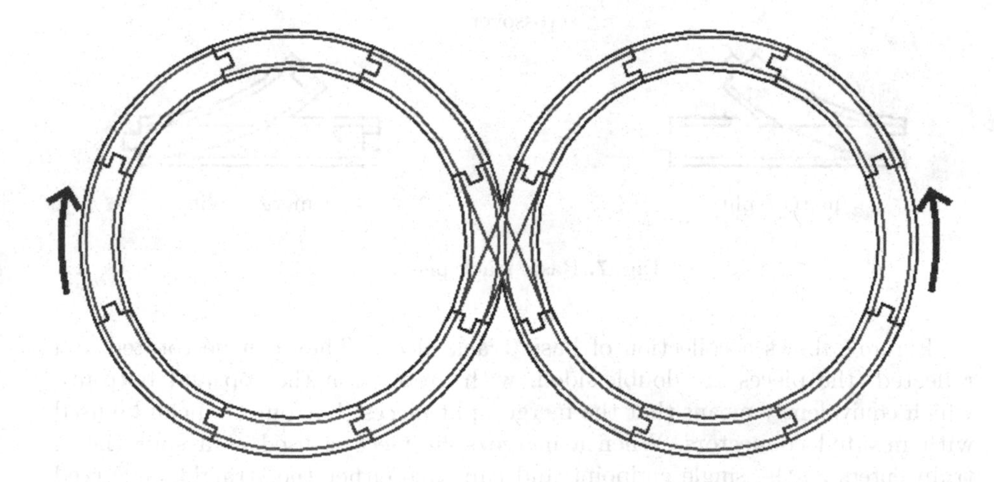

Fig. 6. A figure eight layout. (Color figure online)

Figure 6 shows a simple figure eight layout consisting of two circles linked by a crossover piece. The blue lines (coloured in the electronic version of the paper) show the guides for the train wheels—in these simple sets, they are grooves rather than raised rails. Notice that there are multiple pathways through the crossover piece. It would be possible for a train to run continuously around one of the circles, but in practice the tendency to follow a straight path means that it always transfers through the crossover piece to the other circle.

It's clear from the diagram that when a train runs on this layout, it runs along each section of track in a consistent direction. If it runs clockwise in the left circle, then it runs anticlockwise in the right circle, and this never changes. Consequently, if two trains run simultaneously on the track, both of them in the correct consistent direction, there can never be a head-on collision. For example, if one train starts clockwise in the left circle, and the other train starts anticlockwise in the right circle, they can never move in opposite directions within the same circle. They might side-swipe each other by entering the crossover section with bad timing, or a faster train might rear-end a slower train, but we will ignore these possibilities and focus on the absence of head-on collisions as the safety property that we want to guarantee.

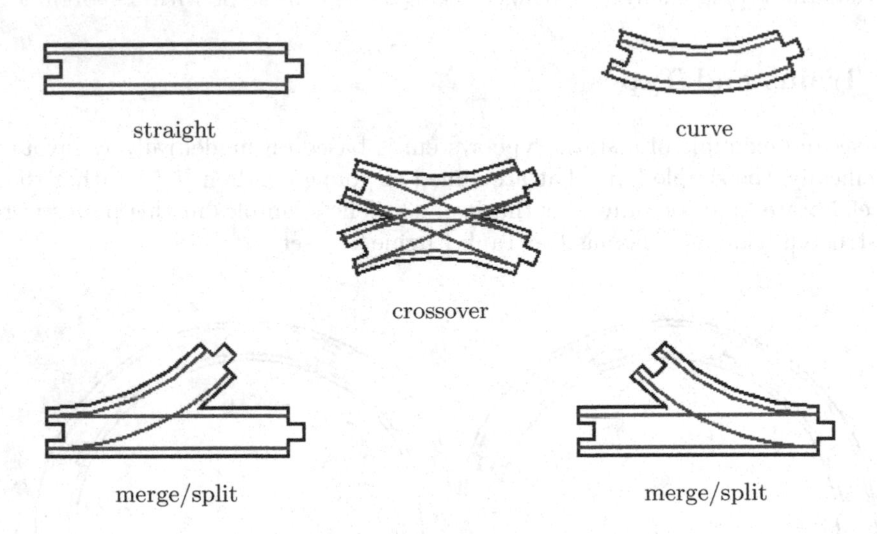

Fig. 7. Basic track pieces.

Figure 7 shows a collection of basic track pieces. They can be rotated and reflected (the pieces are double-sided, with grooves on the top and bottom), which equivalently means that the merge/split pieces (bottom row) can be used with inverted connectors. When a merge/split piece is used as a split (i.e. a train enters at the single endpoint and can take either the straight or curved branch), there is a lever that can be set to determine the choice of branch. We will ignore this feature, because we are interested in the safety of layouts under the assumption that any physically possible route can be taken.

The pieces in Fig. 7 can be used to construct the figure eight layout (Fig. 6) as well as more elaborate layouts such as the one in Fig. 8. It is easy to see that the layout of Fig. 8 has the same "no head-on collisions" property as the figure eight layout.

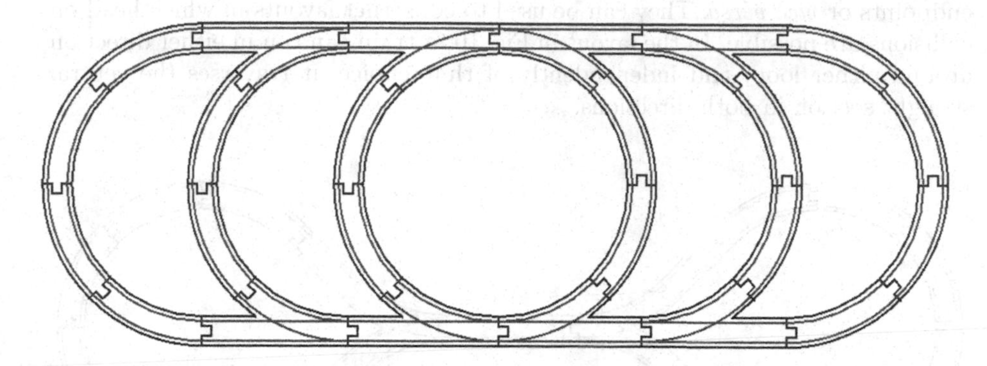

Fig. 8. A layout with multiple paths.

Each track piece has a number of endpoints, where it can be connected to other pieces. We will refer to each endpoint as either positive (the protruding connector) or negative (the hole). The pieces in Fig. 7 have the property that if a train enters from a negative endpoint, it must leave from a positive endpoint. This property is preserved inductively when track pieces are joined together, and also when a closed (no unconnected endpoints) layout is formed. This inductively-preserved invariant is the essence of reasoning with a type system, if we consider the type of a track piece or layout to be the collection of polarities of its endpoints. If we imagine an arrow from negative to positive endpoints in each piece, the whole layout is oriented so that there are never two arrowheads pointing towards each other. This is exactly the "no head-on collisions" property. It is possible to use the same argument in the opposite direction, with trains running from positive to negative endpoints, to safely orient the layout in the opposite sense.

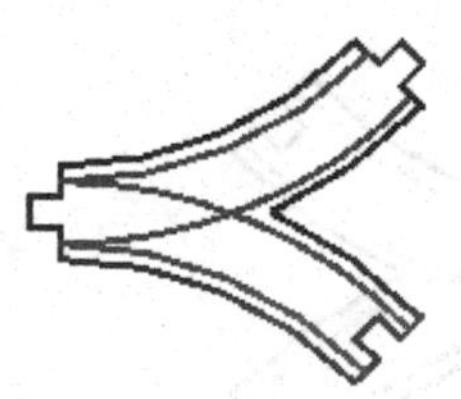

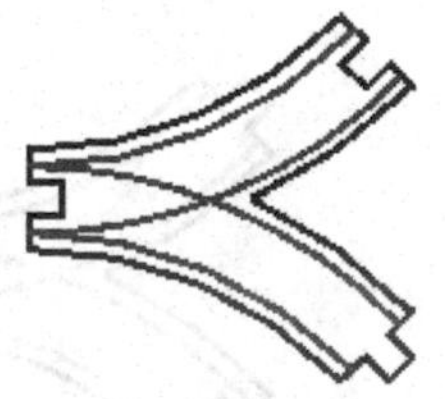

Fig. 9. The Y pieces.

This argument could be formalised by defining a syntax for track layouts in the language of traced monoidal categories [9,11] or compact closed categories [6,10] and associating a directed graph with every track piece and layout.

The track pieces in Fig. 7 are not the only ones. Figure 9 shows the Y pieces, which violate the property that trains run consistently from negative to positive endpoints or *vice versa*. They can be used to construct layouts in which head-on collisions are possible. In the layout in Fig. 10, a train can run in either direction around either loop, and independently of that choice, it traverses the central straight section in both directions.

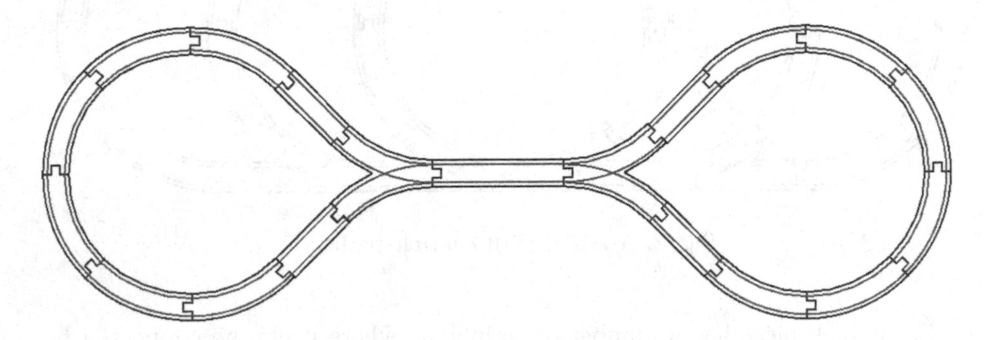

Fig. 10. An unsafe layout using Y pieces.

It is possible to build safe layouts that contain Y pieces. Joining two Y pieces as in Fig. 11 gives a structure that is similar to the crossover piece (Fig. 7) except that the polarities of the endpoints are different. This “Y crossover” can be used as the basis for a safe figure eight (Fig. 12). However, safety of this layout cannot be proved by using the type system. If a train runs clockwise in the circle on the right, following the direction from negative to positive endpoints, then its anticlockwise journey around the circle on the left goes against the polarities. To prove safety of this layout, we can introduce the concept of logical polarities, which can be different from the physical polarities. In the circle on the left, assign logical polarities so that the protruding connectors are negative and the holes are positive, and then the original proof works.

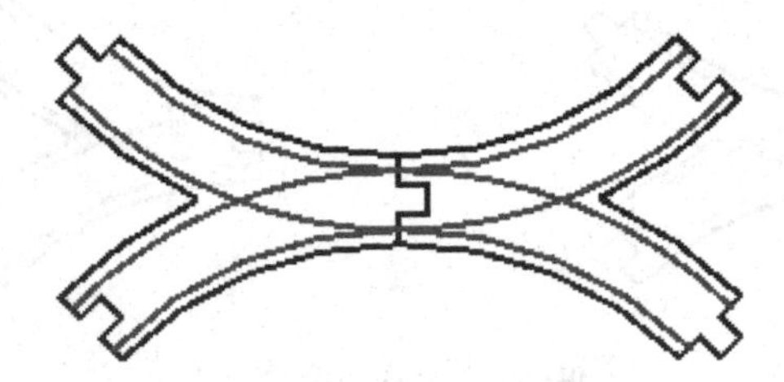

Fig. 11. Joining Y pieces to form a crossover.

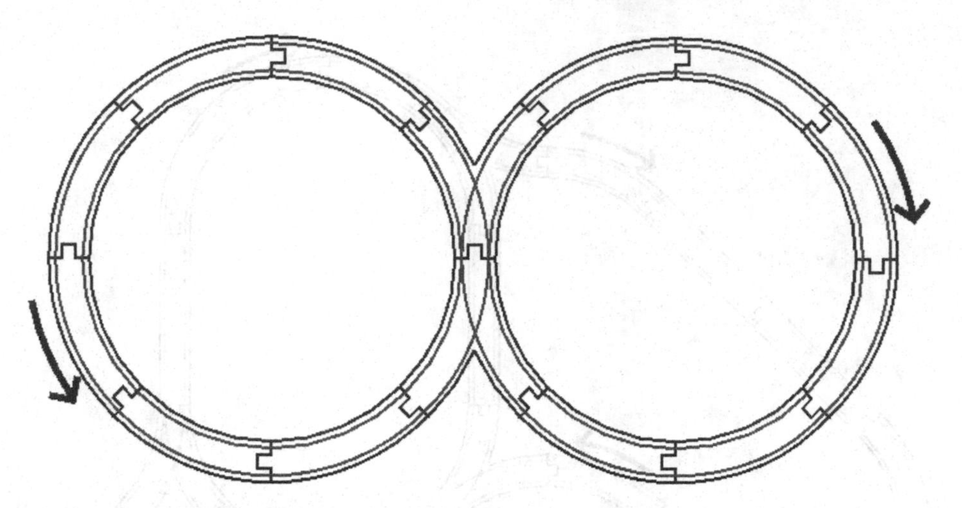

Fig. 12. A safe figure eight using Y pieces. In the circle on the right, the direction of travel follows the physical polarity, but in the circle on the left, the direction of travel is against the physical polarity. To prove safety, assign logical polarities in the circle on the left, which are opposite to the physical polarities.

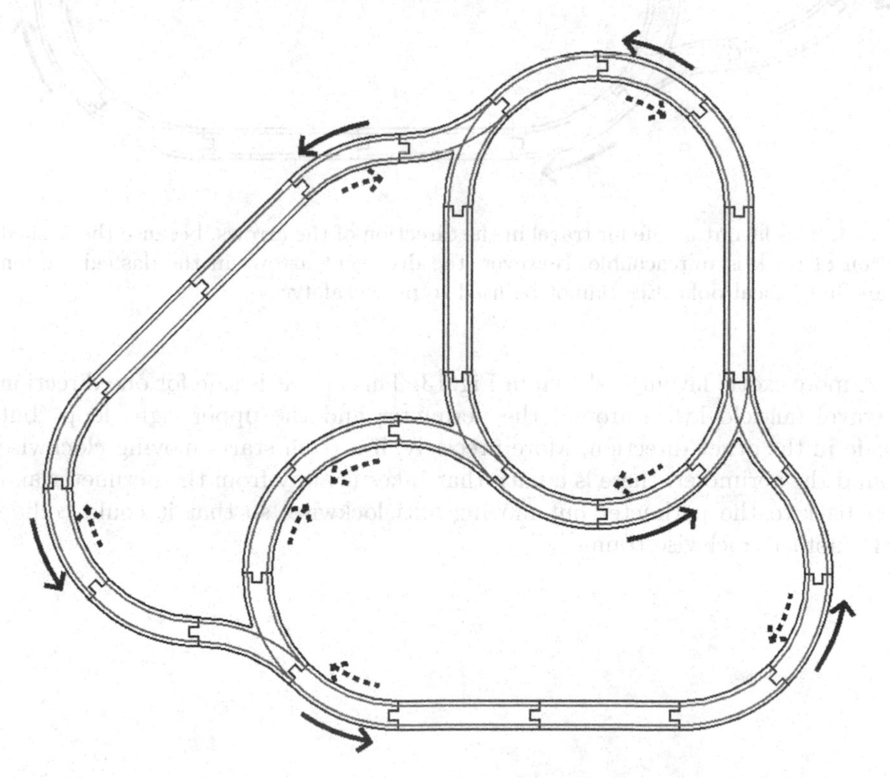

Fig. 13. A layout using Y pieces that is safe in one direction (solid arrows) but not the other (dashed arrows).

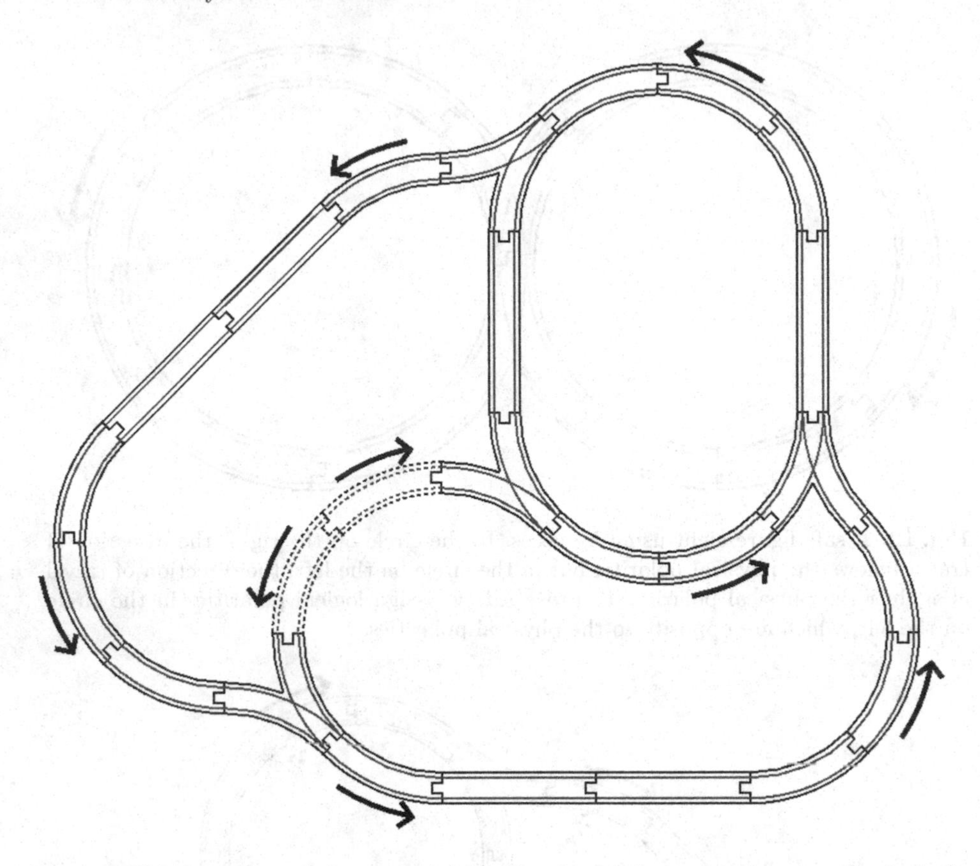

Fig. 14. This layout is safe for travel in the direction of the arrows, because the dashed section of track is unreachable. However, the divergent arrows in the dashed section mean that logical polarities cannot be used to prove safety.

A more exotic layout is shown in Fig. 13. This layout is safe for one direction of travel (anticlockwise around the perimeter and the upper right loop) but unsafe in the other direction. More precisely, if a train starts moving clockwise around the perimeter, there is a path that takes it away from the perimeter and then back to the perimeter but moving anticlockwise, so that it could collide with another clockwise train.

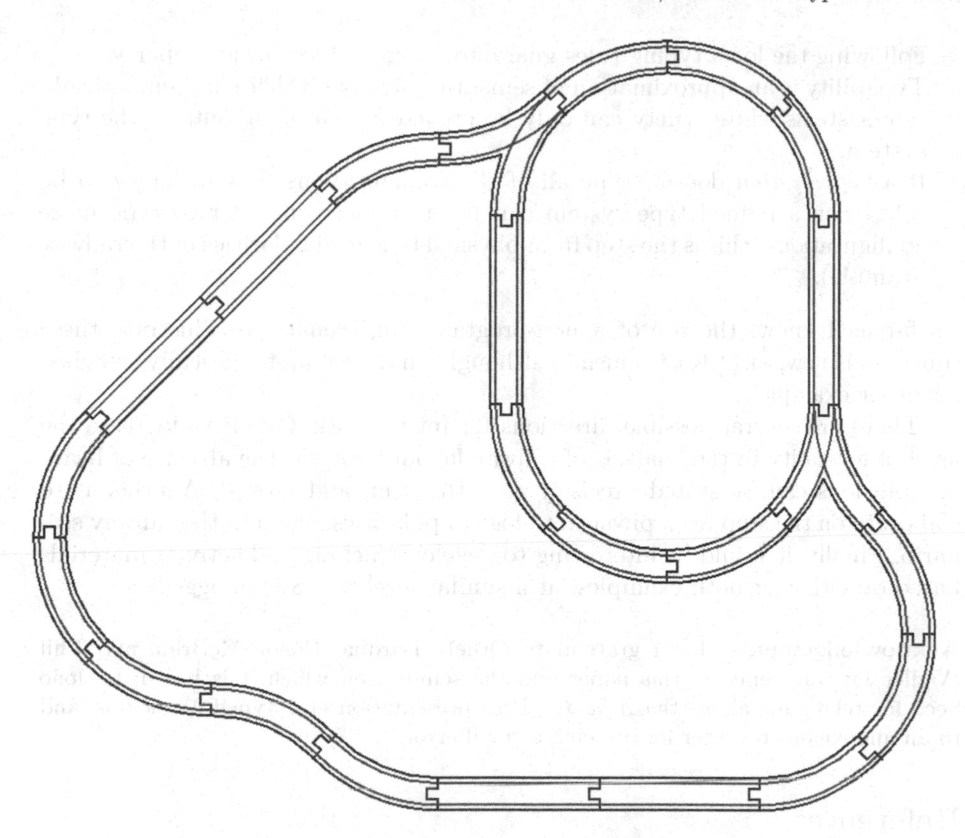

Fig. 15. The layout of Fig. 13 with the problematic section of track removed. This layout is safe in both directions. For clockwise travel around the perimeter, following the physical polarities, logical polarities are assigned to the inner loop.

Safety of the anticlockwise direction cannot be proved by physical polarities, because of the Y pieces. Figure 14 shows that it cannot be proved even by using logical polarities. This is because the section with dashed lines, where the arrows diverge, would require a connection between two logically negative endpoints. To prove safety we can observe that for the safe direction of travel, the section with dashed lines is unreachable. Therefore we can remove it (Fig. 15) to give an equivalent layout in which safety can be proved by logical polarities. In fact the layout of Fig. 15 is safe in both directions.

4 Conclusion

I have illustrated the ideas of static type systems in two non-programming domains: serial cables, and model railways. The examples demonstrate the following concepts.

– Typing rules impose local constraints on how components can be connected.

- Following the local typing rules guarantees a global semantic property.
- Typability is an approximation of semantic safety, and there are semantically safe systems whose safety can only be proved by reasoning outside the type system.
- If a type system doesn't type all of the configurations that we know to be safe, then a refined type system can be introduced in order to type more configurations (this is the step from physical to logical polarities in the railway example).

As far as I know, the use of a non-programming scenario to illustrate these concepts is new, or at least unusual, although I have not systematically searched for other examples.

There are several possible directions for future work. One is to increase the level of formality in the analysis of railway layouts, so that the absence of head-on collisions can be stated precisely as a theorem, and proved. Another is to elaborate on the step from physical to logical polarities, again in the railway scenario. Finally, it would be interesting to develop teaching and activity materials based on either or both examples, at a similar level to CS Unplugged.

Acknowledgements. I am grateful to Ornela Dardha, Conor McBride and Phil Wadler for comments on this paper and the seminar on which it is based; to João Seco for telling me about the *Alligator Eggs* presentation of untyped λ-calculus; and to an anonymous reviewer for noticing a small error.

References

1. Brio. www.brio.uk
2. CS Unplugged. csunplugged.org
3. Hornby. www.hornby.com
4. Theoretical Computer Science. www.journals.elsevier.com/theoretical-computer-science
5. Thomas & Friends. www.thomasandfriends.com
6. Abramsky, S., Gay, S.J., Nagarajan, R.: Interaction categories and the foundations of typed concurrent programming. In: Broy, M. (ed.) Proceedings of the NATO Advanced Study Institute on Deductive Program Design, pp. 35–113 (1996)
7. Awdrey, W.: Thomas the tank engine (1946)
8. Girard, J.-Y.: Linear logic. Theoret. Comput. Sci. **50**, 1–102 (1987)
9. Joyal, A., Street, R., Verity, D.: Traced monoidal categories. Math. Proc. Cambridge Philos. Soc. **119**(3), 447–468 (1996)
10. Kelly, G.M., Laplaza, M.L.: Coherence for compact closed categories. J. Pure Appl. Algebra **19**, 193–213 (1980)
11. Ştefănescu, G.: Network Algebra. Springer, Heidelberg (2000). https://doi.org/10.1007/978-1-4471-0479-7
12. Victor, B.: Alligator eggs. worrydream.com/AlligatorEggs

Cathoristic Logic

A Logic for Capturing Inferences Between Atomic Sentences

Richard Evans[1](✉) and Martin Berger[2]

[1] Imperial College, London, UK
richardevans@google.com
[2] University of Sussex, Brighton, UK
M.F.Berger@sussex.ac.uk

Abstract. Cathoristic logic is a multi-modal logic where negation is replaced by a novel operator allowing the expression of incompatible sentences. We present the syntax and semantics of the logic including complete proof rules, and establish a number of results such as compactness, a semantic characterisation of elementary equivalence, the existence of a quadratic-time decision procedure, and Brandom's incompatibility semantics property. We demonstrate the usefulness of the logic as a language for knowledge representation.

Keywords: Modal logic · Hennessy-Milner logic · Transition systems · Negation · Exclusion · Elementary equivalence · Incompatibility semantics · Knowledge representation · Philosophy of language

1 Introduction

Natural language is full of incompatible alternatives. If Pierre is the current king of France, then nobody else can simultaneously fill that role. A traffic light can be green, amber or red - but it cannot be more than one colour at a time. Mutual exclusion is a natural and ubiquitous concept.

First-order logic can represent mutually exclusive alternatives, of course. To say that Pierre is the only king of France, we can write, following Russell:

$$\mathit{king}(\mathit{france}, \mathit{pierre}) \wedge \forall x.(\mathit{king}(\mathit{france}, x) \rightarrow x = \mathit{pierre}).$$

To say that a particular traffic light, tl, is red - and red is its only colour - we could write:

$$\mathit{colour}(\mathit{tl}, \mathit{red}) \wedge \forall x.\mathit{colour}(\mathit{tl}, x) \rightarrow x = \mathit{red}.$$

In this approach, incompatibility is a *derived* concept, reduced to a combination of universal quantification and identity. First-order logic, in other words, uses relatively complex machinery to express a simple concept:

A. Di Pierro et al. (Eds.): Festschrift Hankin, LNCS 12065, pp. 17–85, 2020.
https://doi.org/10.1007/978-3-030-41103-9_2

- Quantification's complexity comes from the rules governing the distinction between free and bound variables[1].
- Identity's complexity comes from the infinite collection of axioms required to formalise the indiscernibility of identicals.

The costs of quantification and identity, such as a larger proof search space, have to be borne every time one expresses a sentence that excludes others - even though incompatibility does not, prima facie, appear to have anything to do with the free/bound variable distinction, or require the full power of the identity relation.

This paper introduces an alternative approach, where exclusion is expressed directly, as a first-class concept. Cathoristic logic[2] is the simplest logic we could find in which incompatible statements can be expressed. It is a multi-modal logic, a variant of Hennessy-Milner logic, that replaces negation with a new logical primitive

$$!A$$

pronounced *tantum*[3] A. Here A is a finite set of alternatives, and $!A$ says that the alternatives in A exhaust all possibilities. For example:

$$!\{green, amber, red\}$$

states that nothing but *green*, *amber* or *red* is possible. Our logic uses modalities to state facts, for example $\langle amber \rangle$ expresses that *amber* is currently the case. The power of the logic comes from the conjunction of modalities and tantum. For example

$$\langle amber \rangle \wedge !\{green, amber, red\}$$

expresses that *amber* is currently the case and *red* as well as *green* are the only two possible alternatives to *amber*. Any statement that exceeds what tantum A allows, like

$$\langle blue \rangle \wedge !\{green, amber, red\},$$

is necessarily false. When the only options are green, amber, or red, then blue is not permissible. Now to say that Pierre is the only king of France, we write:

$$\langle king \rangle \langle france \rangle (\langle pierre \rangle \wedge !\{pierre\}).$$

Crucially, cathoristic logic's representation involves no universal quantifier and no identity relation. It is a purely propositional formulation. To say that the traffic light is currently red, and red is its only colour, we write:

$$\langle tl \rangle \langle colour \rangle (\langle red \rangle \wedge !\{red\}).$$

---

[1] Efficient handling of free/bound variables is an active field of research, e.g. nominal approaches to logic [23]. The problem was put in focus in recent years with the rising interest in the computational cost of syntax manipulation in languages with binders.

[2] "Cathoristic" comes from the Greek καθορίζειν: to impose narrow boundaries. We are grateful to Tim Whitmarsh for suggesting this word.

[3] "Tantum" is Latin for "only".

This is simpler, both in terms of representation length and computational complexity, than the formulation in first-order logic given on the previous page. Properties changing over time can be expressed by adding extra modalities that can be understood as time-stamps. To say that the traffic light was red at time t_1 and amber at time t_2, we can write:

$$\langle tl \rangle \langle colour \rangle (\langle t_1 \rangle (\langle red \rangle \wedge !\{red\}) \wedge \langle t_2 \rangle (\langle amber \rangle \wedge !\{amber\}))$$

Change over time can be expressed in first-order logic with bounded quantification - but modalities are succinct and avoid introducing bound variables.

Having claimed that incompatibility is a natural logical concept, not easily expressed in first-order logic[4], we will now argue the following:

- Incompatibility is conceptually prior to negation.
- Negation arises as the weakest form of incompatibility.

1.1 Material Incompatibility and Negation

Every English speaker knows that

"Jack is male" is incompatible with "Jack is female"

But *why* are these sentences incompatible? The orthodox position is that these sentences are incompatible because of the following general law:

If someone is male, then it is not the case that they are female

Recast in first-order logic:

$$\forall x.(male(x) \rightarrow \neg female(x)).$$

In other words, according to the orthodox position, the incompatibility between the two particular sentences depends on a general law involving universal quantification, implication and negation.

Brandom [7] follows Sellars in proposing an alternative explanation: "Jack is male" is incompatible with "Jack is female" because "is male" and "is female" are *materially incompatible* predicates. They claim we can understand incompatible predicates even if we do not understand universal quantification or negation. Material incompatibility is conceptually prior to logical negation.

Imagine, to make this vivid, a primitive people speaking a primordial language of atomic sentences. These people can express sentences that *are* incompatible. But they cannot express *that* they are incompatible. They recognise when atomic sentences are incompatible, and see that one sentence entails another - but their behaviour outreaches their ability to articulate it.

Over time, these people *may* advance to a more sophisticated language where incompatibilities are made explicit, using a negation operator - but this is a later (and optional) development:

[4] We will precisify this claim in later sections; (1) first-order logic's representation of incompatibility is longer in terms of formula length than cathoristic logic's (see Sect. 4.2); and (2) logic programs in cathoristic logic can be optimised to run significantly faster than their equivalent in first-order logic (see Sect. 5.3).

> [If negation is added to the language], it lets one say that two claims are materially incompatible: "If a monochromatic patch is red, then it is not blue." That is, negation lets one make explicit in the form of claims - something that can be said and (so) thought - a relation that otherwise remained implicit in what one practically did, namely treat two claims as materially incompatible[5].

But before making this optional explicating step, our primitive people understand incompatibility without understanding negation. If this picture of our primordial language is coherent, then material incompatibility is conceptually independent of logical negation.

Now imagine a modification of our primitive linguistic practice in which no sentences are ever treated as incompatible. If one person says "Jack is male" and another says "Jack is female", nobody counts these claims as *conflicting*. The native speakers never disagree, back down, retract their claims, or justify them. They just say things. Without an understanding of incompatibility, and the variety of behaviour that it engenders, we submit (following Brandom) that there is insufficient richness in the linguistic practice for their sounds to count as assertions. Without material incompatibility, their sounds are just *barks*.

> Suppose the reporter's differential responsive dispositions to call things red are matched by those of a parrot trained to utter the same noises under the same stimulation. What practical capacities of the human distinguish the reporter from the parrot? What, besides the exercise of regular differential responsive dispositions, must one be able to *do*, in order to count as having or grasping *concepts*? ... To grasp or understand a concept is, according to Sellars, to have practical mastery over the inferences it is involved in... The parrot does not treat "That's red" as incompatible with "That's green"[6].

If this claim is also accepted, then material incompatibility is not just conceptually *independent* of logical negation, but conceptually *prior*.

1.2 Negation as the Minimal Incompatible

In [7] and [8], Brandom describes logical negation as a limiting form of material incompatibility:

> Incompatible sentences are Aristotelian *contraries*. A sentence and its negation are *contradictories*. What is the relation between these? Well, the contradictory is a contrary: any sentence is incompatible with its negation. What distinguishes the contradictory of a sentence from all the rest of its contraries? The contradictory is the *minimal* contrary: the one that is entailed by all the rest. Thus every contrary of "Plane figure f is a circle" - for instance "f is a triangle", "f is an octagon", and so on - entails "f is *not* a circle".

[5] [8] pp. 47–48.

[6] [7] pp. 88–89, our emphasis.

If someone asserts that it is not the case that Pierre is the (only) King of France, we have said very little. There are so many different ways in which it could be true:

- The King of France might be Jacques
- The King of France might be Louis
- ...
- There may be no King of France at all
- There may be no country denoted by the word "France"

Each of these concrete propositions is incompatible with Pierre being the King of France. To say "It is not the case that the King of France is Pierre" is just to claim that one of these indefinitely many concrete possibilities is true. Negation is just the logically weakest form of incompatibility.

In the rest of this paper, we assume - without further argument - that material incompatibility is conceptually prior to logical negation. We develop a simple modal logic to articulate Brandom's intuition: a language, without negation, in which we can nevertheless make incompatible claims.

1.3 Inferences Between Atomic Sentences

So far, we have justified the claim that incompatibility is a fundamental logical concept by arguing that incompatibility is conceptually prior to negation. Now incompatibility is an inferential relation between *atomic sentences*. In this subsection, we shall describe *other* inferential relations between atomic sentences - inferential relations that first-order logic cannot articulate (or can only do so awkwardly), but that cathoristic logic handles naturally.

The *atomic sentences* of a natural language can be characterised as the sentences which do not contain any other sentences as constituent parts[7]. According to this criterion, the following are atomic:

- Jack is male
- Jack loves Jill

The following is not atomic:

Jack is male and Jill is female

because it contains the complete sentence "Jack is male" as a syntactic constituent. Note that, according to this criterion, the following *is* atomic, despite using "and":

Jack loves Jill and Joan

[7] Compare Russell [24] p. 117: "A sentence is of atomic form when it contains no logical words and no subordinate sentence". We use a broader notion of atomicity by focusing solely on whether or not it contains a subordinate sentence, allowing logical words such as "and" *as long as they are conjoining noun-phrases* and not sentences.

Here, "Jack loves Jill" is not a syntactic constituent[8].

There are many types of inferential relations between atomic sentences of a natural language. For example:

- "Jack is male" is incompatible with "Jack is female"
- "Jack loves Jill" implies "Jack loves"
- "Jack walks slowly" implies "Jack walks"
- "Jack loves Jill and Joan" implies "Jack loves Jill"
- "Jack is wet and cold" implies "Jack is cold"

The first of these examples involves an incompatibility relation, while the others involve entailment relations. A key question this paper seeks to answer is: what is the simplest logic that can capture these inferential relations between atomic sentences?

1.4 Wittgenstein's Vision of a Logic of Elementary Propositions

In the *Tractatus* [34], Wittgenstein claims that the world is a set of atomic sentences in an idealised logical language. Each atomic sentence was supposed to be *logically independent* of every other, so that they could be combined together in every possible permutation, without worrying about their mutual compatibility. But already there were doubts and problem cases. He was aware that certain statements seemed atomic, but did not seem logically independent:

> For two colours, e.g., to be at one place in the visual field is impossible, and indeed logically impossible, for it is excluded by the logical structure of colour. (6.3751)

At the time of writing the *Tractatus*, he hoped that further analysis would reveal that these statements were not really atomic.

Later, in the *Philosophical Remarks* [33], he renounced the thesis of the logical independence of atomic propositions. In §76, talking about incompatible colour predicates, he writes:

> That makes it look as if *a construction might be possible within the elementary proposition.* That is to say, as if there were a construction in logic which didn't work by means of truth functions. What's more, it also seems that these constructions have an effect on one proposition's following logically from another. For, if different degrees exclude one another it follows from the presence of one that the other is not present. In that case, *two elementary propositions can contradict one another.*

Here, he is clearly imagining a logical language in which there are incompatibilities between atomic propositions. In §82:

[8] To see that "Jack loves Jill" is not a constituent of "Jack loves Jill and Joan", observe that "and" conjoins constituents of the *same syntactic type*. But "Jack loves Jill" is a sentence, while "Joan" is a noun. Hence the correct parsing is "Jack (loves (Jill and Joan))", rather than "(Jack loves Jill) and Joan".

> This is how it is, what I said in the Tractatus doesn't exhaust the grammatical rules for 'and', 'not', 'or', etc.; *there are rules for the truth functions which also deal with the elementary part of the proposition.* The fact that one measurement is right *automatically* excludes all others.

Wittgenstein does not, unfortunately, show us what this language would look like. In this paper, we present cathoristic logic as one way of formalising inferences between atomic sentences.

1.5 Outline

The rest of this paper is organised as follows: The next section briefly recapitulates the mathematical background of our work. Section 3 introduces the syntax and semantics of cathoristic logic with examples. Section 4 discusses how cathoristic logic can be used to model inferences between atomic sentences. Section 5 describes informally how our logic is useful as a knowledge representation language. Section 6 presents core results of the paper, in particular a semantic characterisation of elementary equivalence and a decision procedure with quadratic time-complexity. The decision procedure has been implemented in Haskell and is available for public use [1] under a liberal open-source license. This section also shows that Brandom's incompatibility semantics condition holds for cathoristic logic. Section 7 presents the proof rules for cathoristic logic and proves completeness. Section 8 provides two translations from cathoristic logic into first-order logic, and proves compactness using one of them. Section 9 investigates a variant of cathoristic logic with an additional negation operator, and provides a decision procedure for this extension that has an exponential time-complexity. Section 10 extends cathoristic logic with first-order quantifiers and sketches the translation of first-order formulae into first-order cathoristic logic. The conclusion surveys related work and lists open problems. Appendix A outlines a different approach to giving the semantics of cathoristic logic, including a characterisation of the corresponding elementary equivalence. The appendix also discusses the question of non-deterministic models. The remaining appendices present routine proof of facts used in the main section.

A reader not interested in mathematical detail is recommended to look only at Sects. 1, 3, 4, 5, the beginning of Sect. 7, and the Conclusion.

2 Mathematical Preliminaries

This section briefly surveys the mathematical background of our paper. A fuller account of order-theory can be found in [9]. Labelled transition systems are explored in [16, 26] and bisimulations in [25]. Finally, [11] is one of many books on first-order logic.

Order-Theory. A *preorder* is a pair $(S, \sqsubseteq)$ where S is a set, and $\sqsubseteq$ is a binary relation on S that is reflexive and transitive. Let $T \subseteq S$ and $x \in S$. We say x is an *upper bound* of T provided $t \sqsubseteq x$ for all $t \in T$. If in addition $x \sqsubseteq y$ for all upper bounds y of T, we say that x is the *least* upper bound of T. The set of all least upper bounds of T is denoted $\bigsqcup T$. *Lower bounds*, *greatest lower bounds* and $\bigsqcap T$ are defined mutatis mutandis. A *partial order* is a preorder $\sqsubseteq$ that is also anti-symmetric. A partial order $(S, \sqsubseteq)$ is a *lattice* if every pair of elements in S has a least upper and a greatest lower bound. A lattice is a *bounded lattice* if it has top and bottom elements $\top$ and $\bot$ such that for all $x \in S$:

$$x \sqcap \bot = \bot \qquad x \sqcup \bot = x \qquad x \sqcap \top = x \qquad x \sqcup \top = \top.$$

If $(S, \sqsubseteq)$ is a preorder, we can turn it into a partial-order by quotienting: let $a \simeq b$ iff $a \sqsubseteq b$ as well as $b \sqsubseteq a$. Clearly $\simeq$ is an equivalence. Let E be the set of all $\simeq$-equivalence classes of S. We get a canonical partial order, denoted $\sqsubseteq_E$, on E by setting: $[a]_\simeq \sqsubseteq_E [b]_\simeq$ whenever $a \sqsubseteq b$. If all relevant upper and lower bounds exist in $(S, \sqsubseteq)$, then $(E, \sqsubseteq_E)$ becomes a bounded lattice by setting

$$[x]_\simeq \sqcap [y]_\simeq = [x \sqcap y]_\simeq \quad [x]_\simeq \sqcup [y]_\simeq = [x \sqcup y]_\simeq \quad \bot_E = [\bot]_\simeq \quad \top_E = [\top]_\simeq.$$

Transition Systems. Let Σ be a set of *actions*. A *labelled transition system over* Σ is a pair $(\mathcal{S}, \rightarrow)$ where $\mathcal{S}$ is a set of *states* and $\rightarrow \subseteq \mathcal{S} \times \Sigma \times \mathcal{S}$ is the *transition relation*. We write $x \xrightarrow{a} y$ to abbreviate $(x, a, y) \in \rightarrow$. We let $s, t, w, w', x, y, z, ...$ range over states, $a, a', b, ...$ range over actions and $\mathcal{L}, \mathcal{L}', ...$ range over labelled transition systems. We usually speak of labelled transition systems when the set of actions is clear from the context. We say $\mathcal{L}$ is *deterministic* if $x \xrightarrow{a} y$ and $x \xrightarrow{a} z$ imply that $y = z$. Otherwise $\mathcal{L}$ is *non-deterministic.* A labelled transition system is *finitely branching* if for each state s, the set $\{t \mid s \xrightarrow{a} t\}$ is finite.

Simulations and Bisimulations. Given two labelled transition systems $\mathcal{L}_i = (S_i, \rightarrow_i)$ over Σ for $i = 1, 2$, a *simulation from* $\mathcal{L}_1$ *to* $\mathcal{L}_2$ is a relation $\mathcal{R} \subseteq S_1 \times S_2$ such that whenever $(s, s') \in \mathcal{R}$: if $s \xrightarrow{a}_1 s'$ then there exists a transition $t \xrightarrow{a}_2 t'$ with $(t, t') \in \mathcal{R}$. We write $s \preceq_{sim} t$ whenever $(s, t) \in \mathcal{R}$ for some simulation $\mathcal{R}$. We say $\mathcal{R}$ is a *bisimulation between* $\mathcal{L}_1$ *and* $\mathcal{L}_2$ if both, $\mathcal{R}$ and $\mathcal{R}^{-1}$ are simulations. Here $\mathcal{R}^{-1} = \{(y, x) \mid (x, y) \in \mathcal{R}\}$. We say two states s, s' are *bisimilar*, written $s \sim s'$ if there is a bisimulation $\mathcal{R}$ with $(s, s') \in \mathcal{R}$.

First-Order Logic. A *many-sorted first-order signature* is specified by the following data:

- a non-empty set of *sorts*
- a set of *function symbols* with associated *arities*, i.e. non-empty list of sorts $\#(f)$ for each function symbol f
- a set of *relation symbols* with associated *arities*, i.e. a list of sorts $\#(R)$ for each relation symbol R

- a set of *constant symbols* with associated *arity*, i.e. a sort $\#(c)$ for each constant symbol c.

We say a function symbol f is *n-ary* if $\#(f)$ has length $n+1$. Likewise, a relation symbol is *n-ary* if $\#(R)$ has length n.

Let $\mathcal{S}$ be a signature. An $\mathcal{S}$*-model* $\mathcal{M}$ is an object with the following components:

- for each sort σ a set U_σ called *universe* of sort σ. The members of U_σ are called σ*-elements* of $\mathcal{M}$
- an element $c^{\mathcal{M}}$ of U_σ for each constant c of sort σ
- a function $f^{\mathcal{M}} : (U_{\sigma_1} \times \cdots \times U_{\sigma_n}) \rightarrow U_\sigma$ for each function symbol f of arity $(\sigma_1, ..., \sigma_n, \sigma)$
- a relation $R^{\mathcal{M}} \subseteq U_{\sigma_1} \times \cdots \times U_{\sigma_n}$ for each relation symbol R of arity $(\sigma_1, ..., \sigma_n)$

Given an infinite set of variables for each sort σ, the *terms* and *first-order formulae* for $\mathcal{S}$ are given by the following grammar

$$\begin{aligned} t &::= x \mid c \mid f(t_1, ..., t_n) \\ \phi &::= t = t' \mid R(t_1, ..., t_n) \mid \neg\phi \mid \phi \wedge \psi \mid \forall x.A \end{aligned}$$

Here x ranges over variables of all sorts, c over constants, R over n-ary relational symbols and f over n-ary function symbols from $\mathcal{S}$. Other logical constructs such as disjunction or existential quantification are given by de Morgan duality, and truth $\top$ is an abbreviation for $x = x$. If $\mathcal{S}$ has just a single sort, we speak of *single-sorted first-order logic* or just *first-order logic.*

Given an $\mathcal{S}$-model $\mathcal{M}$, an *environment*, ranged over by σ, is a partial function from variables to $\mathcal{M}$'s universes. We write $x \mapsto u$ for the environment that maps x to u and is undefined for all other variables. Moreover, $\sigma, x \mapsto u$ is the environment that is exactly like σ, except that it also maps x to u, assuming that x is not in the domain of σ. The *interpretation* $[\![t]\!]_{\mathcal{M},\sigma}$ of a term t w.r.t. $\mathcal{M}$ and σ is given by the following clauses, assuming that the domain of σ contains all free variables of t:

- $[\![x]\!]_{\mathcal{M},\sigma} = \sigma(x)$.
- $[\![c]\!]_{\mathcal{M},\sigma} = c^{\mathcal{M}}$.
- $[\![f(t_1, ..., t_n)]\!]_{\mathcal{M},\sigma} = f^{\mathcal{M}}([\![t_1]\!]_{\mathcal{M},\sigma}, ..., [\![t_n]\!]_{\mathcal{M},\sigma})$.

The *satisfaction relation* $\mathcal{M} \models_\sigma \phi$ is given by the following clauses, this time assuming that the domain of σ contains all free variables of ϕ:

- $\mathcal{M} \models_\sigma t = t'$ iff $[\![t]\!]_{\mathcal{M},\sigma} = [\![t']\!]_{\mathcal{M},\sigma}$.
- $\mathcal{M} \models_\sigma R(t_1, ..., t_n)$ iff $R^{\mathcal{M}}([\![t_1]\!]_{\mathcal{M},\sigma}, ..., [\![t_n]\!]_{\mathcal{M},\sigma})$.
- $\mathcal{M} \models_\sigma \neg\phi$ iff $\mathcal{M} \not\models_\sigma \phi$.
- $\mathcal{M} \models_\sigma \phi \wedge \psi$ iff $\mathcal{M} \models_\sigma \phi$ and $\mathcal{M} \models_\sigma \psi$.
- $\mathcal{M} \models_\sigma \forall x.\phi$ iff for all u in the universe of $\mathcal{M}$ we have $\mathcal{M} \models_{\sigma, x \mapsto v} \phi$.

Note that if σ and σ' agree on the free variables of t, then $[\![t]\!]_{\mathcal{M},\sigma} = [\![t]\!]_{\mathcal{M},\sigma'}$. Likewise $\mathcal{M} \models_\sigma \phi$ if and only if $\mathcal{M} \models_{\sigma'} \phi$, provided σ and σ' agree on the free variables of ϕ.

The *theory* of a model $\mathcal{M}$, written $\mathsf{Th}(\mathcal{M})$, is the set of all formulae made true by $\mathcal{M}$, i.e. $\mathsf{Th}(\mathcal{M}) = \{\phi \mid \mathcal{M} \models \phi\}$. We say two models $\mathcal{M}$ and $\mathcal{N}$ are *elementary equivalent* if $\mathsf{Th}(\mathcal{M}) = \mathsf{Th}(\mathcal{N})$.

3 Cathoristic Logic

In this section we introduce the syntax and semantics of cathoristic logic.

3.1 Syntax

Syntactically, cathoristic logic is a multi-modal logic with one new operator.

Definition 1. *Let Σ be a non-empty set of* actions. *Actions are ranged over by $a, a', a_1, b, ...$, and A ranges over finite subsets of Σ. The* formulae *of cathoristic logic, ranged over by $\phi, \psi, \xi ...$, are given by the following grammar.*

$$\phi ::= \top \mid \phi \wedge \psi \mid \langle a \rangle \phi \mid !A$$

The first three forms of ϕ are standard from Hennessy-Milner logic [17]: $\top$ is logical truth, $\wedge$ is conjunction, and $\langle a \rangle \phi$ means that the current state can transition via action a to a new state at which ϕ holds. Tantum A, written $!A$, is the key novelty of cathoristic logic. Asserting $!A$ means: in the current state at most the modalities $\langle a \rangle$ that satisfy $a \in A$ are permissible.

We assume that $\langle a \rangle \phi$ binds more tightly than conjunction, so $\langle a \rangle \phi \wedge \psi$ is short for $(\langle a \rangle \phi) \wedge \psi$. We often abbreviate $\langle a \rangle \top$ to $\langle a \rangle$. We define falsity $\bot$ as $!\emptyset \wedge \langle a \rangle$ where a is an arbitrary action in Σ. Hence, Σ must be non-empty. Note that, in the absence of negation, we cannot readily define disjunction, implication, or $[a]$ modalities by de Morgan duality.

Convention 1. *From now on we assume a fixed set Σ of actions, except where stated otherwise.*

3.2 Semantics

The semantics of cathoristic logic is close to Hennessy-Milner logic, but uses deterministic transition systems augmented with labels on states.

Definition 2. *A* cathoristic transition system *is a triple $\mathcal{L} = (S, \rightarrow, \lambda)$, where $(S, \rightarrow)$ is a deterministic labelled transition system over Σ, and λ is a function from states to sets of actions (not necessarily finite), subject to the following constraints:*

- *For all states $s \in S$ it is the case that $\{a \mid \exists t\ s \xrightarrow{a} t\} \subseteq \lambda(s)$. We call this condition* admissibility.

- *For all states $s \in S$, $\lambda(s)$ is either finite or Σ. We call this condition* well-sizedness.

The intended interpretation is that $\lambda(w)$ is the set of allowed actions emanating from w. The λ function is the semantic counterpart of the ! operator. The admissibility restriction is in place because transitions $s \xrightarrow{a} t$ where $a \notin \lambda(s)$ would be saying that an a action is possible at s but at the same time prohibited at s. Well-sizedness is not a fundamental restriction but rather a convenient trick. Cathoristic transition systems have two kinds of states:

- States s without restrictions on outgoing transitions. Those are labelled with $\lambda(s) = \Sigma$.
- States s with restriction on outgoing transitions. Those are labelled by a finite set $\lambda(s)$ of actions.

Defining λ on all states and not just on those with restrictions makes some definitions and proofs slightly easier.

As with other modal logics, satisfaction of formulae is defined relative to a particular state in the transition system, giving rise to the following definition.

Definition 3. *A* cathoristic model, *ranged over by $\mathfrak{M}, \mathfrak{M}', \ldots$, is a pair $(\mathcal{L}, s)$, where $\mathcal{L}$ is a cathoristic transition system $(S, \rightarrow, \lambda)$, and s is a state from S. We call s the* start state *of the model. An cathoristic model is a* tree *if the underlying transition system is a tree whose root is the start state.*

Satisfaction of a formula is defined relative to a cathoristic model.

Definition 4. *The* satisfaction relation *$\mathfrak{M} \models \phi$ is defined inductively by the following clauses, where we assume that $\mathfrak{M} = (\mathcal{L}, s)$ and $\mathcal{L} = (S, \rightarrow, \lambda)$.*

$$
\begin{array}{ll}
\mathfrak{M} \models \top & \\
\mathfrak{M} \models \phi \wedge \psi & \textit{iff } \mathfrak{M} \models \phi \textit{ and } \mathfrak{M} \models \psi \\
\mathfrak{M} \models \langle a \rangle \phi & \textit{iff there is a transition } s \xrightarrow{a} t \textit{ such that } (\mathcal{L}, t) \models \phi \\
\mathfrak{M} \models !A & \textit{iff } \lambda(s) \subseteq A
\end{array}
$$

The first three clauses are standard. The last clause enforces the intended meaning of $!A$: the permissible modalities in the model are *at least as constrained* as required by $!A$. They may even be more constrained if the inclusion $\lambda(s) \subseteq A$ is proper. For infinite sets Σ of actions, allowing $\lambda(s)$ to return arbitrary infinite sets in addition to Σ does not make a difference because A is finite by construction, so $\lambda(s) \subseteq A$ can never hold anyway for infinite $\lambda(s)$.

We continue with concrete examples. The model in Fig. 1 satisfies all the following formulae, amongst others.

$$
\begin{array}{lllll}
\langle a \rangle & \langle a \rangle \langle b \rangle & \langle a \rangle !\{b, c\} & \langle a \rangle !\{b, c, d\} & \langle c \rangle \\
\langle c \rangle !\emptyset & \langle c \rangle !\{a\} & \langle c \rangle !\{a, b\} & \langle a \rangle \wedge \langle c \rangle & \langle a \rangle (\langle b \rangle \wedge !\{b, c\})
\end{array}
$$

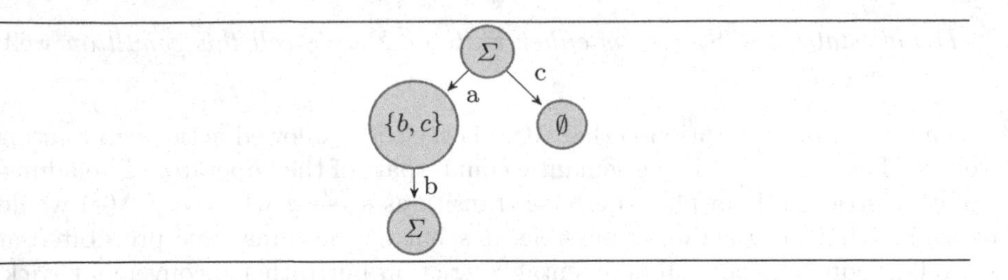

Fig. 1. Example model.

Here we assume, as we do with all subsequent figures, that the top state is the start state. The same model does not satisfy any of the following formulae.

$$\langle b\rangle \qquad !\{a\} \qquad !\{a,c\} \qquad \langle a\rangle!\{b\} \qquad \langle a\rangle\langle c\rangle \qquad \langle a\rangle\langle b\rangle!\{c\}$$

Figure 2 shows various models of $\langle a\rangle\langle b\rangle$ and Fig. 3 shows one model that does, and one that does not, satisfy the formula $!\{a, b\}$. Both models validate $!\{a, b, c\}$.

Cathoristic logic does not have the operators $\neg$, $\vee$, or $\rightarrow$. This has the following two significant consequences. First, every satisfiable formula has a unique (up to isomorphism) simplest model. In Fig. 2, the left model is the unique simplest model satisfying $\langle a\rangle\langle b\rangle$. We will clarify below that model simplicity is closely related to the process theoretic concept of similarity, and use the existence of unique simplest models in our quadratic-time decision procedure.

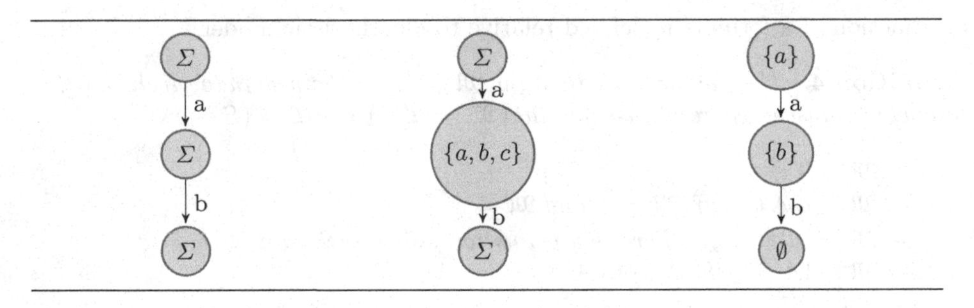

Fig. 2. Three models of $\langle a\rangle\langle b\rangle\top$

Secondly, cathoristic logic is different from other logics in that there is an asymmetry between tautologies and contradictories: logics with conventional negation have an infinite number of non-trivial tautologies, as well as an infinite number of contradictories. In contrast, because cathoristic logic has no negation or disjunction operator, it is expressively limited in the tautologies it can express: $\top$ and conjunctions of $\top$ are its sole tautologies. On the other hand, the tantum operator enables an infinite number of contradictories to be expressed. For example:

$$\langle a\rangle \wedge !\emptyset \qquad \langle a\rangle \wedge !\{b\} \qquad \langle a\rangle \wedge !\{b,c\} \qquad \langle b\rangle \wedge !\emptyset$$

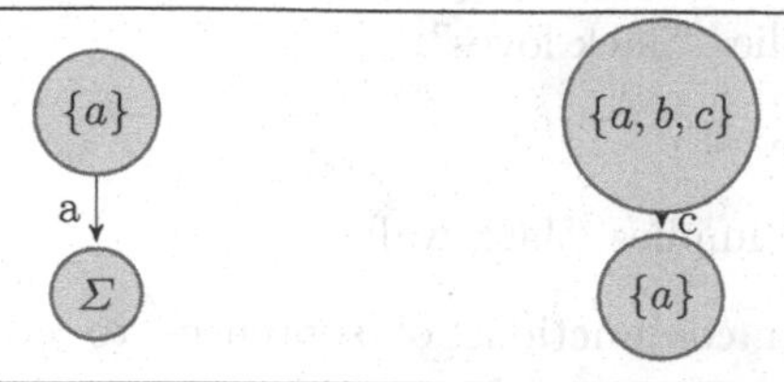

Fig. 3. The model on the left validates $!\{a, b\}$ while the model on the right does not.

Next, we present the semantic consequence relation.

Definition 5. *We say the formula* ϕ semantically implies ψ, *written* $\phi \models \psi$, *provided for all cathoristic models* $\mathfrak{M}$ *if* $\mathfrak{M} \models \phi$ *then also* $\mathfrak{M} \models \psi$. *We sometimes write* $\models \phi$ *as a shorthand for* $\top \models \phi$.

Cathoristic logic shares with other (multi)-modal logics the following implications:

$$\langle a \rangle \langle b \rangle \models \langle a \rangle \qquad \langle a \rangle (\langle b \rangle \wedge \langle c \rangle) \models \langle a \rangle \langle b \rangle$$

As cathoristic logic is restricted to deterministic models, it also validates the following formula:

$$\langle a \rangle \langle b \rangle \wedge \langle a \rangle \langle c \rangle \models \langle a \rangle (\langle b \rangle \wedge \langle c \rangle)$$

Cathoristic logic also validates all implications in which the set of constraints is relaxed from left to right. For example:

$$!\{c\} \models !\{a, b, c\} \qquad !\emptyset \models !\{a, b\}$$

4 Inferences Between Atomic Sentences

Cathoristic logic arose in part as an attempt to answer the question: what is the simplest logic that can capture inferences between atomic sentences of natural language? In this section, we give examples of such inferences, and then show how cathoristic logic handles them. We also compare our approach with attempts at expressing the inferences in first-order logic.

4.1 Intra-atomic Inferences in Cathoristic Logic

Natural language admits many types of inference between atomic sentences. First, exclusion:

"Jack is male" is incompatible with "Jack is female".

Second, entailment inferences from dyadic to monadic predicates:

"Jack loves Jill" implies "Jack loves".

Third, adverbial inferences:

"Jack walks quickly" implies "Jack walks".

Fourth, inferences from conjunctions of sentences to conjunctions of noun-phrases (and vice-versa):

"Jack loves Jill" and "Jack loves Joan" together imply that "Jack loves Jill and Joan".

Fifth, inferences from conjunctions of sentences to conjunction of predicates[9] (and vice-versa):

"Jack is bruised" and "Jack is humiliated" together imply that "Jack is bruised and humiliated".

They all can be handled directly and naturally in cathoristic logic, as we shall now show.

Incompatibility, such as that between "Jack is male" and "Jack is female", is translated into cathoristic logic as the pair of incompatible sentences:

$$\langle jack\rangle\langle sex\rangle(\langle male\rangle \wedge !\{male\}) \qquad \langle jack\rangle\langle sex\rangle(\langle female\rangle \wedge !\{female\}).$$

Cathoristic logic handles entailments from dyadic to monadic predicates[10]. "Jack loves Jill" is translated into cathoristic logic as:

$$\langle jack\rangle\langle loves\rangle\langle jill\rangle.$$

The semantics of modalities ensures that this directly entails:

$$\langle jack\rangle\langle loves\rangle.$$

Similarly, cathoristic logic supports inferences from triadic to dyadic predicates:

[9] See [28] p. 282 for a spirited defence of predicate conjunction against Fregean regimentation.

[10] Although natural languages are full of examples of inferences from dyadic to monadic predicates, there are certain supposed counterexamples to the general rule that a dyadic predicate always implies a monadic one. For example, "Jack explodes the device" does not, on its most natural reading, imply that "Jack explodes". Our response to cases like this is to distinguish between two distinct monadic predicates $explodes_1$ and $explodes_2$:

- $X explodes_1$ iff X is an object that undergoes an explosion
- $X explodes_2$ iff X is an agent that initiates an explosion

Now "Jack explodes the device" does imply that "Jack $explodes_2$" but does not imply that "Jack $explodes_1$". There is no deep problem here - just another case where natural language overloads the same word in different situation to have different meanings.

"Jack passed the biscuit to Mary" implies "Jack passed the biscuit".

This can be expressed directly in cathoristic logic as:

$$\langle \mathit{jack} \rangle \langle \mathit{passed} \rangle \langle \mathit{biscuit} \rangle \langle \mathit{to} \rangle (\langle \mathit{mary} \rangle \land ! \{\mathit{mary}\}) \models \langle \mathit{jack} \rangle \langle \mathit{passed} \rangle \langle \mathit{biscuit} \rangle.$$

Adverbial inferences is captured in cathoristic logic as follows.

$$\langle \mathit{jack} \rangle \langle \mathit{walks} \rangle \langle \mathit{quickly} \rangle$$

entails:

$$\langle \mathit{jack} \rangle \langle \mathit{walks} \rangle.$$

Cathoristic logic directly supports inferences from conjunctions of sentences to conjunctions of noun-phrases. As our models are deterministic, we have the general rule that $\langle a \rangle \langle b \rangle \land \langle a \rangle \langle c \rangle \models \langle a \rangle (\langle b \rangle \land \langle c \rangle)$ from which it follows that

$$\langle \mathit{jack} \rangle \langle \mathit{loves} \rangle \langle \mathit{jill} \rangle \qquad \text{and} \qquad \langle \mathit{jack} \rangle \langle \mathit{loves} \rangle \langle \mathit{joan} \rangle$$

together imply

$$\langle \mathit{jack} \rangle \langle \mathit{loves} \rangle (\langle \mathit{jill} \rangle \land \langle \mathit{joan} \rangle).$$

Using the same rule, we can infer that

$$\langle \mathit{jack} \rangle \langle \mathit{bruised} \rangle \land \langle \mathit{jack} \rangle \langle \mathit{humiliated} \rangle$$

together imply

$$\langle \mathit{jack} \rangle (\langle \mathit{bruised} \rangle \land \langle \mathit{humiliated} \rangle).$$

4.2 Intra-atomic Inferences in First-Order Logic

Next, we look at how these inferences are handled in first-order logic.

Incompatible Predicates in First-Order Logic. How are incompatible predicates represented in first-order logic? Brachman and Levesque [6] introduce the topic by remarking:

> We would consider it quite "obvious" in this domain that if it were asserted that $John$ were a Man, then we should answer "no" to the query $Woman(John)$.

They propose adding an extra axiom to express the incompatibility:

$$\forall x.(Man(x) \rightarrow \neg Woman(x))$$

This proposal imposes a burden on the knowledge-representer: an extra axiom must be added for every pair of incompatible predicates. This is burdensome for large sets of incompatible predicates. For example, suppose there are 50 football teams, and a person can only support one team at a time. We would need to add $\binom{50}{2}$ axioms, which is unwieldy.

$$\forall x.\neg(\mathit{SupportsArsenal}(x) \wedge \mathit{SupportsLiverpool}(x))$$
$$\forall x.\neg(\mathit{SupportsArsenal}(x) \wedge \mathit{SupportsManUtd}(x))$$
$$\forall x.\neg(\mathit{SupportsLiverpool}(x) \wedge \mathit{SupportsManUtd}(x))$$
$$\vdots$$

Or, if we treat the football-teams as objects, and have a two-place *Supports* relation between people and teams, we could have:

$$\forall xyz.(\mathit{Supports}(x, y) \wedge y \neq z \rightarrow \neg \mathit{Supports}(x, z)).$$

If we also assume that each football team is distinct from all others, this certainly captures the desired uniqueness condition. But it does so by using relatively complex logical machinery.

Inferences from Dyadic to Monadic Predicates in First-Order Logic. If we want to capture the inference from "Jack loves Jill" to "Jack loves" in first-order logic, we can use a non-logical axiom:

$$\forall x.y.(\mathit{Loves}_2(x, y) \rightarrow \mathit{Loves}_1(x))$$

We would have to add an extra axiom like this for every n-place predicate. This is cumbersome at best. In cathoristic logic, by contrast, we do not need to introduce any non-logical machinery to capture these inferences because they all follow from the general rule that $\langle a \rangle \langle b \rangle \models \langle a \rangle$.

Adverbial Inferences in First-Order Logic. How can we represent verbs in traditional first-order logic so as to support adverbial inference? Davidson [10] proposes that every n-place action verb be analysed as an $n{+}1$-place predicate, with an additional slot representing an event. For example, he analyses "I flew my spaceship to the Morning Star" as

$$\exists x.(\mathit{Flew}(I, \mathit{MySpaceship}, x) \wedge \mathit{To}(x, \mathit{TheMorningStar}))$$

Here, x ranges over events. This implies

$$\exists x.\mathit{Flew}(I, \mathit{MySpaceship}, x)$$

This captures the inference from "I flew my spaceship to the Morning Star" to "I flew my spaceship".

First-order logic cannot support logical inferences between atomic sentences. If it is going to support inferences from adverbial sentences, it *cannot* treat them as atomic and must instead *reinterpret* them as logically complex propositions.

The cost of Davidson's proposal is that a seemingly simple sentence - such as "Jones walks" - turns out, on closer inspection, not to be atomic at all - but to involve existential quantification:

$$\exists x. \mathit{Walks}(\mathit{Jones}, x)$$

First-order logic *can* handle such inferences - but only by reinterpreting the sentences as logically-complex compound propositions.

5 Cathoristic Logic as a Language for Knowledge Representation

Cathoristic logic has been used as the representation language for a large, complex, dynamic multi-agent simulation [13]. This is an industrial-sized application, involving tens of thousands of rules and facts[11]. In this simulation, the entire world state is stored as a cathoristic model.

We found that cathoristic logic has two distinct advantages as a language for knowledge representation. First, it is ergonomic: ubiquitous concepts (such as uniqueness) can be expressed directly. Second, it is efficient: the tantum operator allows certain sorts of optimisation that would not otherwise be available. We shall consider these in turn.

5.1 Representing Facts in Cathoristic Logic

A sentence involving a one-place predicate of the form $p(a)$ is expressed in cathoristic logic as

$$\langle a \rangle \langle p \rangle$$

A sentence involving a many-to-many two-place relation of the form $r(a, b)$ is expressed in cathoristic logic as

$$\langle a \rangle \langle r \rangle \langle b \rangle$$

But a sentence involving a many-to-one two-place relation of the form $r(a, b)$ is expressed as:

$$\langle a \rangle \langle r \rangle (\langle b \rangle \wedge !\{b\})$$

So, for example, to say that "Jack likes Jill" (where "likes" is, of course, a many-many relation), we would write:

$$\langle \mathit{jack} \rangle \langle \mathit{likes} \rangle \langle \mathit{jill} \rangle$$

---

[11] The application had thousands of paying users, and was available for download on the App Store for the iPad [12].

But to say that "Jack is married to Joan" (where "is-married-to" is a many-one relation), we would write:

$$\langle jack\rangle\langle married\rangle(\langle joan\rangle \wedge !\{joan\})$$

Colloquially, we might say that "Jack is married to Joan - and only Joan". Note that the relations are placed in infix position, so that the facts about an object are "contained" within the object. One reason for this particular way of structuring the data will be explained below.

Consider the following facts about a gentleman named Brown:

$$\langle brown\rangle \begin{pmatrix} \langle sex\rangle(\langle male\rangle \wedge !\{male\}) \\ \wedge \\ \langle friends\rangle(\langle lucy\rangle \wedge \langle elizabeth\rangle) \end{pmatrix}$$

All facts starting with the prefix $\langle brown\rangle$ form a sub-tree of the entire database. And all facts which start with the prefix $\langle brown\rangle\langle friends\rangle$ form a sub-tree of that tree. A sub-tree can be treated as an individual via its prefix. A sub-tree of formulae is the cathoristic logic equivalent of an *object* in an object-oriented programming language.

To model change over time, we assert and retract statements from the database, using a non-monotonic update mechanism. If a fact is inserted into the database that involves a state-labelling restricting the permissible transitions emanating from that state, then all transitions out of that state that are incompatible with the restriction are removed. So, for example, if the database currently contains the fact that the traffic light is amber, and then we update the database to assert the traffic light is red:

$$\langle tl\rangle\langle colour\rangle(\langle red\rangle \wedge !\{red\})$$

Now the restriction on the state (that red is the only transition) means that the previous transition from that state (the transition labelled with amber) is automatically removed.

The tree-structure of formulae allows us to express the *life-time of data* in a natural way. If we wish a piece of data d to exist for just the duration of a proposition t, then we make t be a sub-expression of d. For example, if we want the friendships of an agent to exist just as long as the agent, then we place the relationships inside the agent:

$$\langle brown\rangle\langle friends\rangle$$

Now, when we remove $\langle brown\rangle$ all the sub-trees, including the data about who he is friends with, will be automatically deleted as well.

Another advantage of our representation is that we get a form of *automatic currying* which simplifies queries. So if, for example, Brown is married to Elizabeth, then the database would contain

$$\langle brown\rangle\langle married\rangle(\langle elizabeth\rangle \wedge !\{elizabeth\})$$

In cathoristic logic, if we want to find out whether Brown is married, we can query the sub-formula directly - we just ask if

$$\langle brown \rangle \langle married \rangle$$

In first-order logic, if *married* is a two-place predicate, then we need to fill in the extra argument place with a free variable - we would need to find out if there exists an x such that $married(brown, x)$ - this is more cumbersome to type and slower to compute.

5.2 Simpler Postconditions

In this section, we contrast the representation in action languages based on first-order logic[12], with our cathoristic logic-based representation. Action definitions are rendered in typewriter font.

When expressing the pre- and postconditions of an action, planners based on first-order logic have to explicitly describe the propositions that are removed when an action is performed:

```
action move(A, X, Y)
    preconditions
        at(A, X)
    postconditions
        add: at(A, Y)
        remove: at(A, X)
```

Here, we need to explicitly state that when A moves from X to Y, A is no longer at X. It might seem obvious to us that if A is now at Y, he is no longer at X - but we need to explicitly tell the system this. This is unnecessary, cumbersome and error-prone. In cathoristic logic, by contrast, the exclusion operator means we do not need to specify the facts that are no longer true:

```
action move (A, X, Y)
    preconditions
        <A><at>(<X> /\ !{X})
    postconditions
        add: <A><at>(<Y> /\ !{Y})
```

The tantum operator ! makes it clear that something can only be at one place at a time, and the non-monotonic update rule described above *automatically* removes the old invalid location data.

5.3 Using Tantum ! to Optimise Preconditions

Suppose, for example, we want to find all married couples who are both Welsh. In Prolog, we might write something like:

[12] E.g. STRIPS [14].

```
welsh_married_couple(X, Y) :-
    welsh(X),
    welsh(Y),
    spouse(X,Y).
```

Rules like this create a large search-space because we need to find all instances of $welsh(X)$ and all instances of $welsh(Y)$ and take the cross-product [27]. If there are n Welsh people, then we will be searching n^2 instances of (X, Y) substitutions.

If we express the rule in cathoristic logic, the compiler is able to use the extra information expressed in the ! operator to reorder the literals to find the result significantly faster. Assuming someone can only have a single spouse at any moment, the rule is expressed in cathoristic logic as:

```
welsh_married_couple(X, Y) :-
    <welsh> <X>,
    <welsh> <Y>,
    <spouse> <X> (<Y> /\ !{Y}).
```

Now the compiler is able to reorder these literals to minimise the search-space. It can see that, once X is instantiated, the following literal can be instantiated without increasing the search-space:

```
<spouse> <X> (<Y> /\ !{Y})
```

The *tantum* operator can be used by the compiler to see that there is at most one Y who is the spouse of X. So the compiler reorders the clauses to produce:

```
welsh_married_couple (X, Y) :-
    <welsh> <X>,
    <spouse> <X> (<Y> /\ !{Y}),
    <welsh> <Y>.
```

Now it is able to find all results by just searching n instances - a significant optimisation. In our application, this optimisation has made a significant difference to the run-time cost of query evaluation.

6 Semantics and Decision Procedure

In this section we provide our key semantic results. We define a partial ordering $\preceq$ on models, and show how the partial ordering can be extended into a bounded lattice. We use the bounded lattice to construct a quadratic-time decision procedure.

6.1 Semantic Characterisation of Elementary Equivalence

Elementary equivalence induces a notion of model equivalence: two models are elementarily equivalent exactly when they make the same formulae true. Elementary equivalence as a concept thus relies on cathoristic logic even for its definition. We now present an alternative characterisation that is purely semantic, using the concept of (mutual) simulation from process theory. Apart from its intrinsic interest, this characterisation will also be crucial for proving completeness of the proof rules.

We first define a pre-order $\preceq$ on models by extending the notion of simulation on labelled transition systems to cathoristic models. Then we prove an alternative characterisation of $\preceq$ in terms of set-inclusion of the theories induced by models. We then show that two models are elementarily equivalent exactly when they are related by $\preceq$ and by $\preceq^{-1}$.

Definition 6. *Let $\mathcal{L}_i = (S_i, \rightarrow_i, \lambda_i)$ be cathoristic transition systems for $i = 1, 2$. A relation $\mathcal{R} \subseteq S_1 \times S_2$ is a* simulation *from $\mathcal{L}_1$ to $\mathcal{L}_2$, provided:*

- *$\mathcal{R}$ is a simulation on the underlying transition systems.*
- *Whenever $(x, y) \in \mathcal{R}$ then also $\lambda_1(x) \supseteq \lambda_2(y)$.*

If $\mathfrak{M}_i = (\mathcal{L}_i, x_i)$ are models, we say $\mathcal{R}$ is a simulation *from $\mathfrak{M}_1$ to $\mathfrak{M}_2$, provided the following hold.*

- *$\mathcal{R}$ is a simulation from $\mathcal{L}_1$ to $\mathcal{L}_2$ as cathoristic transition systems.*
- *$(x_1, x_2) \in \mathcal{R}$.*

Note that the only difference from the usual definition of simulation is the additional requirement on the state labelling functions λ_1 and λ_2.

Definition 7. *The largest simulation from $\mathfrak{M}_1$ to $\mathfrak{M}_2$ is denoted $\mathfrak{M}_1 \preceq_{sim} \mathfrak{M}_2$. It is easy to see that $\preceq_{sim}$ is itself a simulation from $\mathfrak{M}_1$ to $\mathfrak{M}_2$, and the union of all such simulations. If $\mathfrak{M}_1 \preceq_{sim} \mathfrak{M}_2$ we say $\mathfrak{M}_2$* simulates *$\mathfrak{M}_1$.*

We write $\simeq$ for $\preceq_{sim} \cap \preceq_{sim}^{-1}$. We call $\simeq$ the mutual simulation *relation.*

We briefly discuss the relationship of $\simeq$ with bisimilarity, a notion of equality well-known from process theory and modal logic. For non-deterministic transition systems $\simeq$ is a strictly coarser relation than bisimilarity.

Definition 8. *We say $\mathcal{R}$ is a* bisimulation *if $\mathcal{R}$ is a simulation from $\mathfrak{M}_1$ to $\mathfrak{M}_2$ and $\mathcal{R}^{-1}$ is a simulation from $\mathfrak{M}_2$ to $\mathfrak{M}_1$. By $\sim$ we denote the largest bisimulation, and we say that $\mathfrak{M}_1$ and $\mathfrak{M}_2$ are* bisimilar *whenever $\mathfrak{M}_1 \sim \mathfrak{M}_2$.*

Lemma 1. *On cathoristic models, $\sim$ and $\simeq$ coincide.*

Proof. Straightforward from the definitions. □

Definition 9. *Let $\mathsf{Th}(\mathfrak{M})$ be the* theory *of $\mathfrak{M}$, i.e. the formulae made true by $\mathfrak{M}$, i.e. $\mathsf{Th}(\mathfrak{M}) = \{\phi \mid \mathfrak{M} \models \phi\}$.*

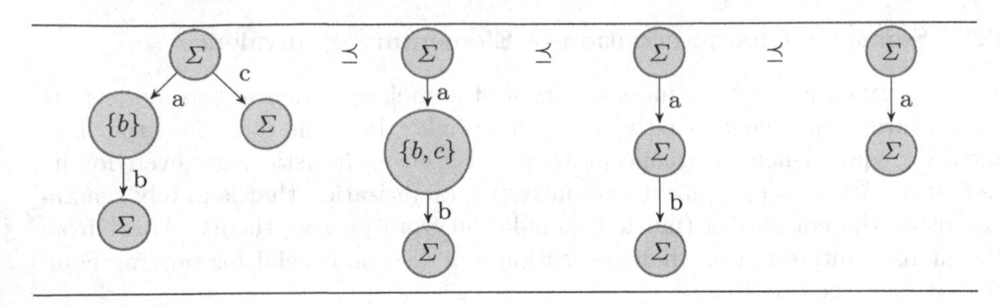

Fig. 4. Examples of $\preceq$

We give an alternative characterisation of $\preceq_{sim}^{-1}$ using theories. In what follows, we will mostly be interested in $\preceq_{sim}^{-1}$, so we give it its own symbol.

Definition 10. *Let* $\preceq$ *be short for* $\preceq_{sim}^{-1}$.

Figure 4 gives some examples of models and how they are related by $\preceq$.

Theorem 1 (Characterisation of elementary equivalence).

1. $\mathfrak{M}' \preceq \mathfrak{M}$ *if and only if* $\mathsf{Th}(\mathfrak{M}) \subseteq \mathsf{Th}(\mathfrak{M}')$.
2. $\mathfrak{M}' \simeq \mathfrak{M}$ *if and only if* $\mathsf{Th}(\mathfrak{M}) = \mathsf{Th}(\mathfrak{M}')$.

Proof. For (1) assume $\mathfrak{M}' \preceq \mathfrak{M}$ and $\mathfrak{M} \models \phi$. We must show $\mathfrak{M}' \models \phi$. Let $\mathfrak{M} = (\mathcal{L}, w)$ and $\mathfrak{M}' = (\mathcal{L}', w')$. The proof proceeds by induction on ϕ. The cases for $\top$ and $\wedge$ are trivial. Assume $\phi = \langle a \rangle \psi$ and assume $(\mathcal{L}, w) \models \langle a \rangle \psi$. Then $w \xrightarrow{a} x$ and $(\mathcal{L}, x) \models \psi$. As $\mathfrak{M}'$ simulates $\mathfrak{M}$, there is an x' such that $(x, x') \in R$ and $w' \xrightarrow{a} x'$. By the induction hypothesis, $(\mathcal{L}', x') \models \psi$. Therefore, by the semantic clause for $\langle\rangle$, $(\mathcal{L}', w') \models \langle a \rangle \psi$. Assume now that $\phi = \,!\, A$, for some finite $A \subseteq \Sigma$, and that $(\mathcal{L}, w) \models \,!\, A$. By the semantic clause for !, $\lambda(w) \subseteq A$. Since $(\mathcal{L}', w') \preceq (\mathcal{L}, w)$, by the definition of simulation of cathoristic transition systems, $\lambda(w) \supseteq \lambda'(w')$. Therefore, $\lambda'(w') \subseteq \lambda(w) \subseteq A$. Therefore, by the semantic clause for !, $(\mathcal{L}', w') \models \,!\, A$.

For the other direction, let $\mathfrak{M} = (\mathcal{L}, w)$ and $\mathfrak{M}' = (\mathcal{L}', w')$. Assume $\mathsf{Th}(\mathfrak{M}) \subseteq \mathsf{Th}(\mathfrak{M}')$. We need to show that $\mathfrak{M}'$ simulates $\mathfrak{M}$. In other words, we need to produce a relation $R \subseteq S \times S'$ where S is the state set of $\mathcal{L}$, S' is the state set for $\mathcal{L}'$ and $(w, w') \in R$ and R is a simulation from $(\mathcal{L}, w)$ to $(\mathcal{L}', w')$. Define $R = \{(x, x') \mid \mathsf{Th}((\mathcal{L}, x)) \subseteq \mathsf{Th}((\mathcal{L}', x'))\}$. Clearly, $(w, w') \in R$, as $\mathsf{Th}((\mathcal{L}, w)) \subseteq \mathsf{Th}((\mathcal{L}', w'))$. To show that R is a simulation, assume $x \xrightarrow{a} y$ in $\mathcal{L}$ and $(x, x') \in R$. We need to provide a y' such that $x' \xrightarrow{a} y'$ in $\mathcal{L}'$ and $(y, y') \in R$. Consider the formula $\langle a \rangle \mathsf{char}((\mathcal{L}, y))$. Now $x \models \langle a \rangle \mathsf{char}((\mathcal{L}, y))$, and since $(x, x') \in R$, $x' \models \langle a \rangle \mathsf{char}((\mathcal{L}, y))$. By the semantic clause for $\langle a \rangle$, if $x' \models \langle a \rangle \mathsf{char}((\mathcal{L}, y))$ then there is a y' such that $y' \models \mathsf{char}((\mathcal{L}, y))$. We need to show $(y, y') \in R$, i.e. that $y \models \phi$ implies $y' \models \phi$ for all ϕ. Assume $y \models \phi$. Then by the definition of $\mathsf{char}()$, $\mathsf{char}((\mathcal{L}, y)) \models \phi$. Since $y' \models \mathsf{char}((\mathcal{L}, y))$, $y' \models \phi$. So $(y, y') \in R$, as required.

Finally, we need to show that whenever $(x, x') \in R$, then $\lambda(x) \supseteq \lambda'(x')$. Assume, first, that $\lambda(x)$ is finite. Then $(\mathcal{L}, x) \models !\ \lambda(x)$. But as $(x, x') \in R$, $\mathsf{Th}((\mathcal{L}, x)) \subseteq \mathsf{Th}((\mathcal{L}', x'))$, so $(\mathcal{L}', x') \models !\ \lambda(x)$. But, by the semantic clause for $!$, $(\mathcal{L}', x') \models !\ \lambda(x)$ iff $\lambda'(x') \subseteq \lambda(x)$. Therefore $\lambda(x) \supseteq \lambda'(x')$. If, on the other hand, $\lambda(x)$ is infinite, then $\lambda(x) = \Sigma$ (because the only infinite state labelling that we allow is Σ). Every state labelling is a subset of Σ, so here too, $\lambda(x) = \Sigma \supseteq \lambda'(x')$.

This establishes (1), and (2) is immediate from the definitions. □

Theorem 1.1 captures one way in which the model theory of classical and cathoristic logic differ. In classical logic the theory of each model is complete, and $\mathsf{Th}(\mathcal{M}) \subseteq \mathsf{Th}(\mathcal{N})$ already implies that $\mathsf{Th}(\mathcal{M}) = \mathsf{Th}(\mathcal{N})$, i.e. $\mathcal{M}$ and $\mathcal{N}$ are elementarily equivalent. Cathoristic logic's lack of negation changes this drastically, and gives $\preceq$ the structure of a non-trivial bounded lattice as we shall demonstrate below.

Theorem 1 has various consequences.

Corollary 1. *1. If ϕ has a model then it has a model whose underlying transition system is a tree, i.e. all states except for the start state have exactly one predecessor, and the start state has no predecessors.*
2. If ϕ has a model then it has a model where every state is reachable from the start state.

Proof. Both are straightforward because $\simeq$ is closed under tree-unfolding as well as under removal of states not reachable from the start state. □

6.2 Quotienting Models

The relation $\preceq$ is not a partial order, only a pre-order. For example

$$\mathfrak{M}_1 = ((\{w\}, \emptyset, \{w \mapsto \Sigma\}), w) \qquad \mathfrak{M}_2 = ((\{v\}, \emptyset, \{v \mapsto \Sigma\}), v)$$

are two distinct models with $\mathfrak{M}_1 \preceq \mathfrak{M}_2$ and $\mathfrak{M}_2 \preceq \mathfrak{M}_1$. The difference between the two models, the name of the unique state, is trivial and not relevant for the formulae they make true: $\mathsf{Th}(\mathfrak{M}_1) = \mathsf{Th}(\mathfrak{M}_2)$. As briefly mentioned in the mathematical preliminaries (Sect. 2), we obtain a proper partial-order by simply quotienting models:

$$\mathfrak{M} \simeq \mathfrak{M}' \quad \text{iff} \quad \mathfrak{M} \preceq \mathfrak{M}' \text{ and } \mathfrak{M}' \preceq \mathfrak{M}$$

and then ordering the $\simeq$-equivalence classes as follows:

$$[\mathfrak{M}]_\simeq \preceq [\mathfrak{M}']_\simeq \quad \text{iff} \quad \mathfrak{M} \preceq \mathfrak{M}'.$$

Greatest lower and least upper bounds can also be computed on representatives:

$$\bigsqcup \{[\mathfrak{M}]_\simeq \mid \mathfrak{M} \in S\} \quad = \quad [\bigsqcup S]_\simeq$$

whenever $\bigsqcup S$ exists, and likewise for the greatest lower bound. We also define

$$[\mathfrak{M}]_\simeq \models \phi \quad \text{iff} \quad \mathfrak{M} \models \phi.$$

It is easy to see that these definitions are independent of the chosen representatives.

In the rest of this text we will usually be sloppy and work with concrete models instead of $\simeq$-equivalence classes of models because the quotienting process is straightforward and not especially interesting. We can do this because all relevant subsequent constructions are also representation independent.

6.3 The Bounded Lattice of Models

It turns out that $\preceq$ on ($\simeq$-equivalence classes of) models is not just a partial order, but a bounded lattice, except that a bottom element is missing.

Definition 11. *We extend the collection of models with a single* bottom *element* $\bot$, *where* $\bot \models \phi$ *for all* ϕ. *We also write* $\bot$ *for* $[\bot]_{\simeq}$. *We extend the relation* $\preceq$ *and stipulate that* $\bot \preceq \mathfrak{M}$ *for all models* $\mathfrak{M}$.

Theorem 2. *The collection of (equivalence classes of) models together with* $\bot$, *and ordered by* $\preceq$ *is a bounded lattice.*

Proof. The topmost element in the lattice is the model $((\{w\}, \emptyset, \{w \mapsto \Sigma\}), w)$ (for some state w): this is the model with no transitions and no transition restrictions. The bottom element is $\bot$. Below, we shall define two functions glb and lub, and show that they satisfy the required properties of $\sqcap$ and $\sqcup$ respectively. □

Cathoristic logic is unusual in that every set of models has a unique (up to isomorphism) least upper bound. Logics with disjunction, negation or implication do not have this property.

Consider propositional logic, for example. Define a model of propositional logic as a set of atomic formulae that are set to true. Then we have a natural ordering on propositional logic models:

$$\mathfrak{M} \leq \mathfrak{M}' \quad \text{iff} \quad \mathfrak{M} \supseteq \mathfrak{M}'$$

Consider all the possible models that satisfy $\phi \vee \psi$:

$$\{\phi\} \qquad \{\psi\} \qquad \{\phi, \psi\} \qquad \{\phi, \psi, \xi\} \qquad \cdots$$

This set of satisfying models has no least upper bound, since $\{\phi\} \nleq \{\psi\}$ and $\{\psi\} \nleq \{\phi\}$. Similarly, the set of models satisfying $\neg(\neg\phi \wedge \neg\psi)$ has no least upper bound.

The fact that cathoristic logic models have unique least upper bounds is used in proving completeness of our inference rules, and implementing the quadratic-time decision procedure.

6.4 Computing the Least Upper Bound of the Models that Satisfy a Formula

In our decision procedure, we will see if $\phi \models \psi$ by constructing the least upper bound of the models satisfying ϕ, and checking whether it satisfies ψ.

In this section, we define a function $\mathsf{simpl}(\phi)$ that satisfies the following condition:

$$\mathsf{simpl}(\phi) = \bigsqcup\{\mathfrak{M} | \mathfrak{M} \models \phi\}$$

Define $\mathsf{simpl}(\phi)$ as:

$$\begin{aligned}
\mathsf{simpl}(\top) &= ((\{v\}, \emptyset, \{v \mapsto \Sigma\}), v) \\
\mathsf{simpl}(!A) &= ((\{v\}, \emptyset, \{v \mapsto A\}), v) \\
\mathsf{simpl}(\phi_1 \wedge \phi_2) &= \mathsf{glb}(\mathsf{simpl}(\phi_1), \mathsf{simpl}(\phi_2)) \\
\mathsf{simpl}(\langle a \rangle \phi) &= ((S \cup \{w'\}, \rightarrow \cup (w' \xrightarrow{a} w), \lambda \cup \{w' \mapsto \Sigma\}]), w') \\
& \quad \text{where } \mathsf{simpl}(\phi) = ((S, \rightarrow, \lambda), w) \text{and } w' \text{ is a new state} \\
& \quad \text{not appearing in } S
\end{aligned}$$

Note that, by our conventions, $\mathsf{simpl}(\phi)$ really returns a $\simeq$-equivalence class of models (Figs. 5, 6, 7 and 8).

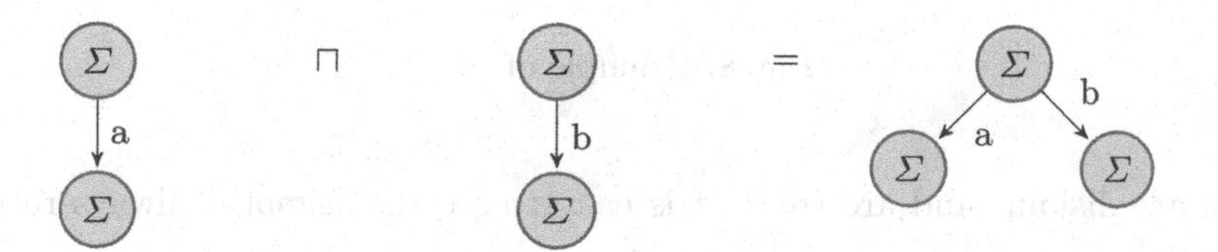

Fig. 5. Example of ⊓.

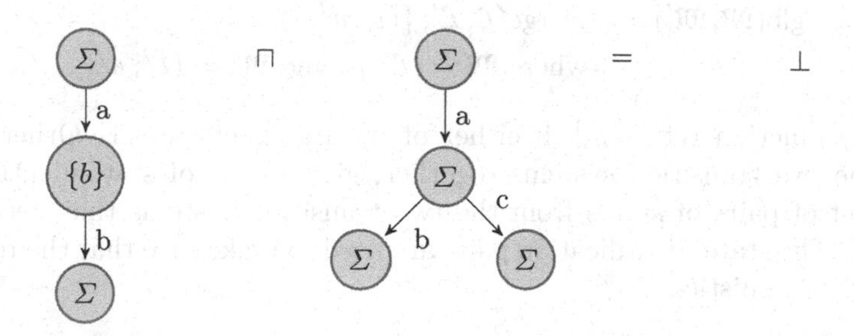

Fig. 6. Example of ⊓.

The only complex case is the clause for $\mathsf{simpl}(\phi_1 \wedge \phi_2)$, which uses the glb function, defined as follows, where we assume that the sets of states in the two

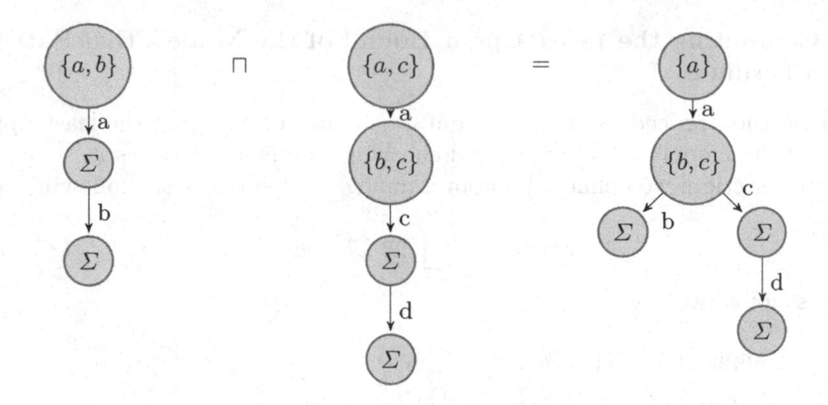

Fig. 7. Example of $\sqcap$.

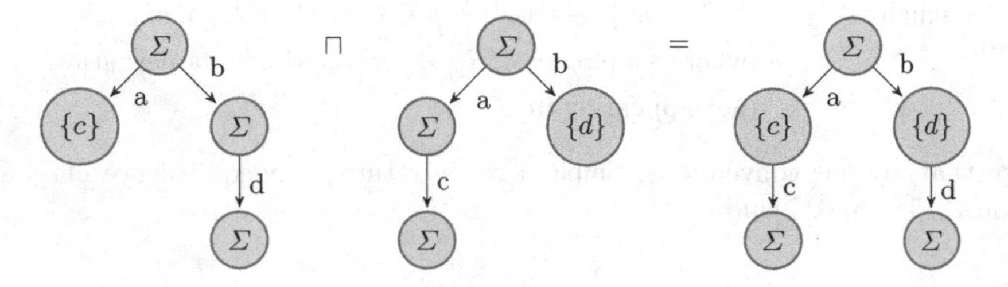

Fig. 8. Example of $\sqcap$.

models are disjoint and are trees. It is easy to see that $\mathsf{simpl}(\cdot)$ always returns tree models.

$$
\begin{aligned}
\mathsf{glb}(\bot, \mathfrak{M}) &= \bot \\
\mathsf{glb}(\mathfrak{M}, \bot) &= \bot \\
\mathsf{glb}(\mathfrak{M}, \mathfrak{M}') &= \mathsf{merge}(\mathcal{L}, \mathcal{L}', \{(w, w')\}) \\
&\quad \text{where } \mathfrak{M} = (\mathcal{L}, w) \text{ and } \mathfrak{M}' = (\mathcal{L}', w')
\end{aligned}
$$

The merge function returns $\bot$ if either of its arguments are $\bot$. Otherwise, it merges the two transition systems together, given a set of state-identification pairs (a set of pairs of states from the two transition systems that need to be identified). The state-identification pairs are used to make sure that the resulting model is deterministic.

$$
\mathsf{merge}(\mathcal{L}, \mathcal{L}', ids) = \begin{cases} \bot & \text{if } \mathsf{inconsistent}(\mathcal{L}, \mathcal{L}', ids) \\ \mathsf{join}(\mathcal{L}, \mathcal{L}') & \text{if } ids = \emptyset \\ \mathsf{merge}(\mathcal{L}, \mathcal{L}'', ids') & \text{else, where } \mathcal{L}'' = \mathsf{applyIds}(ids, \mathcal{L}') \\ & \text{and } ids' = \mathsf{getIds}(\mathcal{L}, \mathcal{L}', ids) \end{cases}
$$

The $\mathsf{inconsistent}$ predicate is true if there is pair of states in the state-identification set such that the out-transitions of one state is incompatible with the state-labelling on the other state:

$$\begin{aligned}&\mathsf{inconsistent}(\mathcal{L}, \mathcal{L}', ids)\\ &\qquad \text{iff } \exists (w, w') \in ids \text{ with } \mathsf{out}(\mathcal{L}, w) \nsubseteq \lambda'(w') \text{ or } \mathsf{out}(\mathcal{L}', w') \nsubseteq \lambda(w).\end{aligned}$$

Here the out function returns all the actions immediately available from the given state w.

$$\mathsf{out}(((S, \rightarrow, \lambda), w)) \quad = \quad \{a \mid \exists w'. w \xrightarrow{a} w'\}$$

The join function takes the union of the two transition systems.

$$\mathsf{join}((S, \rightarrow, \lambda), (S', \rightarrow', \lambda')) \quad = \quad (S \cup S', \rightarrow \cup \rightarrow', \lambda'')$$

Here λ'' takes the constraints arising from both, λ and λ' into account:

$$\lambda''(s) \quad = \quad \begin{array}{l}\{\lambda(s) \cap \lambda'(s) \mid s \in S \cup S'\}\\ \cup \{\lambda(s) \mid s \in S \setminus S'\}\\ \cup \{\lambda(s) \mid s \in S' \setminus S\}.\end{array}$$

The $\mathsf{applyIds}$ function applies all the state-identification pairs as substitutions to the Labelled Transition System:

$$\mathsf{applyIds}(ids, (S, \rightarrow, \lambda)) \quad = \quad (S', \rightarrow', \lambda')$$

where

$$\begin{aligned} S' &= S\,[w/w' \mid (w, w') \in ids]\\ \rightarrow' &= \rightarrow\,[w/w' \mid (w, w') \in ids]\\ \lambda' &= \lambda\,[w/w' \mid (w, w') \in ids]\end{aligned}$$

Here $[w/w' \mid (w, w') \in ids]$ means the simultaneous substitution of w for w' for all pairs (w, w') in ids. The getIds function returns the set of extra state-identification pairs that need to be added to respect determinism:

$$\mathsf{getIds}(\mathcal{L}, \mathcal{L}', ids) \quad = \quad \{(x, x') \mid (w, w') \in ids, \exists a \,.\, w \xrightarrow{a} x, w' \xrightarrow{a} x'\}$$

The function $\mathsf{simpl}(\cdot)$ has the expected properties, as the next lemma shows.

Lemma 2. $\mathsf{simpl}(\phi) \models \phi$.

Proof. By induction on ϕ. □

Lemma 3. glb *as defined is the greatest lower bound*

We will show that:

- $\mathsf{glb}(\mathfrak{M}, \mathfrak{M}') \preceq \mathfrak{M}$ and $\mathsf{glb}(\mathfrak{M}, \mathfrak{M}') \preceq \mathfrak{M}'$
- If $\mathfrak{N} \preceq \mathfrak{M}$ and $\mathfrak{N} \preceq \mathfrak{M}'$, then $\mathfrak{N} \preceq \mathsf{glb}(\mathfrak{M}, \mathfrak{M}')$

If $\mathfrak{M}, \mathfrak{M}'$ or $\mathsf{glb}(\mathfrak{M}, \mathfrak{M}')$ are equal to $\bot$, then we just apply the rule that $\bot \preceq m$ for all models m. So let us assume that $\mathsf{consistent}(\mathfrak{M}, \mathfrak{M}')$ and that $\mathsf{glb}(\mathfrak{M}, \mathfrak{M}') \neq \bot$.

Proof. To show $\mathsf{glb}(\mathfrak{M}, \mathfrak{M}') \preceq \mathfrak{M}$, we need to provide a simulation $\mathcal{R}$ from $\mathfrak{M}$ to $\mathsf{glb}(\mathfrak{M}, \mathfrak{M}')$. If $\mathfrak{M} = ((S, \rightarrow, \lambda), w)$, then define $\mathcal{R}$ as the identity relation on the states of S:

$$\mathcal{R} = \{(x, x) \mid x \in S\}$$

It is straightforward to show that $\mathcal{R}$ as defined is a simulation from $\mathfrak{M}$ to $\mathsf{glb}(\mathfrak{M}, \mathfrak{M}')$. If there is a transition $x \xrightarrow{a} y$ in $\mathfrak{M}$, then by the construction of merge, there is also a transition $x \xrightarrow{a} y$ in $\mathsf{glb}(\mathfrak{M}, \mathfrak{M}')$. We also need to show that $\lambda_{\mathfrak{M}}(x) \supseteq \lambda_{\mathsf{glb}(\mathfrak{M}, \mathfrak{M}')}(x)$ for all states x in $\mathfrak{M}$. This is immediate from the construction of merge. □

Proof. To show that $\mathfrak{N} \preceq \mathfrak{M}$ and $\mathfrak{N} \preceq \mathfrak{M}'$ imply $\mathfrak{N} \preceq \mathsf{glb}(\mathfrak{M}, \mathfrak{M}')$, assume there is a simulation $\mathcal{R}$ from $\mathfrak{M}$ to $\mathfrak{N}$ and there is a simulation $\mathcal{R}'$ from $\mathfrak{M}'$ to $\mathfrak{N}$. We need to provide a simulation $\mathcal{R}*$ from $\mathsf{glb}(\mathfrak{M}, \mathfrak{M}')$ to $\mathfrak{N}$.

Assume the states of $\mathfrak{M}$ and $\mathfrak{M}'$ are disjoint. Define:

$$\mathcal{R}* = \mathcal{R} \cup \mathcal{R}'$$

We need to show that $\mathcal{R}*$ as defined is a simulation from $\mathsf{glb}(\mathfrak{M}, \mathfrak{M}')$ to $\mathfrak{N}$.

Suppose $x \xrightarrow{a} y$ in $\mathsf{glb}(\mathfrak{M}, \mathfrak{M}')$ and that $(x, x_2) \in \mathcal{R} \cup \mathcal{R}'$. We need to provide a y_2 such that $x_2 \xrightarrow{a} y_2$ in $\mathfrak{N}$ and $(y, y_2) \in \mathcal{R} \cup \mathcal{R}'$. If $x \xrightarrow{a} y$ in $\mathsf{glb}(\mathfrak{M}, \mathfrak{M}')$, then, from the definition of merge, either $x \xrightarrow{a} y$ in $\mathfrak{M}$ or $x \xrightarrow{a} y$ in $\mathfrak{M}'$. If the former, and given that $\mathcal{R}$ is a simulation from $\mathfrak{M}$ to $\mathfrak{N}$, then there is a y_2 such that $(y, y_2) \in \mathcal{R}$ and $x_2 \xrightarrow{a} y_2$ in $\mathfrak{N}$. But, if $(y, y_2) \in \mathcal{R}$, then also $(y, y_2) \in \mathcal{R} \cup \mathcal{R}'$.

Finally, we need to show that if $(x, y) \in \mathcal{R} \cup \mathcal{R}'$ then

$$\lambda_{\mathsf{glb}(\mathfrak{M}, \mathfrak{M}')}(x) \supseteq \lambda_{\mathfrak{N}}(y)$$

If $(x, y) \in \mathcal{R} \cup \mathcal{R}'$ then either $(x, y) \in \mathcal{R}$ or $(x, y) \in \mathcal{R}'$. Assume the former. Given that $\mathcal{R}$ is a simulation from $\mathfrak{M}$ to $\mathfrak{N}$, we know that if $(x, y) \in \mathcal{R}$, then

$$\lambda_{\mathfrak{M}}(x) \supseteq \lambda_{\mathfrak{N}}(y)$$

Let $\mathfrak{M} = ((S, \rightarrow, \lambda), w)$. If $x \neq w$ (i.e. x is some state other than the start state), then, from the definition of merge, $\lambda_{\mathsf{glb}(\mathfrak{M}, \mathfrak{M}')}(x) = \lambda_{\mathfrak{M}}(x)$. So, given $\lambda_{\mathfrak{M}} \supseteq \lambda_{\mathfrak{N}}(y)$, $\lambda_{\mathsf{glb}(\mathfrak{M}, \mathfrak{M}')}(x) \supseteq \lambda_{\mathfrak{N}}(y)$. If, on the other hand, $x = w$ (i.e. x is the start state of our cathoristic model $\mathfrak{M}$), then, from the definition of merge:

$$\lambda_{\mathsf{glb}(\mathfrak{M}, \mathfrak{M}')}(w) = \lambda_{\mathfrak{M}}(w) \cap \lambda_{\mathfrak{M}'}(w')$$

where w' is the start state of $\mathfrak{M}'$. In this case, given $\lambda_{\mathfrak{M}}(w) \supseteq \lambda_{\mathfrak{N}}(y)$ and $\lambda_{\mathfrak{M}'}(w') \supseteq \lambda_{\mathfrak{N}}(y)$, it follows that $\lambda_{\mathfrak{M}}(w) \cap \lambda_{\mathfrak{M}'}(w') \supseteq \lambda_{\mathfrak{N}}(y)$ and hence

$$\lambda_{\mathsf{glb}(\mathfrak{M}, \mathfrak{M}')}(w) \supseteq \lambda_{\mathfrak{N}}(y)$$

□

Next, define the least upper bound (lub) of two models as:

$$\begin{aligned}\mathsf{lub}(\mathfrak{M}, \bot) &= \mathfrak{M} \\ \mathsf{lub}(\bot, \mathfrak{M}) &= \mathfrak{M} \\ \mathsf{lub}((\mathcal{L}, w), (\mathcal{L}', w')) &= \mathsf{lub}_2(\mathcal{L}, \mathcal{L}', (\mathfrak{M}_\top, z), \{(w, w', z)\})\end{aligned}$$

where $\mathfrak{M}_\top$ is the topmost model ($\mathcal{W} = \{z\}, \rightarrow = \emptyset, \lambda = \{z \mapsto \Sigma\}$) for some state z. lub_2 takes four parameters: the two cathoristic transition systems $\mathcal{L}$ and $\mathcal{L}'$, an accumulator representing the constructed result so far, and a list of state triples (each triple contains one state from each of the two input models plus the state of the accumulated result) to consider next. It is defined as:

$$\begin{aligned}\mathsf{lub}_2(\mathcal{L}, \mathcal{L}', \mathfrak{M}, \emptyset) &= \mathfrak{M} \\ \mathsf{lub}_2(\mathcal{L}, \mathcal{L}', ((\mathcal{W}, \rightarrow, \lambda), y), \{(w, w', x)\} \cup R) &= \mathsf{lub}_2(\mathcal{L}, \mathcal{L}', ((\mathcal{W} \cup \mathcal{W}', \rightarrow \cup \rightarrow', \lambda'), y), R' \cup R\}\end{aligned}$$

where:

$$\begin{aligned}\{(a_i, w_i, w_i') \mid i = 1...n\} &= \mathsf{sharedT}((\mathcal{L}, w), (\mathcal{L}', w')) \\ \mathcal{W}' &= \{x_i \mid i = 1...n\} \\ \rightarrow' &= \{(x, a_i, x_i) \mid i = 1...n\} \\ \lambda' &= \lambda[x \mapsto \lambda(w) \cup \lambda(w)'] \\ R' &= \{(w_i, w_i', x_i) \mid i = 1...n\}\end{aligned}$$

Here $\lambda[x \mapsto S]$ is the state labelling function that is exactly like λ, except that it maps x to S. Moreover, $\mathsf{sharedT}$ returns the shared transitions between two models, and is defined as:

$$\mathsf{sharedT}(((\mathcal{W}, \rightarrow, \lambda), w)((\mathcal{W}', \rightarrow', \lambda'), w')) = \{(a, x, x') \mid w \xrightarrow{a} x \land w' \xrightarrow{a}{}' x'\}$$

If $((S*, \rightarrow *, \lambda*), w*) = ((S, \rightarrow, \lambda), w) \sqcup ((S', \rightarrow', \lambda'), w')$ then define the set $\mathsf{triples}_{\mathsf{lub}}$ as the set of triples $(x, x', x*) \mid x \in S, x' \in S', x* \in S*$ that were used during the construction of lub above. So $\mathsf{triples}_{\mathsf{lub}}$ stores the associations between states in $\mathfrak{M}$, $\mathfrak{M}'$ and $\mathfrak{M} \sqcup \mathfrak{M}'$ (Figs. 9, 10 and 11).

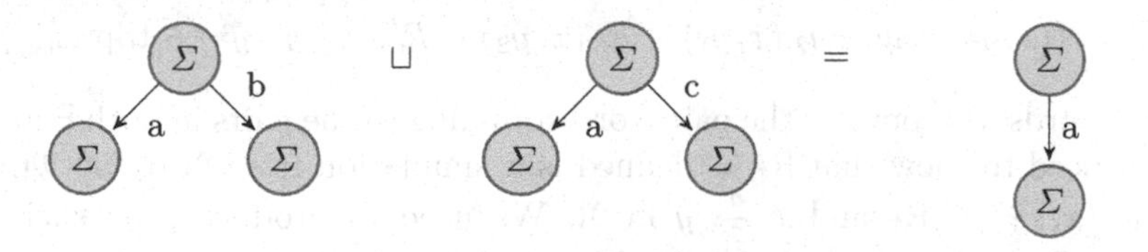

Fig. 9. Example of $\sqcup$

Lemma 4. lub *as defined is the least upper bound*

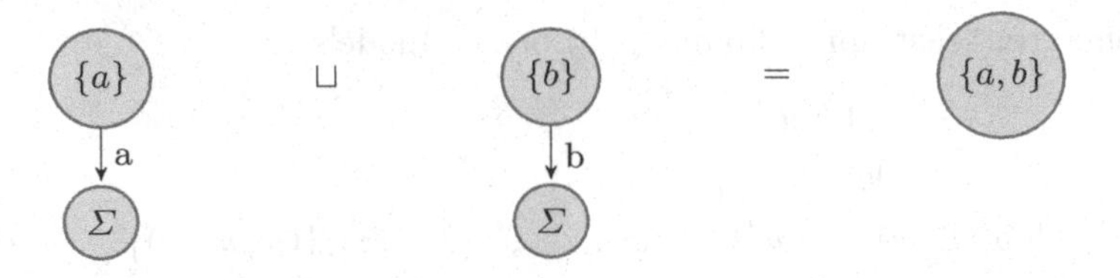

Fig. 10. Example of $\sqcup$

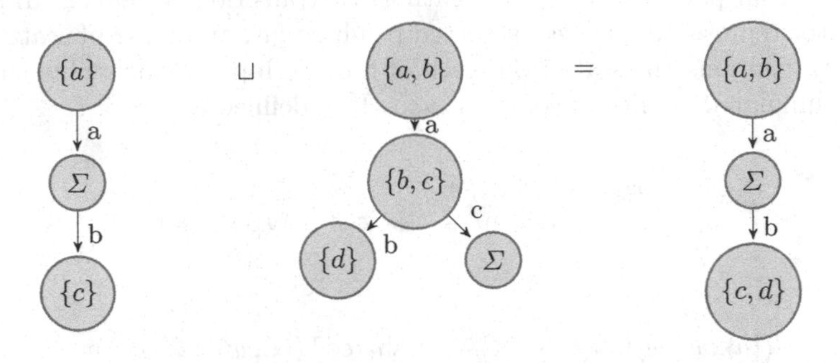

Fig. 11. Example of $\sqcup$

We will show that:

- $\mathfrak{M} \preceq \mathsf{lub}(\mathfrak{M}, \mathfrak{M}')$ and $\mathfrak{M}' \preceq \mathsf{lub}(\mathfrak{M}, \mathfrak{M}')$
- If $\mathfrak{M} \preceq \mathfrak{N}$ and $\mathfrak{M}' \preceq \mathfrak{N}$, then $\mathsf{lub}(\mathfrak{M}, \mathfrak{M}') \preceq \mathfrak{N}$

If $\mathfrak{M}$ or $\mathfrak{M}'$ are equal to $\bot$, then we just apply the rule that $\bot \preceq m$ for all models m. So let us assume that neither $\mathfrak{M}$ not $\mathfrak{M}'$ are $\bot$.

Proof. To see that $\mathfrak{M} \preceq \mathsf{lub}(\mathfrak{M}, \mathfrak{M}')$, observe that, by construction of lub above, every transition in $\mathsf{lub}(\mathfrak{M}, \mathfrak{M}')$ has a matching transition in $\mathfrak{M}$, and every state label in $\mathsf{lub}(\mathfrak{M}, \mathfrak{M}')$ is a superset of the corresponding state label in $\mathfrak{M}$.

To show that $\mathfrak{M} \preceq \mathfrak{N}$ and $\mathfrak{M}' \preceq \mathfrak{N}$ together imply $\mathsf{lub}(\mathfrak{M}, \mathfrak{M}') \preceq \mathfrak{N}$, assume a simulation $\mathcal{R}$ from $\mathfrak{N}$ to $\mathfrak{M}$ and a simulation $\mathcal{R}'$ from $\mathfrak{N}$ to $\mathfrak{M}'$. We need to produce a simulation relation $\mathcal{R}*$ from $\mathfrak{N}$ to $\mathsf{lub}(\mathfrak{M}, \mathfrak{M}')$. Define

$$\mathcal{R}* = \{(x, y*) \mid \exists y_1 . \exists y_2 . (x, y_1) \in \mathcal{R}, (x, y_2) \in \mathcal{R}', (y_1, y_2, y*) \in \mathsf{triples}_{\mathsf{lub}}\}$$

In other words, $\mathsf{R}*$ contains the pairs corresponding to the pairs in both R and R'. We just need to show that $\mathsf{R}*$ as defined is a simulation from $\mathfrak{N}$ to $\mathsf{lub}(\mathfrak{M}, \mathfrak{M}')$. Assume $(x, x*) \in \mathsf{R}*$ and $x \xrightarrow{a} y$ in $\mathfrak{N}$. We need to produce a $y*$ such that $(x*, y*) \in \mathsf{R}*$ and $x* \xrightarrow{a} y*$ in $\mathsf{lub}(\mathfrak{M}, \mathfrak{M}')$. Given that $\mathcal{R}$ is a simulation from $\mathfrak{N}$ to $\mathfrak{M}$, and that $\mathcal{R}'$ is a simulation from $\mathfrak{N}$ to $\mathfrak{M}'$, we know that there is a pair of states x_1, y_1 in $\mathfrak{M}$ and a pair of states x_2, y_2 in $\mathfrak{M}'$ such that $(x, x_1) \in \mathsf{R}$ and $(x, x_2) \in \mathsf{R}'$ and $x_1 \xrightarrow{a} y_1$ in $\mathfrak{M}$ and $x_2 \xrightarrow{a} y_2$ in $\mathfrak{M}'$. Now, from the construction of lub above, there is a triple $(y_1, y_2, y*) \in \mathsf{triples}_{\mathsf{lub}}$. Now, from the construction of $\mathsf{R}*$ above, $(x*, y*) \in \mathsf{R}*$.

Finally, we need to show that for all states x and y, if $(x, y) \in \mathsf{R}*, \lambda_{\mathfrak{N}}(x) \supseteq \lambda_{\mathsf{lub}(\mathfrak{M},\mathfrak{M}')}(y)$. Given that $\mathcal{R}$ is a simulation from $\mathfrak{N}$ to $\mathfrak{M}$, and that $\mathcal{R}'$ is a simulation from $\mathfrak{N}$ to $\mathfrak{M}'$, we know that if $(x, y_1) \in \mathsf{R}$, then $\lambda_{\mathfrak{N}}(x) \supseteq \lambda_{\mathfrak{M}}(y_1)$. Similarly, if $(x, y_2) \in \mathsf{R}$, then $\lambda_{\mathfrak{N}}(x) \supseteq \lambda'_{\mathfrak{M}}(y_2)$. Now, from the construction of lub, $\lambda_{\mathsf{lub}(\mathfrak{M},\mathfrak{M}')}(y*) = \lambda_{\mathfrak{M}}(y_1) \cup \lambda_{\mathfrak{M}}(y_2)$ for all triples $(y_1, y_2, y*) \in \mathsf{triples}_{\mathsf{lub}}$. So $\lambda_{\mathfrak{N}}(x) \supseteq \lambda_{\mathsf{lub}(\mathfrak{M},\mathfrak{M}')}(y)$, as required. □

6.5 A Decision Procedure for cathoristic logic

We use the semantic constructions above to provide a quadratic-time decision procedure. The complexity of the decision procedure is an indication that cathoristic logic is useful as a query language in knowledge representation.

Cathoristic logic's lack of connectives for negation, disjunction or implication is the key reason for the efficiency of the decision procedure. Although any satisfiable formula has an infinite number of models, we have shown that the satisfying models form a bounded lattice with a least upper bound. The $\mathsf{simpl}()$ function defined above gives us the least upper bound of all models satisfying an expression. Using this least upper bound, we can calculate entailment by checking a *single model*. To decide whether $\phi \models \psi$, we use the following algorithm.

1. Compute $\mathsf{simpl}(\phi)$.
2. Check if $\mathsf{simpl}(\phi) \models \psi$.

The correctness of this algorithm is given by the follow theorem.

Theorem 3. *The following are equivalent:*

1. *For all cathoristic models $\mathfrak{M}$, $\mathfrak{M} \models \phi$ implies $\mathfrak{M} \models \psi$.*
2. $\mathsf{simpl}(\phi) \models \psi$.

Proof. The implication from (1) to (2) is trivial because $\mathsf{simpl}(\phi) \models \phi$ by construction.

For the reverse direction, we make use of the following lemma (proved in the Appendix):

Lemma 5. *If $\mathfrak{M} \models \phi$ then $\mathfrak{M} \preceq \mathsf{simpl}(\phi)$.*

With Lemma 5 in hand, the proof of Theorem 3 is straightforward. Assume $\mathfrak{M} \models \phi$. We need to show $\mathfrak{M} \models \psi$. Now if $\mathfrak{M} \models \phi$ then $\mathfrak{M} \preceq \mathsf{simpl}(\phi)$ (by Lemma 5). Further, if $\mathfrak{M}' \models \xi$ and $\mathfrak{M} \preceq \mathfrak{M}'$ then $\mathfrak{M} \models \xi$ by Theorem 1. So, substituting ψ for ξ and $\mathsf{simpl}(\phi)$ for $\mathfrak{M}'$, it follows that $\mathfrak{M} \models \psi$. □

Construction of $\mathsf{simpl}(\phi)$ is quadratic in the size of ϕ, and computing whether a model satisfies ψ is of order $|\psi| \times |\phi|$, so computing whether $\phi \models \psi$ is quadratic time.

6.6 Incompatibility Semantics

One of cathoristic logic's unusual features is that it satisfies Brandom's incompatibility semantics constraint, *even though it has no negation operator*. In this section, we formalise what this means, and prove it.

Define the *incompatibility set of* ϕ as:

$$\mathcal{I}(\phi) = \{\psi \mid \forall \mathfrak{M}.\mathfrak{M} \not\models \phi \land \psi\}$$

The reason why Brandom introduces the incompatibility set [13] is that he wants to use it define *semantic content*:

> Here is a semantic suggestion: represent the propositional content expressed by a sentence with the set of sentences that express propositions incompatible with it[14].

Now if the propositional content of a claim determines its logical consequences, and the propositional content is identified with the incompatibility set, then the incompatibility set must determine the logical consequences. A logic satisfies Brandom's *incompatibility semantics constraint* if

$$\phi \models \psi \quad \text{iff} \quad \mathcal{I}(\psi) \subseteq \mathcal{I}(\phi)$$

Not all logics satisfy this property. Brandom has shown that first-order logic and the modal logic S5 satisfy the incompatibility semantics property. Hennessy-Milner logic satisfies it, but Hennessy-Milner logic without negation does not. Cathoristic logic is the simplest logic we have found that satisfies the property.

To establish the incompatibility semantics constraint for cathoristic logic, we need to define a related incompatibility function on models. $\mathcal{J}(\mathfrak{M})$ is the set of models that are incompatible with $\mathfrak{M}$:

$$\mathcal{J}(\mathfrak{M}) = \{\mathfrak{M}_2 \mid \mathfrak{M} \sqcap \mathfrak{M}_2 = \bot\}$$

We shall make use of two lemmas, proved in Appendix B:

Lemma 6. *If* $\phi \models \psi$ *then* $\mathsf{simpl}(\phi) \preceq \mathsf{simpl}(\psi)$.

Lemma 7. $\mathcal{I}(\psi) \subseteq \mathcal{I}(\phi)$ *implies* $\mathcal{J}(\mathsf{simpl}(\psi)) \subseteq \mathcal{J}(\mathsf{simpl}(\phi))$.

Theorem 4. $\phi \models \psi$ *iff* $\mathcal{I}(\psi) \subseteq \mathcal{I}(\phi)$.

Proof. Left to right: Assume $\phi \models \psi$ and $\xi \in \mathcal{I}(\psi)$. We need to show $\xi \in \mathcal{I}(\phi)$. By the definition of $\mathcal{I}$, if $\xi \in \mathcal{I}(\psi)$ then $\mathsf{simpl}(\xi) \sqcap \mathsf{simpl}(\psi) = \bot$. If $\mathsf{simpl}(\xi) \sqcap \mathsf{simpl}(\psi) = \bot$, then either

[13] Brandom [8] defines incompatibility slightly differently: he defines the set of *sets* of formulae which are incompatible with a *set* of formulae. But in cathoristic logic, if a set of formulae is incompatible, then there is an incompatible subset of that set with exactly two members. So we can work with the simpler definition in the text above.

[14] [8] p. 123.

- $\mathsf{simpl}(\xi) = \bot$
- $\mathsf{simpl}(\psi) = \bot$
- Neither $\mathsf{simpl}(\xi)$ nor $\mathsf{simpl}(\psi)$ are $\bot$, but $\mathsf{simpl}(\xi) \sqcap \mathsf{simpl}(\psi) = \bot$.

If $\mathsf{simpl}(\xi) = \bot$, then $\mathsf{simpl}(\xi) \sqcap \mathsf{simpl}(\phi) = \bot$ and we are done. If $\mathsf{simpl}(\psi) = \bot$, then as $\phi \models \psi$, by Lemma 6, $\mathsf{simpl}(\phi) \preceq \mathsf{simpl}(\psi)$. Now the only model that is $\preceq \bot$ is $\bot$ itself, so $\mathsf{simpl}(\phi) = \bot$. Hence $\mathsf{simpl}(\xi) \sqcap \mathsf{simpl}(\phi) = \bot$, and we are done. The interesting case is when neither $\mathsf{simpl}(\xi)$ nor $\mathsf{simpl}(\psi)$ are $\bot$, but $\mathsf{simpl}(\xi) \sqcap \mathsf{simpl}(\psi) = \bot$. Then (by the definition of $\mathsf{consistent}$ in Sect. 6.4), either $\mathsf{out}(\mathsf{simpl}(\xi)) \nsubseteq \lambda(\mathsf{simpl}(\psi))$ or $\mathsf{out}(\mathsf{simpl}(\psi)) \nsubseteq \lambda(\mathsf{simpl}(\xi))$. In the first sub-case, if $\mathsf{out}(\mathsf{simpl}(\xi)) \nsubseteq \lambda(\mathsf{simpl}(\psi))$, then there is some action a such that $\xi \models \langle a \rangle \top$ and $a \notin \lambda(\mathsf{simpl}(\psi))$. If $a \notin \lambda(\mathsf{simpl}(\psi))$ then $\psi \models !A$ where $a \notin A$. Now $\phi \models \psi$, so $\phi \models !A$. In other words, ϕ also entails the A-restriction that rules out the a transition. So $\mathsf{simpl}(\xi) \sqcap \mathsf{simpl}(\phi) = \bot$ and $\xi \in \mathcal{I}(\phi)$. In the second sub-case, $\mathsf{out}(\mathsf{simpl}(\psi)) \nsubseteq \lambda(\mathsf{simpl}(\xi))$. Then there is some action a such that $\psi \models \langle a \rangle \top$ and $a \notin \lambda(\mathsf{simpl}(\xi))$. If $a \notin \lambda(\mathsf{simpl}(\xi))$ then $\xi \models !A$ where $a \notin A$. But if $\psi \models \langle a \rangle \top$ and $\phi \models \psi$, then $\phi \models \langle a \rangle \top$ and ϕ is also incompatible with ξ's A-restriction. So $\mathsf{simpl}(\xi) \sqcap \mathsf{simpl}(\phi) = \bot$ and $\xi \in \mathcal{I}(\phi)$.

Right to left: assume, for reductio, that $\mathfrak{M} \models \phi$ and $\mathfrak{M} \not\models \psi$. we will show that $\mathcal{I}(\psi) \nsubseteq \mathcal{I}(\phi)$. Assume $\mathfrak{M} \models \phi$ and $\mathfrak{M} \not\models \psi$. We will construct another model $\mathfrak{M}_2$ such that $\mathfrak{M}_2 \in \mathcal{J}(\mathsf{simpl}(\psi))$ but $\mathfrak{M}_2 \notin \mathcal{J}(\mathsf{simpl}(\phi))$. This will entail, via Lemma 7, that $\mathcal{I}(\psi) \nsubseteq \mathcal{I}(\phi)$.

If $\mathfrak{M} \not\models \psi$, then there is a formula ψ' that does not contain $\wedge$ such that $\psi \models \psi'$ and $\mathfrak{M} \not\models \psi'$. ψ' must be either of the form (i) $\langle a_1 \rangle ... \langle a_n \rangle \top$ (for $n > 0$) or (ii) of the form $\langle a_1 \rangle ... \langle a_n \rangle\ !\{A\}$ where $A \subseteq \mathcal{S}$ and $n >= 0$.

In case (i), there must be an i between 0 and n such that $\mathfrak{M} \models \langle a_1 \rangle ... \langle a_i \rangle \top$ but $\mathfrak{M} \not\models \langle a_1 \rangle ... \langle a_{i+1} \rangle \top$. We need to construct another model $\mathfrak{M}_2$ such that $\mathfrak{M}_2 \sqcap \mathsf{simpl}(\psi) = \bot$, but $\mathfrak{M}_2 \sqcap \mathsf{simpl}(\phi) \neq \bot$. Letting $\mathfrak{M} = ((\mathcal{W}, \rightarrow, \lambda), w)$, then $\mathfrak{M} \models \langle a_1 \rangle ... \langle a_i \rangle \top$ implies that there is at least one sequence of states of the form $w, w_1, ..., w_i$ such that $w \xrightarrow{a_1} w_1 \rightarrow ... \xrightarrow{a_i} w_i$. Now let $\mathfrak{M}_2$ be just like $\mathfrak{M}$ but with additional transition-restrictions on each w_i that it not include a_{i+1}. In other words, $\lambda_{\mathfrak{M}_2}(w_i) = \lambda_{\mathfrak{M}}(w_i) - \{a_{i+1}\}$ for all w_i in sequences of the form $w \xrightarrow{a_1} w_1 \rightarrow ... \xrightarrow{a_i} w_i$. Now $\mathfrak{M}_2 \sqcap \mathsf{simpl}(\psi) = \bot$ because of the additional transition restriction we added to $\mathfrak{M}_2$, which rules out $\langle a_1 \rangle ... \langle a_{i+1} \rangle \top$, and a-fortiori ψ. But $\mathfrak{M}_2 \sqcap \mathsf{simpl}(\phi) \neq \bot$, because $\mathfrak{M} \models \phi$ and $\mathfrak{M}_2 \preceq \mathfrak{M}$ together imply $\mathfrak{M}_2 \models \phi$. So $\mathfrak{M}_2$ is indeed the model we were looking for, that is incompatible with $\mathsf{simpl}(\psi)$ while being compatible with $\mathsf{simpl}(\phi)$.

In case (ii), $\mathfrak{M} \models \langle a_1 \rangle ... \langle a_n \rangle \top$ but $\mathfrak{M} \not\models \langle a_1 \rangle ... \langle a_n \rangle !A$ for some $A \subset \mathcal{S}$. We need to produce a model $\mathfrak{M}_2$ that is incompatible with $\mathsf{simpl}(\psi)$ but not with $\mathsf{simpl}(\phi)$. Given that $\mathfrak{M} \models \langle a_1 \rangle ... \langle a_n \rangle \top$, there is a sequence of states $w, w_1, ..., w_n$ such that $w \xrightarrow{a_1} w_1 \rightarrow ... \xrightarrow{a_i} w_n$. Let $\mathfrak{M}_2$ be the model just like $\mathfrak{M}$ except it has an additional transition from each such w_n with an action $a \notin A$. Clearly, $\mathfrak{M}_2 \sqcap \mathsf{simpl}(\psi') = \bot$ because of the additional a-transition, and given that $\psi \models \psi'$, it follows that $\mathfrak{M}_2 \sqcap \mathsf{simpl}(\psi) = \bot$. Also, $\mathfrak{M}_2 \sqcap \mathsf{simpl}(\phi) \neq \bot$, because $\mathfrak{M}_2 \preceq \mathfrak{M}$ and $\mathfrak{M} \models \phi$. □

7 Inference Rules

We now present the inference rules for cathoristic logic. There are no axioms.

$$\frac{-}{\phi \vdash \phi}\ \textsc{Id} \qquad \frac{-}{\phi \vdash \top}\ \top\textsc{-Right} \qquad \frac{-}{\bot \vdash \phi}\ \bot\textsc{-Left} \qquad \frac{\phi \vdash \psi \quad \psi \vdash \xi}{\phi \vdash \xi}\ \textsc{Trans}$$

$$\frac{\phi \vdash \psi}{\phi \land \xi \vdash \psi}\ \land\textsc{-Left 1} \qquad \frac{\phi \vdash \psi}{\xi \land \phi \vdash \psi}\ \land\textsc{-Left 2} \qquad \frac{\phi \vdash \psi \quad \phi \vdash \xi}{\phi \vdash \psi \land \xi}\ \land\textsc{-Right}$$

$$\frac{a \notin A}{!A \land \langle a \rangle \phi \vdash \bot}\ \bot\textsc{-Right 1} \qquad \frac{-}{\langle a \rangle \bot \vdash \bot}\ \bot\textsc{-Right 2} \qquad \frac{\phi \vdash !A \quad A \subseteq A'}{\phi \vdash !A'}\ !\textsc{-Right 1}$$

$$\frac{\phi \vdash !A \quad \phi \vdash !B}{\phi \vdash !(A \cap B)}\ !\textsc{-Right 2} \qquad \frac{\phi \vdash \psi}{\langle a \rangle \phi \vdash \langle a \rangle \psi}\ \textsc{Normal} \qquad \frac{\phi \vdash \langle a \rangle \psi \land \langle a \rangle \xi}{\phi \vdash \langle a \rangle (\psi \land \xi)}\ \textsc{Det}$$

Fig. 12. Proof rules.

Definition 12. *Judgements are of the following form.*

$$\phi \vdash \psi.$$

We also write $\vdash \phi$ *as a shorthand for* $\top \vdash \phi$. *Figure 12 presents all proof rules.*

Note that ϕ and ψ are single formulae, not sequents. By using single formulae, we can avoid structural inference rules. The proof rules can be grouped in two parts: standard rules and rules unique to cathoristic logic. Standard rules are [Id], [⊤-Right], [⊥-Left], [Trans], [∧-Left 1], [∧-Left 2] and [∧-Right]. They hardly need explanation as they are variants of familiar rules for propositional logic, see e.g. [30,32]. We now explain the rules that givc cathoristic logic its distinctive properties.

The rule [⊥-Right 1] captures the core exclusion property of the tantum !: for example if $A = \{male, female\}$ then $\langle orange \rangle \phi$ is incompatible with $!A$. Thus $!A \land \langle orange \rangle \phi$ must be false.

The rule [⊥-Right 2] expresses that falsity is 'global' and cannot be suppressed by the modalities. For example $\langle orange \rangle \bot$ is false, simply because $\bot$ is already false.

[Normal] enables us to prefix an inference with a may-modality. This rule can also be stated in the following more general form:

$$\frac{\phi_1 \land ... \land \phi_n \vdash \psi}{\langle a \rangle \phi_1 \land ... \land \langle a \rangle \phi_n \vdash \langle a \rangle \psi}\ \textsc{Normal-Multi}$$

But it is not necessary because [Normal-Multi] is derivable from [Normal] as we show in the examples below.

7.1 Example Inferences

We prove that we can use $\phi \wedge \psi \vdash \xi$ to derive $\langle a \rangle \phi \wedge \langle a \rangle \psi \vdash \langle a \rangle \xi$:

$$\dfrac{\dfrac{\langle a \rangle \phi \wedge \langle a \rangle \psi \vdash \langle a \rangle \phi \wedge \langle a \rangle \psi}{\langle a \rangle \phi \wedge \langle a \rangle \psi \vdash \langle a \rangle (\phi \wedge \psi)}\ \textsc{Det} \qquad \dfrac{\phi \wedge \psi \vdash \xi}{\langle a \rangle (\phi \wedge \psi) \vdash \langle a \rangle \xi}\ \textsc{Normal}}{\langle a \rangle \phi \wedge \langle a \rangle \psi \vdash \langle a \rangle \xi}\ \textsc{Trans}$$

Figure 13 demonstrates how to infer $\langle a \rangle !\{b, c\} \wedge \langle a \rangle !\{c, d\} \vdash \langle a \rangle !\{c\}$ and $\langle a \rangle !\{b\} \wedge \langle a \rangle \langle c \rangle \top \vdash \langle d \rangle \top$.

7.2 !-Left and !-Right

The rules [!-Right 1, !-Right 2] jointly express how the subset relation $\subseteq$ on sets of actions relates to provability. Why don't we need a corresponding rule !-Left for strengthening ! on the left hand side?

$$\dfrac{\phi \wedge !A \vdash \psi \qquad A' \subseteq A}{\phi \wedge !A' \vdash \psi}\ \textsc{!-Left}$$

The reason is that [!-Left] can be derived as follows.

$$\dfrac{\dfrac{\phi \wedge !A' \vdash \phi \wedge !A' \qquad A' \subseteq A}{\phi \wedge !A' \vdash \phi \wedge !A}\ \textsc{!-Right 1} \qquad \phi \wedge !A \vdash \psi}{\phi \wedge !A' \vdash \psi}\ \textsc{Trans}$$

Readers familiar with object-oriented programming will recognise [!-Left] as contravariant and [!-Right 1] as covariant subtyping. Honda [18] develops a full theory of subtyping based on similar ideas. All three rules embody the intuition that whenever $A \subseteq A'$ then asserting that $!A'$ is as strong as, or a stronger statement than $!A$. [!-Left] simply states that we can always strengthen our premise, while [!-right 1] allows us to weaken the conclusion.

7.3 Characteristic Formulae

In order to prove completeness, below, we need the notion of a *characteristic formula* of a model. The function $\mathsf{simpl}(\cdot)$ takes a formula as argument and returns the least upper bound of the satisfying models. Characteristic formulae go the other way: given a model $\mathfrak{M}$, $\mathsf{char}(\mathfrak{M})$ is the logically weakest formula that describes that model.

Definition 13. *Let $\mathfrak{M}$ be a cathoristic model that is a tree.*

$$\begin{aligned} \mathsf{char}(\bot) &= \langle a \rangle \top \wedge !\emptyset \text{ for some fixed action } a \in \Sigma \\ \mathsf{char}(\mathfrak{M}, w) &= \mathsf{bang}(\mathfrak{M}, w) \wedge \bigwedge_{w \xrightarrow{a} w'} \langle a \rangle \mathsf{char}(\mathfrak{M}, w') \end{aligned}$$

$$\dfrac{\dfrac{\dfrac{\dfrac{!\{b,c\} \vdash !\{b,c\}}{!\{b,c\} \wedge !\{c,d\} \vdash !\{b,c\}}\ \wedge\,\textsc{Left 1} \qquad \dfrac{!\{c,d\} \vdash !\{c,d\}}{!\{b,c\} \wedge !\{c,d\} \vdash !\{c,d\}}\ \wedge\,\textsc{Left 2}}{!\{b,c\} \wedge !\{c,d\} \vdash !\{c\}}\ !\,\textsc{Right 2}}{\langle a\rangle(!\{b,c\} \wedge !\{c,d\}) \vdash \langle a\rangle !\{c\}}\ \textsc{Normal} \qquad \dfrac{\langle a\rangle !\{b,c\} \wedge \langle a\rangle !\{c,d\} \vdash \langle a\rangle !\{b,c\} \wedge \langle a\rangle !\{c,d\}}{\langle a\rangle !\{b,c\} \wedge \langle a\rangle !\{c,d\} \vdash \langle a\rangle(!\{b,c\} \wedge !\{c,d\})}\ \textsc{Det}}{\langle a\rangle !\{b,c\} \wedge \langle a\rangle !\{c,d\} \vdash \langle a\rangle !\{c\}}\ \textsc{Trans}$$

$$\dfrac{\dfrac{\dfrac{\dfrac{!\{b\} \wedge \langle c\rangle\top \vdash \bot}{\langle a\rangle(!\{b\} \wedge \langle c\rangle\top) \vdash \langle a\rangle\bot}\ \textsc{Normal} \qquad \dfrac{\langle a\rangle !\{b\} \wedge \langle a\rangle\langle c\rangle\top \vdash \langle a\rangle !\{b\} \wedge \langle a\rangle\langle c\rangle\top}{\langle a\rangle !\{b\} \wedge \langle a\rangle\langle c\rangle\top \vdash \langle a\rangle(!\{b\} \wedge \langle c\rangle\top)}\ \textsc{Det}}{\langle a\rangle !\{b\} \wedge \langle a\rangle\langle c\rangle\top \vdash \langle a\rangle\bot}\ \textsc{Trans} \qquad \langle a\rangle\bot \vdash \bot}{\langle a\rangle !\{b\} \wedge \langle a\rangle\langle c\rangle\top \vdash \bot}\ \textsc{Trans} \qquad \bot \vdash \langle d\rangle\top}{\langle a\rangle !\{b\} \wedge \langle a\rangle\langle c\rangle\top \vdash \langle d\rangle\top}\ \textsc{Trans}$$

Fig. 13. Derivations of $\langle a\rangle !\{b,c\} \wedge \langle a\rangle !\{c,d\} \vdash \langle a\rangle !\{c\}$ (top) and $\langle a\rangle !\{b\} \wedge \langle a\rangle\langle c\rangle\top \vdash \langle d\rangle\top$ (bottom).

Note that $\bot$ requires a particular action $a \in \Sigma$. This is why we required, in Sect. 3.1, that Σ is non-empty.

The functions $\mathsf{bang}(\cdot)$ on models are given by the following clauses.

$$\mathsf{bang}((S, \rightarrow, \lambda), w) \;=\; \begin{cases} \top & \text{if } \lambda(w) = \Sigma \\ !\,\lambda(w) & \text{otherwise} \end{cases}$$

Note that $\mathsf{char}(\mathfrak{M})$ is finite if $\mathfrak{M}$ contains no cycles and if $\lambda(x)$ is either Σ or finite for all states x. We state without proof that $\mathsf{simpl}(\cdot)$ and $\mathsf{char}(\cdot)$ are inverses of each other (for tree models $\mathfrak{M}$) in that:

- $\mathsf{simpl}(\mathsf{char}(\mathfrak{M})) \;\simeq\; \mathfrak{M}$.
- $\models \mathsf{char}(\mathsf{simpl}(\phi))$ iff $\models \phi$.

7.4 Soundness and Completeness

Theorem 5. *The rules in Fig. 12 are sound and complete:*

1. *(Soundness)* $\phi \vdash \psi$ *implies* $\phi \models \psi$.
2. *(Completeness)* $\phi \models \psi$ *implies* $\phi \vdash \psi$.

Soundness is immediate from the definitions. To prove completeness we will show that $\phi \models \psi$ implies there is a derivation of $\phi \vdash \psi$. Our proof will make use of two key facts (proved in Sect. 7.5 below):

Lemma 8. *If* $\mathfrak{M} \models \phi$ *then* $\mathsf{char}(\mathfrak{M}) \vdash \phi$.

Lemma 9. *For all formulae* ϕ, *we can derive* $\phi \vdash \mathsf{char}(\mathsf{simpl}(\phi))$.

Lemma 8 states that, if ϕ is satisfied by a model, then there is a proof that the characteristic formula describing that model entails ϕ. In Lemma 9, $\mathsf{simpl}(\phi)$ is the simplest model satisfying ϕ, and $\mathsf{char}(\mathfrak{M})$ is the simplest formula describing m, so $\mathsf{char}(\mathsf{simpl}(\phi))$ is a simplified form of ϕ. This lemma states that cathoristic logic has the inferential capacity to transform any proposition into its simplified form.

With these two lemmas in hand, the proof of completeness is straightforward. Assume $\phi \models \psi$. Then all models which satisfy ϕ also satisfy ψ. In particular, $\mathsf{simpl}(\phi) \models \psi$. Then $\mathsf{char}(\mathsf{simpl}(\phi)) \vdash \psi$ by Lemma 8. But we also have, by Lemma 9, $\phi \vdash \mathsf{char}(\mathsf{simpl}(\phi))$. So by transitivity, we have $\phi \vdash \psi$.

7.5 Proofs of Lemmas 8, 9 and 10

Proof of Lemma 8. If $\mathfrak{M} \models \phi$ then $\mathsf{char}(\mathfrak{M}) \vdash \phi$.
We proceed by induction on ϕ.

Case ϕ is $\top$. Then we can prove $\mathsf{char}(\mathfrak{M}) \vdash \phi$ immediately using axiom [$\top$ RIGHT.

Case ϕ is $\psi \wedge \psi'$. By the induction hypothesis, $\mathsf{char}(\mathfrak{M}) \vdash \psi$ and $\mathsf{char}(\mathfrak{M}) \vdash \psi'$. The proof of $\mathsf{char}(\mathfrak{M}) \vdash \psi \wedge \psi'$ follows immediately using [$\wedge$ Right].

Case ϕ is $\langle\ a\ \rangle\psi$. If $\mathfrak{M} \models \langle a \rangle \psi$, then either $\mathfrak{M} = \bot$ or $\mathfrak{M}$ is a model of the form $(\mathcal{L}, w)$.

Subcase $\mathfrak{M} = \bot$. In this case, $\mathsf{char}(\mathfrak{M}) = \mathsf{char}(\bot) = \bot$. (Recall, that we are overloading $\bot$ to mean both the model at the bottom of our lattice and a formula (such as $\langle a \rangle \top \wedge !\emptyset$) which is always false). In this case, $\mathsf{char}(\bot) \vdash \langle a \rangle \psi$ using [$\bot$ Left].

Subcase m is a model of the form $(\mathcal{L}, w)$. Given $\mathfrak{M} \models \langle a \rangle \psi$, and that $\mathfrak{M}$ is a model of the form $(\mathcal{L}, w)$, we know that:

$$(\mathcal{L}, w) \models \langle a \rangle \psi$$

From the satisfaction clause for $\langle a \rangle$, it follows that:

$$\exists w' \text{ such that } w \xrightarrow{a} w' \text{ and } (\mathcal{L}, w') \models \psi$$

By the induction hypothesis:

$$\mathsf{char}((\mathcal{L}, w')) \vdash \psi$$

Now by [Normal]:

$$\langle a \rangle \mathsf{char}((\mathcal{L}, w')) \vdash \langle a \rangle \psi$$

Using repeated application of [$\wedge$ Left], we can show:

$$\mathsf{char}((\mathcal{L}, w)) \vdash \langle a \rangle \mathsf{char}((\mathcal{L}, w'))$$

Finally, using [Trans], we derive:

$$\mathsf{char}((\mathcal{L}, w)) \vdash \langle a \rangle \psi$$

Case ϕ is $!\psi$. If $(\mathcal{L}, w) \models !A$, then $\lambda(w) \subseteq A$. Then $\mathsf{char}((\mathcal{L}, w)) = !\ \lambda(w) \wedge \phi$. Now we can prove $!\ \lambda(w) \wedge \phi \vdash !A$ using [! Right 1] and repeated applications of [$\wedge$ Left].

Proof of Lemma 9. Now we prove Lemma 9: for all formulae ϕ, we can derive $\phi \vdash \mathsf{char}(\mathsf{simpl}(\phi))$.

Proof. Induction on ϕ.

Case ϕ is $\top$. Then we can prove $\top \vdash \top$ using either [$\top$ Right] or [Id].

Case ϕ is $\psi \wedge \psi'$. By the induction hypothesis, $\psi \vdash \mathsf{char}(\mathsf{simpl}(\psi))$ and $\psi' \vdash \mathsf{char}(\mathsf{simpl}(\psi'))$. Using [$\wedge$ Left] and [$\wedge$ Right], we can show:

$$\psi \wedge \psi' \vdash \mathsf{char}(\mathsf{simpl}(\psi)) \wedge \mathsf{char}(\mathsf{simpl}(\psi'))$$

In order to continue the proof, we need the following lemma, proven in the next subsection.

Lemma 10. *For all cathoristic models $\mathfrak{M}$ and $\mathfrak{M}_2$ that are trees,* $\mathsf{char}(\mathfrak{M}) \wedge \mathsf{char}(\mathfrak{M}_2) \vdash \mathsf{char}(\mathfrak{M} \sqcap \mathfrak{M}_2)$.

From Lemma 10 (substituting $\mathsf{simpl}(\psi)$ for $\mathfrak{M}$ and $\mathsf{simpl}(\psi')$ for $\mathfrak{M}_2$, and noting that $\mathsf{simpl}()$ always produces acyclic models), it follows that:

$$\mathsf{char}(\mathsf{simpl}(\psi)) \wedge \mathsf{char}(\mathsf{simpl}(\psi')) \vdash \mathsf{char}(\mathsf{simpl}(\psi \wedge \psi'))$$

Our desired result follows using [TRANS].

Case ϕ is $\langle a \rangle \psi$. By the induction hypothesis, $\psi \vdash \mathsf{char}(\mathsf{simpl}(\psi))$. Now there are two sub-cases to consider, depending on whether or not $\mathsf{char}(\mathsf{simpl}(\psi)) = \bot$.

Subcase $\mathsf{char}(\mathsf{simpl}(\psi)) = \bot$. In this case, $\mathsf{char}(\mathsf{simpl}(\langle a \rangle \psi))$ also equals $\bot$. By the induction hypothesis:

$$\psi \vdash \bot$$

By [NORMAL]:

$$\langle a \rangle \psi \vdash \langle a \rangle \bot$$

By [$\bot$ RIGHT 2]:

$$\langle a \rangle \bot \vdash \bot$$

The desired proof that:

$$\langle a \rangle \psi \vdash \bot$$

follows by [TRANS].

Subcase $\mathsf{char}(\mathsf{simpl}(\psi)) \neq \bot$. By the induction hypothesis, $\psi \vdash \mathsf{char}(\mathsf{simpl}(\psi))$. So, by [NORMAL]:

$$\langle a \rangle \psi \vdash \langle a \rangle \mathsf{char}(\mathsf{simpl}(\psi))$$

The desired conclusion follows from noting that:

$$\langle a \rangle \mathsf{char}(\mathsf{simpl}(\psi)) = \mathsf{char}(\mathsf{simpl}(\langle a \rangle \psi))$$

Case ϕ is $!A$. If ϕ is $!A$, then $\mathsf{char}(\mathsf{simpl}(\phi))$ is $!A \wedge \top$. We can prove $!A \vdash !A \wedge \top$ using [$\wedge$ RIGHT], [$\top$ RIGHT] and [ID]. □

Proof of Lemma 10. We can now finish the proof of Lemma 9 by giving the missing proof of Lemma 10.

Proof. There are two cases to consider, depending on whether or not $(\mathfrak{M} \sqcap \mathfrak{M}_2) = \bot$.

Case $(\mathfrak{M} \sqcap \mathfrak{M}_2) = \bot$. If $(\mathfrak{M} \sqcap \mathfrak{M}_2) = \bot$, there are three possibilities:

- $\mathfrak{M} = \bot$
- $\mathfrak{M}_2 = \bot$
- Neither $\mathfrak{M}$ nor $\mathfrak{M}_2$ are $\bot$, but together they are incompatible.

If either $\mathfrak{M}$ or $\mathfrak{M}_2$ is $\bot$, then the proof is a simple application of [ID] followed by [$\wedge$ LEFT].

Next, let us consider the case where neither $\mathfrak{M}$ nor $\mathfrak{M}_2$ are $\bot$, but together they are incompatible. Let $\mathfrak{M} = (\mathcal{L}, w_1)$ and $\mathfrak{M}' = (\mathcal{L}', w_1')$. If $\mathfrak{M} \sqcap \mathfrak{M}_2 = \bot$, then there is a finite sequence of actions $a_1, ..., a_{n-1}$ such that both $\mathfrak{M}$ and $\mathfrak{M}'$ satisfy $\langle a_1 \rangle ... \langle a_{n-1} \rangle \top$, but they disagree about the state-labelling on the final state of this chain. In other words, there is a b-transition from the final state in $\mathfrak{M}$ which is ruled-out by the λ' state-labelling in $\mathfrak{M}'$. So there is a set of states $w_1, ..., w_1', ...$ and a finite set X of actions such that:

- $w_1 \xrightarrow{a_1} w_2 \xrightarrow{a_2} \dots \xrightarrow{a_{n-1}} w_n$.
- $w'_1 \xrightarrow{a_1} w'_2 \xrightarrow{a_2} \dots \xrightarrow{a_{n-1}} w'_n$.
- $w_n \xrightarrow{b} w_{n+1}$.
- $\lambda'(w'_n) = X$ with $b \notin X$.

Now it is easy to show, using [$\wedge$ Left], that

$$\mathsf{char}(\mathfrak{M}) \vdash \langle a_1 \rangle ... \langle a_{n-1} \rangle \langle b \rangle \top$$
$$\mathsf{char}(\mathfrak{M}') \vdash \langle a_1 \rangle ... \langle a_{n-1} \rangle !X$$

Now using [$\wedge$ Left] and [$\wedge$ Right]:

$$\mathsf{char}(\mathfrak{M}) \wedge \mathsf{char}(\mathfrak{M}') \vdash \langle a_1 \rangle ... \langle a_{n-1} \rangle \langle b \rangle \top \wedge \vdash \langle a_1 \rangle ... \langle a_{n-1} \rangle !X$$

Now using [Det]:

$$\mathsf{char}(\mathfrak{M}) \wedge \mathsf{char}(\mathfrak{M}') \vdash \langle a_1 \rangle ... \langle a_{n-1} \rangle (\langle b \rangle \top \wedge !X)$$

Now, using [$\bot$ Right 1]:

$$\langle b \rangle \top \wedge !X \vdash \bot$$

Using $n-1$ applications of [$\bot$ Right 2]:

$$\langle a_1 \rangle ... \langle a_{n-1} \rangle (\langle b \rangle \top \wedge !X) \vdash \bot$$

Finally, using [Trans], we derive:

$$\mathsf{char}(\mathfrak{M}) \wedge \mathsf{char}(\mathfrak{M}') \vdash \bot$$

Case $(\mathfrak{M} \sqcap \mathfrak{M}_2) \neq \bot$. From the construction of merge, if $\mathfrak{M}$ and $\mathfrak{M}'$ are acyclic, then $\mathfrak{M} \sqcap \mathfrak{M}'$ is also acyclic. If $\mathfrak{M} \sqcap \mathfrak{M}'$ is acyclic, then $\mathsf{char}(\mathfrak{M} \sqcap \mathfrak{M}')$ is equivalent to a set Γ of sentences of one of two forms:

$$\langle a_1 \rangle ... \langle a_n \rangle \top \qquad\qquad \langle a_1 \rangle ... \langle a_n \rangle !X$$

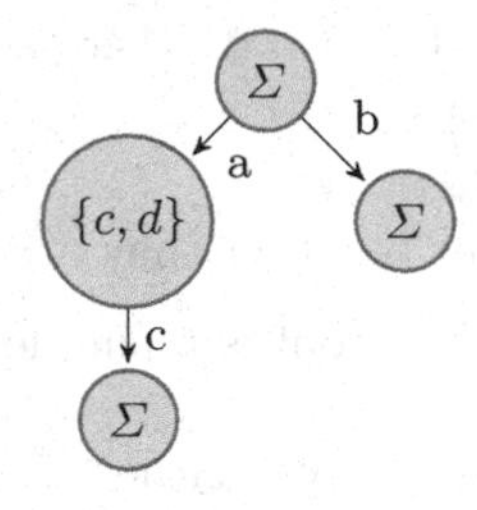

Fig. 14. Example of $\sqcap$

For example, if $\mathfrak{M} \sqcap \mathfrak{M}'$ is as in Fig. 14, then

$$\mathsf{char}(\mathfrak{M} \sqcap \mathfrak{M}') = \langle a \rangle (!\{c, d\} \wedge \langle c \rangle \top) \wedge \langle b \rangle \top$$

This is equivalent to the set Γ of sentences:

$$\langle a\rangle\langle c\rangle\top \qquad \langle b\rangle\top \qquad \langle a\rangle!\{c,d\}$$

Now using [$\wedge$ RIGHT] and [DET] we can show that

$$\bigwedge_{\phi\in\Gamma}\phi \vdash \mathsf{char}(\mathfrak{M}\sqcap\mathfrak{M}')$$

We know that for all $\phi \in \Gamma$

$$\mathfrak{M}\sqcap\mathfrak{M}' \models \phi$$

We just need to show that:

$$\mathsf{char}(\mathfrak{M}) \wedge \mathsf{char}(\mathfrak{M}') \vdash \phi$$

Take any $\phi \in \Gamma$ of the form $\langle a_1\rangle...\langle a_n\rangle!X$ for some finite $X \subseteq \Sigma$. (The case where ϕ is of the form $\langle a_1\rangle...\langle a_n\rangle\top$ is very similar, but simpler). If $\mathfrak{M}\sqcap\mathfrak{M}' \models \langle a_1\rangle...\langle a_n\rangle!X$ then either:

1. $\mathfrak{M} \models \langle a_1\rangle...\langle a_n\rangle!X$ but $\mathfrak{M}' \not\models \langle a_1\rangle...\langle a_n\rangle\top$
2. $\mathfrak{M}' \models \langle a_1\rangle...\langle a_n\rangle!X$ but $\mathfrak{M} \not\models \langle a_1\rangle...\langle a_n\rangle\top$
3. $\mathfrak{M} \models \langle a_1\rangle...\langle a_n\rangle!X_1$ and $\mathfrak{M}' \models \langle a_1\rangle...\langle a_n\rangle!X_2$ and $X_1 \cap X_2 \subseteq X$

In the first two cases, showing $\mathsf{char}(\mathfrak{M})\wedge\mathsf{char}(\mathfrak{M}') \vdash \phi$ is just a matter of repeated application of [$\wedge$ LEFT] and [$\wedge$ RIGHT]. In the third case, let $\mathfrak{M} = (\mathcal{L}, w_1)$ and $\mathfrak{M}' = (\mathcal{L}', w_1')$. If $\mathfrak{M} \models \langle a_1\rangle...\langle a_n\rangle!X_1$ and $\mathfrak{M}' \models \langle a_1\rangle...\langle a_n\rangle!X_2$ then there exists sequences $w_1, ..., w_{n+1}$ and $w_1', ..., w_{n+1}'$ of states such that

- $w_1 \xrightarrow{a_1} ... \xrightarrow{a_n} w_{n+1}$.
- $w_1' \xrightarrow{a_1} ... \xrightarrow{a_n} w_{n+1}'$.
- $\lambda(w_{n+1}) \subseteq X_1$.
- $\lambda'(w_{n+1}') \subseteq X_2$.

Now from the definition of $\mathsf{char}()$:

$$\mathsf{char}((\mathcal{L}, w_{n_1})) \vdash !X_1 \qquad \mathsf{char}((\mathcal{L}', w_{n_1}')) \vdash !X_2$$

Now using [!RIGHT 2]:

$$\mathsf{char}((\mathcal{L}, w_{n_1})) \wedge \mathsf{char}((\mathcal{L}', w_{n_1}')) \vdash !(X_1 \cap X_2)$$

Using [!RIGHT 1]:

$$\mathsf{char}((\mathcal{L}, w_{n_1})) \wedge \mathsf{char}((\mathcal{L}', w_{n_1}')) \vdash !X$$

Using n applications of [NORMAL]:

$$\langle a_1\rangle...\langle a_n\rangle(\mathsf{char}((\mathcal{L}, w_{n_1})) \wedge \mathsf{char}((\mathcal{L}', w_{n_1}'))) \vdash \langle a_1\rangle...\langle a_n\rangle!X$$

Finally, using n applications of [DET]:

$$\mathsf{char}((\mathcal{L}, w_1)) \wedge \mathsf{char}((\mathcal{L}', w_1')) \vdash \langle a_1\rangle...\langle a_n\rangle(\mathsf{char}((\mathcal{L}, w_{n_1})) \wedge \mathsf{char}((\mathcal{L}', w_{n_1}')))$$

So, by [TRANS]

$$\mathsf{char}(\mathfrak{M}) \wedge \mathsf{char}(\mathfrak{M}') \vdash \langle a_1\rangle...\langle a_n\rangle!X$$

$\square$

8 Compactness and the Standard Translation to First-Order Logic

This section studies two embeddings of cathoristic logic into first-order logic. The second embedding is used to prove that cathoristic logic satisfies compactness.

8.1 Translating from Cathoristic to First-Order Logic

The study of how a logic embeds into other logics is interesting in parts because it casts a new light on the logic that is the target of the embedding. A good example is the standard translation of modal into first-order logic. The translation produces various fragments: the finite variable fragments, the fragment closed under bisimulation, guarded fragments. These fragments have been investigated deeply, and found to have unusual properties not shared by the whole of first-order logic. Translations also enable us to push techniques, constructions and results between logics. In this section, we translate cathoristic logic into first-order logic.

Definition 14. *The first-order signature $\mathcal{S}$ has a nullary predicate $\top$, a family of unary predicates $\mathsf{Restrict}_A(\cdot)$, one for each finite subset $A \subseteq \Sigma$, and a family of binary predicates $\mathsf{Arrow}_a(x, y)$, one for each action $a \in \Sigma$.*

The intended interpretation is as follows.

- The universe is composed of states.
- The predicate $\top$ is true everywhere.
- For each finite $A \subseteq \Sigma$ and each state s, $\mathsf{Restrict}_A(s)$ is true if $\lambda(x) \subseteq A$.
- A set of two-place predicates $\mathsf{Arrow}_a(x, y)$, one for each $a \in \Sigma$, where x and y range over states. $\mathsf{Arrow}_a(x, y)$ is true if $x \xrightarrow{a} y$.

If Σ is infinite, then $\mathsf{Restrict}_A(\cdot)$ and $\mathsf{Arrow}_a(\cdot, \cdot)$ are infinite families of relations.

Definition 15. *Choose two fixed variables x, y, let a range over actions in Σ, and A over finite subsets of Σ. Then the restricted fragment of first-order logic that is the target of our translation is given by the following grammar, where w, z range over x, y.*

$$\phi ::= \top \mid \mathsf{Arrow}_a(w, z) \mid \mathsf{Restrict}_A(z) \mid \phi \wedge \psi \mid \exists x.\phi$$

This fragment has no negation, disjunction, implication, or universal quantification.

Definition 16. *The translations $[\![\phi]\!]_x$ and $[\![\phi]\!]_y$ of cathoristic formula ϕ are given relative to a state, denoted by either x or y.*

$$\begin{aligned} [\![\top]\!]_x &= \top & [\![\top]\!]_y &= \top \\ [\![\phi \wedge \psi]\!]_x &= [\![\phi]\!]_x \wedge [\![\psi]\!]_x & [\![\phi \wedge \psi]\!]_y &= [\![\phi]\!]_y \wedge [\![\psi]\!]_y \\ [\![\langle a \rangle \phi]\!]_x &= \exists y.(\mathsf{Arrow}_a(x, y) \wedge [\![\phi]\!]_y) & [\![\langle a \rangle \phi]\!]_y &= \exists x.(\mathsf{Arrow}_a(y, x) \wedge [\![\phi]\!]_x) \\ [\![!A]\!]_x &= \mathsf{Restrict}_A(x) & [\![!A]\!]_y &= \mathsf{Restrict}_A(y) \end{aligned}$$

The translations on the left and right are identical, except for switching x and y. Here is an example translation.

$$[\![\langle a\rangle\top\wedge!\{a\}]\!]_x = \exists y.(\mathsf{Arrow}_a(x,y)\wedge\top)\wedge\mathsf{Restrict}_{\{a\}}(x)$$

We now establish the correctness of the encoding. The key issue is that not every first-order model of our first-order signature corresponds to a cathoristic model because determinism, well-sizedness and admissibility are not enforced by our signature alone. In other words, models may contain 'junk'. We deal with this problem following ideas from modal logic [5]: we add a translation $[\![\mathcal{L}]\!]$ for cathoristic transition systems, and then prove the following theorem.

Theorem 6 (correspondence theorem). *Let ϕ be a cathoristic logic formula and $\mathfrak{M} = (\mathcal{L}, s)$ a cathoristic model.*

$$\mathfrak{M}\models\phi \quad \textit{iff} \quad [\![\mathcal{L}]\!]\models_{x\mapsto s}[\![\phi]\!]_x.$$

And likewise for $[\![\phi]\!]_y$.

The definition of $[\![\mathcal{L}]\!]$ is simple.

Definition 17. *Let $\mathcal{L} = (S, \rightarrow, \lambda)$ be a cathoristic transition system. Clearly $\mathcal{L}$ gives rise to an $\mathcal{S}$-model $[\![\mathcal{L}]\!]$ as follows.*

- *The universe is the set S of states.*
- *The relation symbols are interpreted as follows.*
  - $\top^{[\![\mathcal{L}]\!]}$ *always holds.*
  - $\mathsf{Restrict}_A^{[\![\mathcal{L}]\!]} = \{s\in S \mid \lambda(s)\subseteq A\}$.
  - $\mathsf{Arrow}^{[\![\mathcal{L}]\!]}{}_a = \{(s,t)\in S\times S \mid s\xrightarrow{a}t\}$.

We are now ready to prove Theorem 6.

Proof. By induction on the structure of ϕ. The cases $\top$ and $\phi_1\wedge\phi_2$ are straightforward. The case $\langle a\rangle\psi$ is handled as follows.

$$\begin{aligned}
&[\![\mathcal{L}]\!]\models_{x\mapsto s}[\![\langle a\rangle\psi]\!]_x \\
&\quad\text{iff}\quad [\![\mathcal{L}]\!]\models_{x\mapsto s}\exists y.(\mathsf{Arrow}_a(x,y)\wedge[\![\psi]\!]_y) \\
&\quad\text{iff}\quad \text{exists } t\in S.[\![\mathcal{L}]\!]\models_{x\mapsto s,y\mapsto t}\mathsf{Arrow}_a(x,y)\wedge[\![\psi]\!]_y \\
&\quad\text{iff}\quad \text{exists } t\in S.[\![\mathcal{L}]\!]\models_{x\mapsto s,y\mapsto t}\mathsf{Arrow}_a(x,y) \text{ and } [\![\mathcal{L}]\!]\models_{x\mapsto s,y\mapsto t}[\![\psi]\!]_y \\
&\quad\text{iff}\quad \text{exists } t\in S.s\xrightarrow{a}t \text{ and } [\![\mathcal{L}]\!]\models_{x\mapsto s,y\mapsto t}[\![\psi]\!]_y \\
&\quad\text{iff}\quad \text{exists } t\in S.s\xrightarrow{a}t \text{ and } [\![\mathcal{L}]\!]\models_{y\mapsto t}[\![\psi]\!]_y \qquad (\text{as } x \text{ is not free in } \psi) \\
&\quad\text{iff}\quad \text{exists } t\in S.s\xrightarrow{a}t \text{ and } \mathfrak{M}\models\psi \\
&\quad\text{iff}\quad \mathfrak{M}\models\langle a\rangle\psi
\end{aligned}$$

Finally, if ϕ is $!A$ the derivation comes straight from the definitions.

$$\begin{aligned}
[\![\mathcal{L}]\!]\models_{x\mapsto s}[\![!A]\!]_x \quad &\text{iff}\quad [\![\mathcal{L}]\!]\models_{x\mapsto s}\mathsf{Restrict}_A(x) \\
&\text{iff}\quad \lambda(s)\subseteq A \\
&\text{iff}\quad \mathfrak{M}\models\,!A.
\end{aligned}$$

□

8.2 Compactness by Translation

First-order logic satisfies *compactness*: a set S of sentences has a model exactly when every finite subset of S does. What about cathoristic logic?

We can prove compactness of modal logics using the standard translation from modal to first-order logic [5]: we start from a set of modal formula such that each finite subset has a model. We translate the modal formulae and models to first-order logic, getting a set of first-order formulae such that each finite subset has a first-order model. By compactness of first-order logic, we obtain a first-order model of the translated modal formulae. Then we translate that first-order model back to modal logic, obtaining a model for the original modal formulae, as required. The last step proceeds without a hitch because the modal and the first-order notions of model are identical, save for details of presentation.

Unfortunately we cannot do the same with the translation from cathoristic logic to first-order logic presented in the previous section. The problem are the first-order models termed 'junk' above. The target language of the translation is not expressive enough to have formulae that can guarantee such constraints. As we have no reason to believe that the first-order model whose existence is guaranteed by compactness isn't 'junk', we cannot prove compactness with the translation. We solve this problem with a second translation, this time into a more expressive first-order fragment where we can constrain first-order models easily using formulae. The fragment we use now lives in two-sorted first-order logic (which can easily be reduced to first-order logic [11]).

Definition 18. *The two-sorted first-order signature $\mathcal{S}'$ is given as follows.*

- *$\mathcal{S}'$ has two sorts, states and actions.*
- *The action constants are given by Σ. There are no state constants.*
- *$\mathcal{S}'$ has a nullary predicate $\top$.*
- *A binary predicate $\mathsf{Allow}(\cdot,\cdot)$. The intended meaning of $\mathsf{Allow}(x,a)$ is that at the state denoted by x we are allowed to do the action a.*
- *A ternary predicate $\mathsf{Arrow}(\cdot,\cdot,\cdot)$ where $\mathsf{Arrow}(x,a,y)$ means that there is a transition from the state denoted by x to the state denoted by y, and that transition is labelled a.*

Definition 19. *The encoding $\langle\!\langle \phi \rangle\!\rangle_x$ of cathoristic logic formulae is given by the following clauses.*

$$\begin{aligned}
\langle\!\langle \top \rangle\!\rangle_x &= \top \\
\langle\!\langle \phi \land \psi \rangle\!\rangle_x &= \langle\!\langle \phi \rangle\!\rangle_x \land \langle\!\langle \psi \rangle\!\rangle_x \\
\langle\!\langle \langle a \rangle \phi \rangle\!\rangle_x &= \exists^{st} y.(\mathsf{Arrow}(x,a,y) \land \langle\!\langle \phi \rangle\!\rangle_y) \\
\langle\!\langle !A \rangle\!\rangle_x &= \forall^{act} a.(\mathsf{Allow}(x,a) \rightarrow a \in A)
\end{aligned}$$

Here we use $\exists^{st}$ to indicate that this existential quantifier ranges over the sort of states, and $\forall^{act}$ for the universal quantifier ranging over actions. The expression $a \in A$ is a shorthand for the first-order formula

$$a = a_1 \lor a = a_2 \lor \cdots \lor a = a_n$$

assuming that $A = \{a_1, ..., a_n\}$. Since by definition, A is always a finite set, this is well-defined. The translation could be restricted to a two-variable fragment. Moreover, the standard reduction from many-sorted to one-sorted first-order logic does not increase the number of variables used (although predicates are added, one per sort). We will not consider this matter further here.

We also translate cathoristic transition systems $\langle\langle \mathcal{L} \rangle\rangle$.

Definition 20. *Let $\mathcal{L} = (S, \rightarrow, \lambda)$ be a cathoristic transition system. $\mathcal{L}$ gives rise to an $\mathcal{S}'$-model $\langle\langle \mathcal{L} \rangle\rangle$ as follows.*

- *The sort of states is interpreted by the set S.*
- *The sort of actions is interpreted by the set Σ.*
- *For each constant $a \in \Sigma$, $a^{\langle\langle \mathcal{L} \rangle\rangle}$ is a itself.*
- *The relation symbols are interpreted as follows.*
  - $\top^{\langle\langle \mathcal{L} \rangle\rangle}$ *always holds.*
  - $\mathsf{Allow}^{\langle\langle \mathcal{L} \rangle\rangle}(s, a)$ *holds whenever* $a \in \lambda(s)$.
  - $\mathsf{Arrow}^{\langle\langle \mathcal{L} \rangle\rangle}(s, a, t)$ *holds whenever* $s \xrightarrow{a} t$.

Theorem 7 (correspondence theorem). *Let ϕ be a cathoristic logic formula and $\mathfrak{M} = (\mathcal{L}, s)$ a cathoristic model.*

$$\mathfrak{M} \models \phi \quad \textit{iff} \quad \langle\langle \mathcal{L} \rangle\rangle \models_{x \mapsto s} \langle\langle \phi \rangle\rangle_x.$$

Proof. The proof proceeds by induction on the structure of ϕ and is similar to that of Theorem 7. The case for the may modality proceeds as follows.

$$\begin{aligned}
\mathfrak{M} \models \langle a \rangle \phi \quad & \text{iff} \quad \text{exists state } t \text{ with } \; s \xrightarrow{a} t \text{ and } (\mathcal{L}, t) \models \phi \\
& \text{iff} \quad \text{exists state } t \text{ with } \; s \xrightarrow{a} t \text{ and } \langle\langle \mathcal{L} \rangle\rangle \models_{y \mapsto t} \langle\langle \phi \rangle\rangle_y \qquad \text{by (IH)} \\
& \text{iff} \quad \langle\langle \mathcal{L} \rangle\rangle \models_{x \mapsto s} \exists^{st} y.(\mathsf{Arrow}(x, a, y) \wedge \langle\langle \phi \rangle\rangle_y) \\
& \text{iff} \quad \langle\langle \mathcal{L} \rangle\rangle \models_{x \mapsto s} \langle\langle \langle a \rangle \phi \rangle\rangle_x
\end{aligned}$$

Finally $!A$.

$$\begin{aligned}
\mathfrak{M} \models !A \quad & \text{iff} \quad \lambda(s) \subseteq A \\
& \text{iff} \quad \text{for all } \; a \in \Sigma . a \in A \\
& \text{iff} \quad \langle\langle \mathcal{L} \rangle\rangle \models_{x \mapsto s} \forall^{act} a.(\mathsf{Allow}(x, a) \rightarrow a \in A) \\
& \text{iff} \quad \langle\langle \mathcal{L} \rangle\rangle \models_{x \mapsto s} \langle\langle !A \rangle\rangle_x
\end{aligned}$$

□

We use the following steps in our compactness proof.

1. Choose a set Γ of cathoristic logic formulae such that each finite subset Γ' of Γ has a cathoristic model $(\mathcal{L}, s)$.
2. The translation gives a set $\langle\langle \Gamma \rangle\rangle = \{\langle\langle \phi \rangle\rangle \mid \phi \in \Gamma\}$ of first-order formulae such that each finite subset has a first-order model $\langle\langle \mathcal{L} \rangle\rangle$.

3. By compactness of (two-sorted) first-order logic, we can find a first-order model $\mathcal{M}$ of $\langle\!\langle \Gamma \rangle\!\rangle$.
4. Convert $\mathcal{M}$ into a cathoristic transition system $\mathcal{M}^\sharp$ such that $(\mathcal{M}^\sharp, s) \models \Gamma$.

The problematic step is (4) - for how would we know that the first-order model $\mathcal{M}$ can be converted back to a cathoristic transition system? What if it contains 'junk' in the sense described above? We solve this by adding formulae to $\langle\!\langle \Gamma \rangle\!\rangle$ that preserve finite satisfiability but force the first-order models to be convertible to cathoristic models. To ensure admissibility we use this formula.

$$\phi_{admis} = \forall^{st} s.\forall^{act} a.\forall^{st} t.(\mathsf{Arrow}(s, a, t) \rightarrow \mathsf{Allow}(s, a))$$

The formula ϕ_{det} ensures model determinism.

$$\phi_{det} = \forall^{st} s.\forall^{act} a.\forall^{st} t.\forall^{st} t'.((\mathsf{Arrow}(s, a, t) \wedge \mathsf{Arrow}(s, a, t')) \rightarrow t = t')$$

Lemma 11. *If $\mathcal{L}$ is a cathoristic transition system then $\langle\!\langle \mathcal{L} \rangle\!\rangle \models \phi_{admis} \wedge \phi_{det}$.*

Proof. Straightforward from the definitions. □

We can now add, without changing satisfiability, $\phi_{admis} \wedge \phi_{det}$ to any set of first-order formulae that has a model that is the translation of a cathoristic model.

We also need to deal with well-sizedness in first-order models, because nothing discussed so far prevents models whose state labels are infinite sets without being Σ. Moreover, a model may interpret the set of actions with a proper superset of Σ. This also prevents conversion to cathoristic models. We solve these problems by simply removing all actions that are not in Σ and all transitions involving such actions. We map all infinite state labels to Σ. It is easy to see that this does not change satisfiability of (translations of) cathoristic formulae.

Definition 21. *Let $\mathcal{L} = (S, \rightarrow, \lambda)$ be a cathoristic transition system and X a set, containing actions. The* restriction of $\mathcal{L}$ to X, *written $\mathcal{L} \setminus X$ is the cathoristic model $(S, \rightarrow', \lambda')$ where $\rightarrow' = \{(s, a, t) \in \rightarrow \mid a \notin X\}$, and for all states s we set:*

$$\lambda'(s) = \begin{cases} \lambda(s) \setminus X & \textit{whenever } \lambda(s) \neq \Sigma \\ \Sigma & \textit{otherwise} \end{cases}$$

Lemma 12. *Let ϕ be a cathoristic logic formula and X be a set such that no action occurring in ϕ is in X. Then:*

$$(\mathcal{L}, s) \models \phi \quad \textit{iff} \quad (\mathcal{L} \setminus X, s) \models \phi.$$

Proof. By straightforward induction on the structure of ϕ, using the fact that by assumption X only contains actions not occurring in ϕ. □

Definition 22. *Let $\mathcal{M}$ be a first-order model for the signature $\mathcal{S}'$. We construct a cathoristic transition system $\mathcal{M}^\sharp = (S, \rightarrow, \lambda)$.*

- *The actions Σ are given by the $\mathcal{M}$ interpretation of actions.*
- *The states S are given by the $\mathcal{M}$ interpretation of states.*
- *The reduction relation $s \xrightarrow{a} t$ holds exactly when $\mathsf{Arrow}^{\mathcal{M}}(s, a, t)$.*
- *The function λ is given by the following clause:*

$$\lambda(s) = \begin{cases} X & \text{whenever } X = \{a \mid \mathsf{Allow}^{\mathcal{M}}(s, a)\} \text{ is finite} \\ \Sigma & \text{otherwise} \end{cases}$$

Lemma 13. *Let $\mathcal{M}$ be a first-order model for $\mathcal{S}'$ such that $\mathcal{M} \models \phi_{admis} \wedge \phi_{det}$. Then $\mathcal{M}^{\sharp}$ is an cathoristic transition system with actions Σ.*

Proof. Immediate from the definitions. □

Theorem 8 (correspondence theorem). *Let $\mathcal{M}$ be a first-order model for the signature $\mathcal{S}'$ such that $\mathcal{M} \models \phi_{admis} \wedge \phi_{det}$. Then we have for all cathoristic logic formulae ϕ with actions from Σ:*

$$\mathcal{M} \models_{x \mapsto s} \langle\!\langle \phi \rangle\!\rangle_x \quad \textit{iff} \quad (\mathcal{M}^{\sharp} \setminus X, s) \models \phi.$$

Here X is the set of all elements in the universe of $\mathcal{M}$ interpreting actions that are not in Σ.

Proof. The proof proceeds by induction on the structure of ϕ. □

Definition 23. *Let Γ be a set of cathoristic formulae, and $\mathfrak{M}$ a cathoristic model. We write $\mathfrak{M} \models T$ provided $\mathfrak{M} \models \phi$ for all $\phi \in T$. We say Γ is* satisfiable *provided $\mathfrak{M} \models T$.*

Theorem 9 (Compactness of cathoristic logic). *A set Γ of cathoristic logic formulae is satisfiable iff each finite subset of Γ is satisfiable.*

Proof. For the non-trivial direction, let Γ be a set of cathoristic logic formulae such that any finite subset has a cathoristic model. Define

$$\langle\!\langle \Gamma \rangle\!\rangle \;=\; \{\langle\!\langle \phi \rangle\!\rangle \mid \phi \in \Gamma\} \qquad\qquad \Gamma^* \;=\; \langle\!\langle \Gamma \rangle\!\rangle \cup \{\phi_{admis} \wedge \phi_{det}\}$$

which both are sets of first-order formulae. Clearly each finite subset Γ' of Γ^* has a first-order model. Why? First consider the subset Γ'_{CL} of Γ' which is given as follows.

$$\Gamma'_{CL} \;=\; \{\phi \in \Gamma \mid \langle\!\langle \phi \rangle\!\rangle \in \Gamma'\}$$

Since Γ'_{CL} is finite, by assumption there is a cathoristic model

$$(\mathcal{L}, s) \models \Gamma'_{CL}$$

which means we can apply Theorem 8 to get

$$\langle\!\langle \mathcal{L} \rangle\!\rangle \models_{x \mapsto s} \langle\!\langle \Gamma'_{CL} \rangle\!\rangle,$$

By construction $\Gamma' \setminus \langle\langle \Gamma'_{CL} \rangle\rangle \subseteq \{\phi_{admis} \wedge \phi_{det}\}$, so all we have to show for Γ' to have a model is that

$$\langle\langle \mathcal{L} \rangle\rangle \models_{x \mapsto s} \{\phi_{admis}\} \cup \{\phi_a \mid a \in \Sigma\},$$

but that is a direct consequence of Lemma 11. That means each finite subset of Γ^* has a model and by appealing to compactness of first-order many-sorted logic (which is an immediate consequence of compactness of one-sorted first-order logic [11]), we know there must be a first-order model $\mathcal{M}$ of Γ^*, i.e.

$$\mathcal{M} \models \Gamma^*.$$

Since $\mathcal{M} \models \phi_{admis} \wedge \phi_{det}$ we can apply Theorem 8 that also

$$(\mathcal{M}^\sharp \setminus X, s) \models \Gamma$$

where X is the set of all actions in $\mathcal{M}^\sharp$ that are not in Σ. Hence Γ is satisfiable. □

9 Cathoristic Logic and Negation

We have presented cathoristic logic as a language that can express incompatible claims without negation. In this section, we briefly consider cathoristic logic enriched with negation.

9.1 Syntax and Semantics

Definition 24. *Given a set Σ of actions, the* formulae of cathoristic logic with negation *are given by the following grammar.*

$$\phi ::= ... \mid \neg\phi$$

We can now define disjunction $\phi \vee \psi$ and implication $\phi \rightarrow \psi$ by de Morgan duality: $\phi \vee \psi$ is short for $\neg(\neg\phi \wedge \neg\psi)$, and $\phi \rightarrow \psi$ abbreviates $\neg\phi \vee \psi$.

The semantics of cathoristic logic with negation is just that of plain cathoristic logic except for the obvious clause for negation.

$$\mathfrak{M} \models \neg\phi \quad \text{iff} \quad \mathfrak{M} \not\models \phi$$

Negation is a core operation of classical logic, and its absence makes cathoristic logic unusual. In order to understand cathoristic logic better, we now investigate how negation can be seen as a definable abbreviation in cathoristic logic with disjunction. The key idea is to use the fact that $\langle a \rangle \phi$ can be false in two ways: either there is no a-labelled action at the current state - or there is, but ϕ is false. Both arms of this disjunction can be expressed in cathoristic logic, the former as $!\Sigma \setminus \{a\}$, the latter as $\langle a \rangle \neg\phi$. Hence, we can see $\neg\langle a \rangle \phi$ as a shorthand for

$$!(\Sigma \setminus \{a\}) \vee \langle a \rangle \neg\phi$$

Negation still occurs in this term, but prefixing a formula of lower complexity.

This leaves the question of negating the tantum. That's easy: when $\neg !A$, then clearly the current state can do an action $a \notin A$. In other words

$$\bigvee_{a \in \Sigma} \langle a \rangle \top$$

When Σ is infinite, then so is the disjunction.

Note that both the negation of the modality and the negation of the tantum involve the set Σ of actions. So far, we have defined negation with respect to the whole (possibly infinite) set Σ. For technical reasons, we generalise negation and define it with respect to a *finite* subset $S \subseteq \Sigma$. We use this finitely-restricted version of negation in the decision procedure below.

Definition 25. *The function $\neg_S(\phi)$ removes negation from ϕ relative to a finite subset $S \subseteq \Sigma$:*

$$\begin{aligned} \neg_S(\top) &= \bot & \neg_S(\bot) &= \top \\ \neg_S(\phi \wedge \psi) &= \neg_S(\phi) \vee \neg_S(\psi) & \neg_S(\phi \vee \psi) &= \neg_S(\phi) \wedge \neg_S(\psi) \\ \neg_S(\langle a \rangle \phi) &= !(S - \{a\}) \vee \langle a \rangle \neg_S(\phi) & \neg_S(!A) &= \bigvee_{a \in S - A} \langle a \rangle \top \end{aligned}$$

9.2 Decision Procedure

We can use the fact that cathoristic logic has a quadratic-time decision procedure to build a super-polynomial time decision procedure for cathoristic logic with negation. Given $\phi \models \psi$, let $S = \mathsf{actions}(\phi) \cup \mathsf{actions}(\psi) \cup \{a\}$, where a is a fresh action. The function $\mathsf{actions}(\cdot)$ returns all actions occurring in a formula, e.g. $\mathsf{actions}(\langle a \rangle \phi) = \{a\} \cup \mathsf{actions}(\phi)$ and $\mathsf{actions}(!A) = A$. The decision procedure executes the following steps.

1. Inductively translate away all negations in ϕ using $\neg_S(\phi)$ as defined above. Let the result be ϕ'.
2. Reduce ϕ' to disjunctive normal form by repeated application of the rewrite rules:

$$\phi \wedge (\psi \vee \xi) \rightsquigarrow (\phi \wedge \psi) \vee (\phi \wedge \xi) \qquad (\phi \vee \psi) \wedge \xi \rightsquigarrow (\phi \wedge \xi) \vee (\psi \wedge \xi).$$

3. Let the resulting disjuncts be $\phi_1, ..., \phi_n$. Note that

$$\phi \models \psi \quad \text{iff} \quad \phi_i \models \psi \text{ for all } i = 1, ..., n.$$

   For each disjunct ϕ_i do the following.
   - Notice that $\phi_i \models \psi$ if and only if all S-extensions (defined below) of $\mathsf{simpl}(\phi_i)$ satisfy ψ. So, to check whether $\phi_i \models \psi$, we enumerate the S-extensions of $\mathsf{simpl}(\phi_i)$ (there are a finite number of such extensions - the exact number is exponential in the size of $\mathsf{simpl}(\phi_i)$) and check for each such S-extension $\mathfrak{M}$ whether $\mathfrak{M} \models \psi$, using the algorithm of Sect. 6.5.

Here is the definition of S-extension.

Definition 26. *Given an cathoristic transition system $\mathcal{L} = (\mathcal{W}, \rightarrow, \lambda)$, and a set S of actions, then $(\mathcal{W}', \rightarrow', \lambda')$ is a S-*extension *of $\mathcal{L}$ if it is a valid cathoristic transition system (recall Definition 2) and for all $(x, a, y) \in \rightarrow'$, either:*

- *$(x, a, y) \in \rightarrow$, or;*
- *$x \in \mathcal{W}$, $a \in S$, $a \in \lambda(x)$, and y is a new state not appearing elsewhere in $\mathcal{W}$ or $\mathcal{W}'$.*

The state-labelling λ' is:

$$\lambda'(x) = \lambda(x) \text{ if } x \in \mathcal{W}$$
$$\lambda'(x) = \Sigma \text{ if } x \notin \mathcal{W}$$

In other words, $\mathfrak{M}'$ is an extension of an annotated model $\mathfrak{M}$, if all its transitions are either from $\mathfrak{M}$ or involve states of $\mathfrak{M}$ transitioning via elements of S to new states not appearing in $\mathfrak{M}$ or $\mathfrak{M}'$. The number of extensions grows quickly. If the model $\mathfrak{M}$ has n states, then the number of possible extensions is:

$$(2^{|S|})^n$$

But recall that we are computing these extensions in order to verify ψ. So we can make a significant optimisation by restricting the height of each tree to $|\psi|$. We state, without proof, that this optimisation preserves correctness. A Haskell implementation of the decision procedure is available [1].

10 Quantified Cathoristic Logic

So far, we have presented cathoristic logic as a propositional modal logic. This section sketches quantified cathoristic logic, primarily to demonstrate that this extension works smoothly.

Definition 27. *Let Σ be a non-empty set of actions, ranged over by $a, a', \ldots$ as before. Given a set $\mathcal{V}$ of* variables, *with $x, x', y, y', \ldots$ ranging over $\mathcal{V}$, the* terms, *ranged over by $t, t', \ldots$ and formulae of quantified cathoristic logic are given by the following grammar:*

$$t ::= x \mid a$$
$$\phi ::= \top \mid \phi \wedge \psi \mid \langle t \rangle \phi \mid !A \mid \exists x.\phi \mid \forall x.\phi$$

Now A ranges over finite subsets of terms. The free variables *of a ϕ, denoted $\mathsf{fv}(\phi)$ is given as expected, e.g. $\mathsf{fv}(\langle t \rangle \phi) = \mathsf{fv}(t) \cup \mathsf{fv}(\phi)$ and $\mathsf{fv}(!A) = \bigcup_{t \in A} \mathsf{fv}(t)$ where $\mathsf{fv}(a) = \emptyset$ and $\mathsf{fv}(x) = \{x\}$.*

Definition 28. *The semantics of quantified cathoristic logic is constructed along conventional lines. An* environment *is a map $\sigma : \mathcal{V} \to \Sigma$ with finite domain. We write $\sigma, x : a$ for the environment that is just like σ, except it also maps x to a, implicitly assuming that x is not in σ's domain. The* denotation *$[\![t]\!]_\sigma$ of a term t under an environment σ is given as follows:*

$$[\![a]\!]_\sigma = a \qquad\qquad [\![x]\!]_\sigma = \sigma(x)$$

where we assume that $\mathsf{fv}(t)$ *is a subset of the domain of* σ.

The satisfaction relation $\mathfrak{M} \models_\sigma \phi$ *is defined whenever* $\mathsf{fv}(\phi)$ *is a subset of σ's domain. It is given by the following clauses, where we assume that $\mathfrak{M} = (\mathcal{L}, s)$ and $\mathcal{L} = (S, \to, \lambda)$.*

$$\begin{array}{lll}
\mathfrak{M} \models_\sigma \top & & \\
\mathfrak{M} \models_\sigma \phi \land \psi & \textit{iff} & \mathfrak{M} \models_\sigma \phi \textit{ and } \mathfrak{M} \models_\sigma \psi \\
\mathfrak{M} \models_\sigma \langle t \rangle \phi & \textit{iff} & \textit{there is transition } s \xrightarrow{[\![t]\!]_\sigma} s' \textit{ such that } (\mathcal{L}, s') \models_\sigma \phi \\
\mathfrak{M} \models_\sigma !A & \textit{iff} & \lambda(s) \subseteq \{[\![t]\!] \mid t \in A\} \\
\mathfrak{M} \models_\sigma \forall x.\phi & \textit{iff} & \textit{for all } a \in \Sigma \textit{ we have } \mathfrak{M} \models_{\sigma,x:a} \phi \\
\mathfrak{M} \models_\sigma \exists x.\phi & \textit{iff} & \textit{there exists } a \in \Sigma \textit{ such that } \mathfrak{M} \models_{\sigma,x:a} \phi
\end{array}$$

In quantified cathoristic logic, we can say that there is exactly one king of France, and he is bald, as:

$$\exists x.(\langle king \rangle \langle france \rangle !\{x\} \land \langle x \rangle \langle bald \rangle)$$

Expressing this in first-order logic is more cumbersome:

$$\exists x.(king(france, x) \land bald(x) \land \forall y.(king(france, y) \to y = x))$$

The first-order logic version uses an extra universal quantifier, and also requires the identity relation with concomitant axioms.

To say that every person has exactly one sex, which is either male or female, we can write in quantified cathoristic logic:

$$\forall x.(\langle x \rangle \langle person \rangle \to \langle x \rangle \langle sex \rangle !\{male, female\} \land \exists y.\langle x \rangle \langle sex \rangle (\langle y \rangle \land !\{y\}))$$

This is more elegant than the equivalent in first-order logic:

$$\forall x.(person(x) \to \exists y. \begin{pmatrix} sex(x, y) \\ \land \\ (y = male \ \lor \ y = female) \\ \land \\ \forall z.sex(x, z) \to y = z \end{pmatrix}$$

To say that every traffic light is coloured either green, amber or red, we can write in quantified cathoristic logic:

$$\forall x.(\langle x \rangle \langle light \rangle \to \langle x \rangle \langle colour \rangle !\{green, amber, red\} \land \exists y.\langle x \rangle \langle colour \rangle (\langle y \rangle \land !\{y\}))$$

Again, this is less verbose than the equivalent in first-order logic:

$$\forall x.(light(x) \rightarrow \exists y. \begin{pmatrix} colour(x,y) \\ \wedge \\ (y = green \vee y = amber \vee y = red) \\ \wedge \\ \forall z.colour(x,z) \rightarrow y = z \end{pmatrix}$$

11 Related Work

This section surveys cathoristic logic's intellectual background, and related approaches.

11.1 Brandom's Incompatibility Semantics

In [8], Chapter 5, Appendix I, Brandom developed a new type of semantics, incompatibility semantics, that takes material incompatibility - rather than truth-assignment - as the semantically primitive notion.

Incompatibility semantics applies to any language, $\mathcal{L}$, given as a set of sentences. Given a predicate $\mathsf{Inc}(X)$ which is true of sets $X \subseteq \mathcal{L}$ that are incompatible, he defines an incompatibility function $\mathcal{I}$ from subsets of $\mathcal{L}$ to sets of subsets of $\mathcal{L}$:

$$X \in \mathcal{I}(Y) \quad \text{iff} \quad \mathsf{Inc}(X \cup Y).$$

We assume that $\mathcal{I}$ satisfies the monotonicity requirement (Brandom calls it "Persistence"):

$$\text{If } X \in \mathcal{I}(Y) \text{ and } X \subseteq X' \text{ then } X' \in \mathcal{I}(Y).$$

Now Brandom defines entailment in terms of the incompatibility function. Given a set $X \subseteq \mathcal{L}$ and an individual sentence $\phi \in \mathcal{L}$:

$$X \models \phi \quad \text{iff} \quad \mathcal{I}(\{\phi\}) \subseteq \mathcal{I}(X).$$

Now, given material incompatibility (as captured by the $\mathcal{I}$ function) and entailment, he introduces logical negation as a *derived* concept via the rule:

$$\{\neg\phi\} \in \mathcal{I}(X) \quad \text{iff} \quad X \models \phi.$$

Brandom goes on to show that the $\neg$ operator, as defined, satisfies the laws of classical negation. He also introduces a modal operator, again defined in terms of material incompatibility, and shows that this operator satisfies the laws of $S5$.

Cathoristic logic was inspired by Brandom's vision that material incompatibility is conceptually prior to logical negation: in other words, it is possible for a community of language users to make incompatible claims, even if that language has no explicit logical operators such as negation. The language users of this simple language may go on to introduce logical operators, in order to make certain inferential properties explicit - but this is an optional further development. The language before that addition was already in order as it is.

The approach taken in this paper takes Brandom's original insight in a different direction. While Brandom defines an unusual (non truth-conditional) semantics that applies to any language, we have defined an unusual logic with a standard (truth-conditional) semantics, and then shown that this logic satisfies the Brandomian connection between incompatibility and entailment.

11.2 Peregrin on Defining a Negation Operator

Peregrin [22] investigates the structural rules that any logic must satisfy if it is to connect incompatibility (Inc) and entailment ($\models$) via the Brandomian incompatibility semantics constraint:

$$X \models \phi \quad \text{iff} \quad \mathcal{I}(\{\phi\}) \subseteq \mathcal{I}(X).$$

The general structural rules are:

$$(\bot) \text{ If } \mathsf{Inc}(X) \text{ and } X \subseteq Y \text{ then } \mathsf{Inc}(Y).$$
$$(\models 1)\ \phi, X \models \phi.$$
$$(\models 2) \text{ If } X, \phi \models \psi \text{ and } Y \models \phi \text{ then } X, Y \models \psi.$$
$$(\bot \models 2) \text{ If } X \models \phi \text{ for all } \phi, \text{ then } \mathsf{Inc}(X).$$
$$(\models \bot 2) \text{ If } \mathsf{Inc}(Y \cup \{\phi\}) \text{ implies } \mathsf{Inc}(Y \cup X) \text{ for all } Y, \text{ then } X \models \phi.$$

Peregrin shows that if a logic satisfied the above laws, then incompatibility and entailment are mutually interdefinable, and the logic satisfies the Brandomian incompatibility semantics constraint.

Next, Peregrin gives a pair of laws for defining negation in terms of Inc and $\models$[15]:

$$(\neg 1)\ \mathsf{Inc}(\{\phi, \neg\phi\}).$$
$$(\neg 2) \text{ If } \mathsf{Inc}(X, \phi) \text{ then } X \models \neg\phi.$$

These laws characterise intuitionistic negation as the *minimal incompatible*[16]. Now, in [8], Brandom defines negation slightly differently. He uses the rule:

$$(\neg B)\ \mathsf{Inc}(X, \neg\phi) \text{ iff } X \models \phi.$$

Using this stronger rule, we can infer the classical law of double-negation: $\neg\neg\phi \models \phi$. Peregrin establishes that Brandom's rule for negation entail ($\neg 1$) and ($\neg 2$) above, but not conversely: Brandom's rule is stronger than Peregrin's minimal laws ($\neg 1$) and ($\neg 2$).

Peregrin concludes that the Brandomian constraint between incompatibility and entailment is satisfied by many different logics. Brandom happened to choose a particular rule for negation that led to classical logic, but the general connection between incompatibility and entailment is satisfied by many different logics, including intuitionistic logic. This paper supports Peregrin's conclusion: we have shown that cathoristic logic also satisfies the Brandomian constraint.

---

15 The converse of ($\neg 2$) follows from ($\neg 1$) and the general structural laws above.

16 ψ is the minimal incompatible of ϕ iff for all ξ, if $\mathsf{Inc}(\{\phi\} \cup \{\xi\})$ then $\xi \models \psi$.

11.3 Peregrin and Turbanti on Defining a Necessity Operator

In [8], Brandom gives a rule for defining necessity in terms of incompatibility and entailment:

$$X \in \mathcal{I}(\{\Box\phi\}) \quad \text{iff} \quad \mathsf{Inc}(X) \vee \exists Y.\, Y \notin \mathcal{I}(X) \wedge Y \nvDash \phi.$$

In other words, X is incompatible with $\Box\phi$ if X is compatible with something that does not entail ϕ.

The trouble is, as Peregrin and Turbanti point out, if ϕ is not tautological, then *every set* $X \subseteq \mathcal{L}$ is incompatible with $\Box\phi$. To show this, take any set $X \subseteq \mathcal{L}$. If $\mathsf{Inc}(X)$, then $X \in \mathcal{I}(\Box\phi)$ by definition. If, on the other hand, $\neg\mathsf{Inc}(X)$, then let $Y = \emptyset$. Now $\neg\mathsf{Inc}(X \cup Y)$ as $Y = \emptyset$, and $Y \nvDash \phi$ as ϕ is not tautological. Hence $X \in \mathcal{I}(\Box\phi)$ for all $X \subseteq \mathcal{L}$. Brandom's rule, then, is only capable of specifying a very specific form of necessity: logical necessity.

In [22] and [31], Peregrin and Turbanti describe alternative ways of defining necessity. These alternative rule sets can be used to characterise modal logics other than S5. For example, Turbanti defines the accessibility relation between worlds in terms of a *compossibility relation*, and then argues that the S4 axiom of transitivity fails because compossibility is not transitive.

We draw two conclusions from this work. The first is, once again, that a commitment to connecting incompatibility and entailment via the Brandomian constraint:

$$X \models \phi \quad \text{iff} \quad \mathcal{I}(\{\phi\}) \subseteq \mathcal{I}(X)$$

does not commit us to any particular logical system. There are a variety of logics that can satisfy this constraint. Second, questions about the structure of the accessibility relation in Kripke semantics - questions that can seem hopelessly abstract and difficult to answer - can be re-cast in terms of concrete questions about the incompatibility relation. Incompatibility semantics can shed light on possible-world semantics [31].

11.4 Linear Logic

Linear logic [15] is a refinement of first-order logic and was introduced by J.-Y. Girard and brings the symmetries of classical logic to constructive logic.

Linear logic splits conjunction into additive and multiplicative parts. The former, additive conjunction $A \& B$, is especially interesting in the context of cathoristic logic. In the terminology of process calculus it can be interpreted as an external choice operation [2]. ('External', because the choice is offered to the environment). This interpretation has been influential in the study of types for process calculus, e.g. [19,20,29]. Implicitly, additive conjunction gives an explicit upper bound on how many different options the environment can choose from. For example $A \& B \& C$ has three options (assuming that none of A, B, C can be decomposed into further additive conjunctions). With this in mind, and simplifying a great deal, a key difference between $!A$ and additive conjunction $A \& B$ is that the individual actions in $!A$ have no continuation, while they do with $A \& B$: the tantum $!\{l, r\}$ says that the only permitted actions are l and r. What happens at later states is not constrained by $!A$. In contrast, $A \& B$ says not only that at this point the only permissible options are A and B, but also that if we choose A, then A holds 'for ever', and likewise for choosing B. To be sure, the alternatives in $A \& B$ may themselves contain further additive conjunctions, and in this way express how exclusion changes 'over time'.

In summary, cathoristic logic and linear logic offer operators that restrict the permissible options. How are they related? Linear logic has an explicit linear negation $(\cdot)^{\perp}$ which, unlike classical negation, is constructive. In contrast, cathoristic logic defines a restricted form of negation using $!A$. Can these two perspectives be fruitfully reconciled?

11.5 Process Calculus

Process calculi are models of concurrent computation. They are based on the idea of message passing between actors running in parallel. Labelled transition systems are often used as models for process calculi, and many concepts used in the development of cathoristic logic - for example, bisimulations and Hennessy-Milner logic - originated in process theory (although some, such as bisimulation, evolved independently in other contexts).

Process calculi typically feature a construct called sum, that is an explicit description of mutually exclusive option:

$$\sum_{i \in I} P_i$$

That is a process that can internally choose, or be chosen externally by the environment to evolve into the process P_i for each i. Once the choice is made, all other options disappear. Sums also relate closely to linear logic's additive conjunction. Is this conceptual proximity a coincidence or indicative of deeper common structure?

11.6 Linguistics

Linguists have also investigated how mutually exclusive alternatives are expressed, often in the context of antonymy [3,4,21], but, to the best of our knowledge have not proposed formal theories of linguistic exclusion.

12 Open Problems

In this paper, we have introduced cathoristic logic and established key metalogical properties. However, many questions are left open.

12.1 Excluded Middle

One area we would like to investigate further is what happens to the law of excluded middle in cathoristic logic. The logical law of excluded middle states that either a proposition or its negation must be true. In cathoristic logic

$$\models \phi \vee \neg_S(\phi)$$

does not hold in general. (The negation operator $\neg_S(\cdot)$ was defined in Sect. 9.) For example, let ϕ be $\langle a\rangle\top$ and $S = \Sigma = \{a, b\}$. Then

$$\phi \vee \neg_S\phi \quad = \quad \langle a\rangle\top \vee !\{b\} \ \vee \ \langle a\rangle\bot$$

Now this will not in general be valid - it will be false for example in the model $((\{x\}, \emptyset, \{(x, \Sigma)\}), x)$, the model having just the start state (labelled Σ) and no transitions. Restricting S to be a proper subset of $\Sigma = \{a, b\}$ is also not enough. For example with $S = \{a\}$ we have

$$\langle a\rangle\top \vee \neg_S(\langle a\rangle\top) \quad = \quad \langle a\rangle\top \vee !\emptyset \vee \langle a\rangle\bot$$

This formula cannot hold in any cathoristic model which contains a b-labelled transition, but no a-transition from the start state.

Is it possible to identify classes of models that nevertheless verify excluded middle? The answer to this question appears to depend on the chosen notion of semantic model.

12.2 Understanding the Expressive Strength of Cathoristic Logic

Comparing Cathoristic Logic and Hennessy-Milner Logic. Section 8.1 investigated the relationship between cathoristic logic and first-order logic. Now we compare cathoristic logic with a logic that is much closer in spirit: Hennessy-Milner logic [17], a multi-modal logic designed to reason about process calculi. Indeed, the present shape of cathoristic logic owes much to Hennessy-Milner logic. We contrast both by translation from the former into the latter. This will reveal, more clearly than the translation into first-order logic, the novelty of cathoristic logic.

Definition 29. *Assume a set Σ of symbols, with s ranging over Σ, the* formulae *of Hennessy-Milner logic are given by the following grammar:*

$$\phi ::= \top \mid \bigwedge_{i\in I} \phi_i \mid \langle s\rangle\phi \mid \neg\phi$$

The index set I in the conjunction can be infinite, and needs to be so for applications in process theory.

Definition 30. Models *of Hennessy-Milner logic are simply pairs $(\mathcal{L}, s)$ where $\mathcal{L} = (S, \rightarrow)$ is a labelled transition system over Σ, and $s \in S$. The* satisfaction relation *$(\mathcal{L}, s) \models \phi$ is given by the following inductive clauses.*

$$\begin{array}{lll}(\mathcal{L}, s) \models \top & & \\ (\mathcal{L}, s) \models \bigwedge_{i \in I} \phi_i & \textit{iff} & \textit{for all } i \in I : (\mathcal{L}, s) \models \phi_i \\ (\mathcal{L}, s) \models \langle a \rangle \phi & \textit{iff} & \textit{there is a } s \xrightarrow{a} s' \textit{ such that } (l, s') \models \phi \\ (\mathcal{L}, s) \models \neg \phi & \textit{iff} & (\mathcal{L}, s) \not\models \phi\end{array}$$

There are two differences between cathoristic logic and Hennessy-Milner logic - one syntactic, the other semantic.

- Syntactically, cathoristic logic has the tantum operator (!) instead of logical negation ($\neg$).
- Semantically, cathoristic models are deterministic, while (typically) models of Hennessy-Milner logic are non-deterministic (although the semantics makes perfect sense for deterministic transition systems, too). Moreover, models of Hennessy-Milner logic lack state labels.

Definition 31. *We translate formulae of cathoristic logic into Hennessy-Milner logic using the function* $[\![\cdot]\!]$*:*

$$\begin{array}{rcl}[\![\top]\!] & = & \top \\ [\![\phi_1 \wedge \phi_2]\!] & = & [\![\phi_1]\!] \wedge [\![\phi_2]\!] \\ [\![\langle a \rangle \phi]\!] & = & \langle a \rangle [\![\phi]\!] \\ [\![!A]\!] & = & \bigwedge_{a \in \Sigma \setminus A} \neg \langle a \rangle \top\end{array}$$

If Σ is an infinite set, then the translation of a !-formula will be an infinitary conjunction. If Σ is finite, then the size of the Hennessy-Milner logic formula will be of the order of $n \cdot |\Sigma|$ larger than the original cathoristic formula, where n is the number of tantum operators occurring in the cathoristic formula). In both logics we use the number of logical operators as a measure of size.

We can also translate cathoristic models by forgetting state-labelling:

$$[\![((S, \rightarrow, \lambda), s)]\!] = ((S, \rightarrow), s)$$

We continue with an obvious consequence of the translation.

Theorem 10. *Let* $\mathfrak{M}$ *be a (deterministic or non-deterministic) cathoristic model. Then* $\mathfrak{M} \models \phi$ *implies* $[\![\mathfrak{M}]\!] \models [\![\phi]\!]$*.*

Proof. Straightforward by induction on ϕ. □

However, note that the following natural extension is *not* true under the translation above:

$$\text{If } \phi \models \psi \text{ then } [\![\phi]\!] \models [\![\psi]\!]$$

To see this, consider an entailment which relies on determinism, such as

$$\langle a \rangle \langle b \rangle \wedge \langle a \rangle \langle c \rangle \models \langle a \rangle (\langle b \rangle \wedge \langle c \rangle)$$

The first entailment is valid in cathoristic logic because of the restriction to deterministic models, but not in Hennessy-Milner logic, where it is invalidated by any model with two outgoing a transitions, one of which satisfies $\langle b \rangle$ and one of which satisfies $\langle c \rangle$.

We can restore the desired connection between cathoristic implication and Hennessy-Milner logic implication in two ways. First we can restrict our attention to deterministic models of Hennessy-Milner logic. The second solution is to add a determinism constraint to our translation. Given a set Γ of cathoristic formulae, closed under sub formulae, that contains actions from the set $A \subseteq \Sigma$, let the determinism constraint for Γ be:

$$\bigwedge_{a \in A, \phi \in \Gamma, \psi \in \Gamma} \neg \left(\langle a \rangle \phi \wedge \langle a \rangle \psi \wedge \neg \langle a \rangle (\phi \wedge \psi) \right)$$

If we add this sentence as part of our translation $[\![\cdot]\!]$, we do get the desired result that

$$\text{If } \phi \models \psi \text{ then } [\![\phi]\!] \models [\![\psi]\!]$$

Comparing Cathoristic Logic with Hennessy-Milner Logic and Propositional Logic. Consider the following six languages (Fig. 15):

Language	Description
PL[∧]	Propositional logic without negation
Hennessy-Milner logic[∧]	Hennessy-Milner logic without negation
CL[∧, !]	Cathoristic logic
PL [∧, ¬]	Full propositional logic
HML [∧, ¬]	Full Hennessy-Milner logic
CL [∧, !, ¬]	Cathoristic logic with negation

The top three languages are simple. In each case: there is no facility for expressing disjunction, every formula that is satisfiable has a simplest satisfying model, and there is a simple quadratic-time decision procedure But there are two ways in which CL[∧, !] is more expressive. Firstly, CL[∧, !], unlike HML[∧], is expressive enough to be able to distinguish between any two models that are not bisimilar, cf. Theorem 11. The second way in which CL[∧, !] is significantly more expressive than both PL[∧] and HML[∧] is in its ability to express incompatibility. No two formulae of PL[∧] or HML[∧] are incompatible[17] with each other. But many pairs of formulae of CL[∧, !] are incompatible. (For example: $\langle a \rangle \top$ and $!\emptyset$). Because CL[∧, !] is expressive enough to be able to make incompatible claims, it satisfies Brandom's incompatibility semantics constraint. CL[∧, !]

[17] The notion of incompatibility applies to all logics: two formulae are incompatible if there is no model which satisfies both.

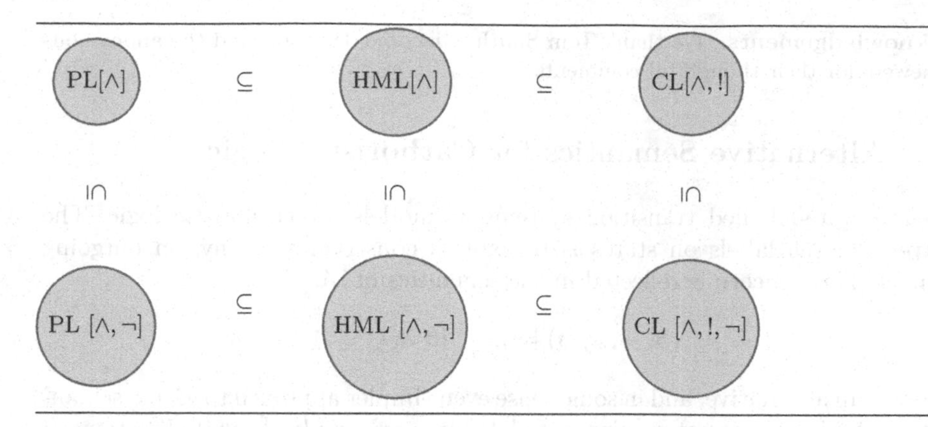

Fig. 15. Conjectured relationships of expressivity between logics. Here $L_1 \subseteq L_2$ means that the logic L_2 is more expressive than L_1. We leave the precise meaning of logical expressivity open.

is the only logic (we are aware of) with a quadratic-time decision procedure that is expressive enough to respect this constraint.

The bottom three language can all be decided in super-polynomial time. We claim that Hennessy-Milner logic is more expressive than PL, and CL$[\wedge, !, \neg]$ is more expressive than full Hennessy-Milner logic. To see that full Hennessy-Milner logic is more expressive than full propositional logic, fix a propositional logic with the nullary operator $\top$ plus an infinite number of propositional atoms $P_{(i,j)}$, indexed by i and j. Now translate each formula of Hennessy-Milner logic via the rules:

$$\llbracket \top \rrbracket = \top \qquad \llbracket \phi \wedge \psi \rrbracket = \llbracket \phi \rrbracket \wedge \llbracket \psi \rrbracket$$
$$\llbracket \neg \phi \rrbracket = \neg \llbracket \phi \rrbracket \qquad \llbracket \langle a_i \rangle \phi_j \rrbracket = P_{(i,j)}$$

We claim Hennessy-Milner logic is more expressive because there are formulae ϕ and ψ of Hennessy-Milner logic such that

$$\phi \models_{\text{HML}} \psi \text{ but } \llbracket \phi \rrbracket \not\models_{\text{PL}} \llbracket \psi \rrbracket$$

For example, let $\phi = \langle a \rangle \langle b \rangle \top$ and $\psi = \langle a \rangle \top$. Clearly, $\phi \models_{\text{HML}} \psi$. But $\llbracket \phi \rrbracket = P_{(i,j)}$ and $\llbracket \psi \rrbracket = P_{(i',j')}$ for some i, j, i', j', and there are no entailments in propositional logic between arbitrary propositional atoms.

We close by stating that CL$[\wedge, !, \neg]$ is more expressive than full Hennessy-Milner logic. As mentioned above, the formula $!A$ of cathoristic logic can be translated into Hennessy-Milner logic as:

$$\bigwedge_{a \in \Sigma - A} \neg \langle a \rangle \top$$

But if Σ is infinite, then this is an infinitary disjunction. Cathoristic logic can express the same proposition in a finite sentence.

Acknowledgements. We thank Tom Smith, Giacomo Turbanti and the anonymous reviewers for their thoughtful comments.

A Alternative Semantics for Cathoristic Logic

We use state-labelled transition systems as models for cathoristic logic. The purpose of the labels on states is to express constraints, if any, on outgoing actions. This concern is reflected in the semantics of $!A$.

$$((S, \rightarrow, \lambda), s) \models !A \quad \text{iff } \lambda(s) \subseteq A$$

There is an alternative, and in some sense even simpler approach to giving semantics to $!A$ which does not require state-labelling: we simply check if all actions of all outgoing transitions at the current state are in A. As the semantics of other formula requires state-labelling in its satisfaction condition, this means we can use plain labelled transition systems (together with a current state) as models. This gives rise to a subtly different theory that we now explore, albeit not in depth.

A.1 Pure Cathoristic Models

Definition 32. *By a* pure cathoristic model, *ranged over by* $\mathfrak{P}, \mathfrak{P}', \ldots$, *we mean a pair* $(\mathcal{L}, s)$ *where* $\mathcal{L} = (S, \rightarrow)$ *is a deterministic labelled transition system and* $s \in S$ *a state.*

Adapting the satisfaction relation to pure cathoristic models is straightforward.

Definition 33. *Using pure cathoristic models, the* satisfaction relation *is defined inductively by the following clauses, where we assume that* $\mathfrak{M} = (\mathcal{L}, s)$ *and* $\mathcal{L} = (S, \rightarrow)$.

$$\begin{array}{lll}
\mathfrak{M} \models \top & & \\
\mathfrak{M} \models \phi \land \psi & \textit{iff} & \mathfrak{M} \models \phi \textit{ and } \mathfrak{M} \models \psi \\
\mathfrak{M} \models \langle a \rangle \phi & \textit{iff} & \textit{there is a } s \xrightarrow{a} t \textit{ such that } (\mathcal{L}, t) \models \phi \\
\mathfrak{M} \models A & \textit{iff} & \{a \mid \exists t.s \xrightarrow{a} t\} \subseteq A
\end{array}$$

Note that all but the last clause are unchanged from Definition 4.

In this interpretation, $!A$ restricts the out-degree of the current state s, i.e. it constraints the 'width' of the graph. It is easy to see that all rules in Fig. 12 are sound with respect to the new semantics. The key advantage pure cathoristic models have is their simplicity: they are unadorned labelled transition systems, the key model of concurrency theory [26]. The connection with concurrency theory is even stronger than that, because, as we show below (Theorem 11), the elementary equivalence on (finitely branching) pure cathoristic models is bisimilarity, one of the more widely used notions of process equivalence. This characterisation even holds if we remove the determinacy restriction in Definition 32.

A.2 Relationship Between Pure and Cathoristic Models

The obvious way of converting an cathoristic model into a pure cathoristic model is by forgetting about the state-labelling:

$$((S, \rightarrow, \lambda), s) \qquad \mapsto \qquad ((S, \rightarrow), s)$$

Let this function be $\mathsf{forget}(\cdot)$. For going the other way, we have two obvious choices:

- $((S, \rightarrow), s) \mapsto ((S, \rightarrow, \lambda), s)$ where $\lambda(t) = \Sigma$ for all states t. Call this map $\mathsf{max}(\cdot)$.
- $((S, \rightarrow), s) \mapsto ((S, \rightarrow, \lambda), s)$ where $\lambda(t) = \{a \mid \exists t'.t \xrightarrow{a} t'\}$ for all states t. Call this map $\mathsf{min}(\cdot)$.

Lemma 14. *Let $\mathfrak{M}$ be an cathoristic model, and $\mathfrak{P}$ a pure cathoristic model.*

1. *$\mathfrak{M} \models \phi$ implies $\mathsf{forget}(\mathfrak{M}) \models \phi$. The reverse implication does not hold.*
2. *$\mathsf{max}(\mathfrak{P}) \models \phi$ implies $\mathfrak{P} \models \phi$. The reverse implication does not hold.*
3. *$\mathsf{min}(\mathfrak{P}) \models \phi$ if and only if $\mathfrak{P} \models \phi$.*

Proof. The implication in (1) is immediate by induction on ϕ. A counterexample for the reverse implication is given by the formula $\phi =!\{a\}$ and the cathoristic model $\mathfrak{M} = (\{s, t\}, s \xrightarrow{a} t, \lambda), s)$ where $\lambda(s) = \{a, b, c\}$: clearly $\mathsf{forget}(\mathfrak{M}) \models \phi$, but $\mathfrak{M} \not\models \phi$.

The implication in (2) is immediate by induction on ϕ. To construct a counterexample for the reverse implication, assume that Σ is a strict superset of $\{a\}$ a. The formula $\phi =!\{a\}$ and the pure cathoristic model $\mathfrak{P} = (\{s, t\}, s \xrightarrow{a} t), s)$ satisfy $\mathfrak{P} \models \phi$, but clearly $\mathsf{max}(\mathfrak{P}) \not\models \phi$.

Finally, (3) is also straightforward by induction on ϕ. □

A.3 Non-determinism and Cathoristic Models

Both, cathoristic models and pure cathoristic models must be deterministic. That is important for the incompatibility semantics. However, formally, the definition of satisfaction makes sense for non-deterministic models as well, pure or otherwise. Such models are important in the theory of concurrent processes. Many of the theorems of the precious section either hold directly, or with small modifications for non-deterministic models. The rules of inference in Fig. 12 are sound except for [DETERMINISM] which cannot hold in properly non-deterministic models. With this omission, they are also complete. Elementary equivalence on non-deterministic cathoristic models also coincides with mutual simulation, while elementary equivalence on non-deterministic pure cathoristic models is bisimilarity. The proofs of both facts follow those of Theorems 1 and 11, respectively. Compactness by translation can be shown following the proof in Sect. 8, except that the constraint ϕ_{det} is unnecessary.

We have experimented with a version of cathoristic logic in which the models are *non-deterministic* labelled-transition systems. Although non-determinism makes some of the constructions simpler, non-deterministic cathoristic logic is unable to express incompatibility properly. Consider, for example, the claim that Jack is married[18] to Jill In standard deterministic cathoristic logic this would be rendered as:

$$\langle jack \rangle \langle married \rangle (\langle jill \rangle \wedge !\{jill\})$$

There are three levels at which this claim can be denied. First, we can claim that Jack is married to someone else - Joan, say:

$$\langle jack \rangle \langle married \rangle (\langle joan \rangle \wedge !\{joan\})$$

Second, we can claim that Jack is unmarried (specifically, that being unmarried is Jack's only property):

$$\langle jack \rangle !\{unmarried\}$$

Third, we can claim that Jack does not exist at all. Bob and Jill, for example, are the only people in our domain:

$$!\{bob, jill\}$$

Now we can assert the same sentences in non-deterministic cathoristic logic, but they are *no longer incompatible with our original sentence.* In non-deterministic cathoristic logic, the following sentences are compatible (as long as there are two separate transitions labelled with *married*, or two separate transitions labelled with *jack*):

$$\langle jack \rangle \langle married \rangle (\langle jill \rangle \wedge !\{jill\}) \qquad \langle jack \rangle \langle married \rangle (\langle joan \rangle \wedge !\{joan\})$$

Similarly, the following sentences are fully compatible as long as there are two separate transitions labelled with *jack*:

$$\langle jack \rangle \langle married \rangle \qquad \langle jack \rangle !\{unmarried\}$$

Relatedly, non-deterministic cathoristic logic does not satisfy Brandom's incompatibility semantics property:

$$\phi \models \psi \text{ iff } \mathcal{I}(\psi) \subseteq \mathcal{I}(\phi)$$

To take a simple counter-example, $\langle a \rangle \langle b \rangle$ implies $\langle a \rangle$, but not conversely. But in non-deterministic cathoristic logic, the set of sentences incompatible with $\langle a \rangle \langle b \rangle$ is identical with the set of sentences incompatible with $\langle a \rangle$.

[18] We assume, in this discussion, that *married* is a many-to-one predicate. We assume that polygamy is one person *attempting* to marry two people (but failing to marry the second).

A.4 Semantic Characterisation of Elementary Equivalence

In Sect. 6.1 we presented a semantic analysis of elementary equivalence, culminating in Theorem 1 which showed that elementary equivalence coincides with $\simeq$, the relation of mutual simulation of models. We shall now carry out a similar analysis for pure cathoristic models, and show that elementary equivalence coincides with bisimilarity, an important concept in process theory and modal logics [25]. Bisimilarity is strictly finer on non-deterministic transition systems than $\simeq$, and more sensitive to branching structure. In the rest of this section, we allow non-deterministic pure models, because the characterisation is more interesting that in the deterministic case.

Definition 34. *A pure cathoristic model $(\mathcal{L}, s)$ is finitely branching if its underlying transition system $\mathcal{L}$ is finitely branching.*

Definition 35. *A binary relation $\mathcal{R}$ is a* bisimulation *between pure cathoristic models $\mathfrak{P}_i = (\mathcal{L}_i), s_i)$ for $i = 1, 2$ provided (1) $\mathcal{R}$ is a bisimulation between $\mathcal{L}_1$ and $\mathcal{L}_2$, and (2) $(s_1, s_2) \in \mathcal{R}$. We say $\mathfrak{P}_1$ and $\mathfrak{P}_2$ are* bisimilar, *written $\mathfrak{P}_1 \sim \mathfrak{P}_2$ if there is a bisimulation between $\mathfrak{P}_1$ and $\mathfrak{P}_2$.*

Definition 36. *The* theory *of $\mathfrak{P}$, written $\mathsf{Th}(\mathfrak{P})$, is the set $\{\phi \mid \mathfrak{P} \models \phi\}$.*

Theorem 11. *Let $\mathfrak{P}$ and $\mathfrak{P}'$ be two finitely branching pure cathoristic models. Then: $\mathfrak{P} \sim \mathfrak{P}'$ if and only if $\mathsf{Th}(\mathfrak{P}) = \mathsf{Th}(\mathfrak{P}')$.*

Proof. Let $\mathfrak{P} = (\mathcal{L}, w)$ and $\mathfrak{P}' = (\mathcal{L}', w')$ be finitely branching, where $\mathcal{L} = (W, \rightarrow)$ and $(W', \rightarrow')$. We first show the left to right direction, so assume that $\mathfrak{P} \sim \mathfrak{P}'$.

The proof is by induction on formulae. The only case which differs from the standard Hennessy-Milner theorem is the case for $!A$, so this is the only case we shall consider. Assume $w \sim w'$ and $w \models !A$. We need to show $w' \models !A$. From the semantic clause for $!$, $w \models !A$ implies $\lambda(w) \subseteq A$. If $w \sim w'$, then $\lambda(w) = \lambda'(w')$. Therefore $\lambda'(w') \subseteq A$, and hence $w' \models !A$.

The proof for the other direction is more involved. For states $x \in W$ and $x' \in W$, we write

$$x \equiv x' \qquad \text{iff} \qquad \mathsf{Th}((\mathcal{L}, x)) = \mathsf{Th}((\mathcal{L}', x')).$$

We define the bisimilarity relation:

$$Z = \{(x, x') \in \mathcal{W} \times \mathcal{W}' \mid x \equiv x'\}$$

To prove $w \sim w'$, we need to show:

- $(w, w') \in Z$. This is immediate from the definition of Z.
- The relation Z respects the transition-restrictions: if $(x, x') \in Z$ then $\lambda(x) = \lambda'(x')$
- The forth condition: if $(x, x') \in Z$ and $x \xrightarrow{a} y$, then there exists a y' such that $x' \xrightarrow{a} y'$ and $(y, y') \in Z$.

– The back condition: if $(x, x') \in Z$ and $x' \xrightarrow{a} y'$, then there exists a y such that $x \xrightarrow{a} y$ and $(y, y') \in Z$.

To show that $(x, x') \in Z$ implies $\lambda(x) = \lambda'(x')$, we will argue by contraposition. Assume $\lambda(x) \neq \lambda'(x')$. Then either $\lambda'(x') \not\subseteq \lambda(x)$ or $\lambda(x) \not\subseteq \lambda'(x')$. If $\lambda'(x') \not\subseteq \lambda(x)$, then $x' \not\models !\lambda(x)$. But $x \models !\lambda(x)$, so x and x' satisfy different sets of propositions and are not equivalent. Similarly, if $\lambda(x) \not\subseteq \lambda'(x')$ then $x \not\models !\lambda'(x')$. But $x' \models !\lambda'(x')$, so again x and x' satisfy different sets of propositions and are not equivalent.

We will show the forth condition in detail. The back condition is similar. To show the forth condition, assume that $x \xrightarrow{a} y$ and that $(x, x') \in Z$ (i.e. $x \equiv x'$). We need to show that $\exists y'$ such that $x' \xrightarrow{a} y'$ and $(y, y') \in Z$ (i.e. $y \equiv y'$).

Consider the set of y'_i such that $x' \xrightarrow{a} y'_i$. Since $x \xrightarrow{a} y$, $x \models \langle a \rangle \top$, and as $x \equiv x'$, $x' \models \langle a \rangle \top$, so we know this set is non-empty. Further, since $(\mathcal{W}', \rightarrow')$ is finitely-branching, there is only a finite set of such y'_i, so we can list them $y'_1, ..., y'_n$, where $n >= 1$.

Now, in the Hennessy-Milner theorem for Hennessy-Milner logic, the proof proceeds as follows: assume, for reductio, that of the $y'_1, ..., y'_n$, there is no y'_i such that $y \equiv y'_i$. Then, by the definition of $\equiv$, there must be formulae $\phi_1, ..., \phi_n$ such that for all i in 1 to n:

$$y'_i \models \phi_i \text{ and } y \not\models \phi_i$$

Now consider the formula:

$$[a](\phi_1 \vee ... \vee \phi_n)$$

As each $y'_i \models \phi_i$, $x' \models [a](\phi_1 \vee ... \vee \phi_n)$, but x does not satisfy this formula, as each ϕ_i is not satisfied at y. Since there is a formula which x and x' do not agree on, x and x' are not equivalent, contradicting our initial assumption.

But this proof cannot be used in cathoristic logic because it relies on a formula $[a](\phi_1 \vee ... \vee \phi_n)$ which cannot be expressed in cathoristic logic: Cathoristic logic does not include the box operator or disjunction, so this formula is ruled out on two accounts. But we can massage it into a form which is more amenable to cathoristic logic's expressive resources:

$$\begin{aligned}[a](\phi_1 \vee ... \vee \phi_n) &= \neg\langle a \rangle \neg(\phi_1 \vee ... \vee \phi_n) \\ &= \neg\langle a \rangle(\neg\phi_1 \wedge ... \wedge \neg\phi_n)\end{aligned}$$

Further, if the original formula $[a](\phi_1 \vee ... \vee \phi_n)$ is true in x' but not in x, then its negation will be true in x but not in x'. So we have the following formula, true in x but not in x':

$$\langle a \rangle(\neg\phi_1 \wedge ... \wedge \neg\phi_n)$$

The reason for massaging the formula in this way is so we can express it in cathoristic logic (which does not have the box operator or disjunction). At this moment, the revised formula is *still* outside cathoristic logic because it uses negation. But we are almost there: the remaining negation is in innermost scope, and innermost scope negation can be simulated in cathoristic logic by the ! operator.

We are assuming, for reductio, that of the $y'_1, ..., y'_n$, there is no y'_i such that $y \equiv y'_i$. But in cathoristic logic without negation, we cannot assume that each y'_i has a formula ϕ_i which is satisfied by y'_i but not by y - it might instead be the other way round: ϕ_i may be satisfied by y but not by y'_i. So, without loss of generality, assume that $y'_1, ..., y'_m$ fail to satisfy formulae $\phi_1, ..., \phi_m$ which y does satisfy, and that $y'_{m+1}, ..., y'_n$ satisfy formulae $\phi_{m+1}, ..., \phi_n$ which y does not:

$$y \models \phi_i \text{ and } y'_i \not\models \phi_i \quad i = 1 \text{ to } m$$
$$y \not\models \phi_j \text{ and } y'_j \models \phi_j \quad j = m+1 \text{ to } n$$

The formula we will use to distinguish between x and x' is:

$$\langle a \rangle (\bigwedge_{i=1}^{m} \phi_i \wedge \bigwedge_{j=m+1}^{n} \mathsf{neg}(y, \phi_j))$$

Here, neg is a meta-language function that, given a state y and a formula ϕ_j, returns a formula that is true in y but incompatible with ϕ_j. We will show that, since $y \not\models \phi_j$, it is always possible to construct $\mathsf{neg}(y, \phi_j)$ using the ! operator.

Consider the possible forms of ϕ_j:

- $\top$: this case cannot occur since all models satisfy $\top$.
- $\phi_1 \wedge \phi_2$: we know $y'_j \models \phi_1 \wedge \phi_2$ and $y \not\models \phi_1 \wedge \phi_2$. There are three possibilities:
  1. $y \not\models \phi_1$ and $y \models \phi_2$. In this case, $\mathsf{neg}(y, \phi_1 \wedge \phi_2) = \mathsf{neg}(y, \phi_1) \wedge \phi_2$.
  2. $y \models \phi_1$ and $y \not\models \phi_2$. In this case, $\mathsf{neg}(y, \phi_1 \wedge \phi_2) = \phi_1 \wedge \mathsf{neg}(y, \phi_2)$.
  3. $y \not\models \phi_1$ and $y \not\models \phi_2$. In this case, $\mathsf{neg}(y, \phi_1 \wedge \phi_2) = \mathsf{neg}(y, \phi_1) \wedge \mathsf{neg}(y, \phi_2)$.
- $!A$: if $y \not\models !A$ and $y'_j \models !A$, then there is an action $a \in \Sigma - A$ such that $y \xrightarrow{a} z$ for some z but there is no such z such that $y'_j \xrightarrow{a} z$. In this case, let $\mathsf{neg}(y, \phi_j) = \langle a \rangle \top$.
- $\langle a \rangle \phi$. There are two possibilities:
  1. $y \models \langle a \rangle \top$. In this case, $\mathsf{neg}(y, \langle a \rangle \phi) = \bigwedge_{y \xrightarrow{a} z} \langle a \rangle \mathsf{neg}(z, \phi)$.
  2. $y \not\models \langle a \rangle \top$. In this case, $\mathsf{neg}(y, \langle a \rangle \phi) = !\{b \mid \exists z. y \xrightarrow{b} z\}$. This set of bs is finite since we are assuming the transition system is finitely-branching.

□

We continue with a worked example of neg. Consider y and y'_j as in Fig. 16. One formula that is true in y'_j but not in y is

$$\langle a \rangle (\langle b \rangle \top \wedge \langle c \rangle \top)$$

Fig. 16. Worked example of neg. Note that the transition system on the left is non-deterministic.

Now:

$$\begin{aligned}
&\mathsf{neg}(y, \langle a\rangle(\langle b\rangle\top \wedge \langle c\rangle\top)) \\
&\qquad = \bigwedge_{y \xrightarrow{a} z} \langle a\rangle \mathsf{neg}(z, \langle b\rangle\top \wedge \langle c\rangle\top) \\
&\qquad = \langle a\rangle \mathsf{neg}(z_1, \langle b\rangle\top \wedge \langle c\rangle\top) \wedge \langle a\rangle \mathsf{neg}(z_2, \langle b\rangle\top \wedge \langle c\rangle\top) \\
&\qquad = \langle a\rangle(\langle b\rangle\top \wedge \mathsf{neg}(z_1, \langle c\rangle\top)) \wedge \langle a\rangle \mathsf{neg}(z_2, \langle b\rangle\top \wedge \langle c\rangle\top) \\
&\qquad = \langle a\rangle(\langle b\rangle\top \wedge \mathsf{neg}(z_1, \langle c\rangle\top)) \wedge \langle a\rangle(\mathsf{neg}(z_2, \langle b\rangle\top) \wedge \langle c\rangle\top) \\
&\qquad = \langle a\rangle(\langle b\rangle\top \wedge !\{b\}) \wedge \langle a\rangle(\mathsf{neg}(z_2, \langle b\rangle\top) \wedge \langle c\rangle\top) \\
&\qquad = \langle a\rangle(\langle b\rangle\top \wedge !\{b\}) \wedge \langle a\rangle(!\{c\} \wedge \langle c\rangle\top)
\end{aligned}$$

The resulting formula is true in y but not in y'_j.

B Omitted Proofs

B.1 Proof of Lemma 5

If $\mathfrak{M} \models \phi$ then $\mathfrak{M} \preceq \mathsf{simpl}(\phi)$.

Proof. We shall show $\mathsf{Th}(\mathsf{simpl}(\phi)) \subseteq \mathsf{Th}(\mathfrak{M})$. The desired result will then follow by applying Theorem 1. We shall show that

$$\text{If } \mathfrak{M} \models \phi \text{ then } \mathsf{Th}(\mathsf{simpl}(\phi)) \subseteq \mathsf{Th}(\mathfrak{M})$$

by induction on ϕ. In all the cases below, let $\mathsf{simpl}(\phi) = (\mathcal{L}, w)$ and let $\mathfrak{M} = (\mathcal{L}', w')$. The case where $\phi = \top$ is trivial. Next, assume $\phi = \langle a\rangle\psi$. We know $\mathfrak{M} \models \langle a\rangle\psi$ and need to show that $\mathsf{Th}(\mathsf{simpl}(\langle a\rangle\psi)) \subseteq \mathsf{Th}(\mathfrak{M})$. Since $(\mathcal{L}', w') \models \langle a\rangle\psi$, there is an x' such that $w' \xrightarrow{a} x'$ and $(\mathcal{L}', x') \models \psi$. Now from the definition of $\mathsf{simpl}()$, $\mathsf{simpl}(\langle a\rangle\psi)$ is a model combining $\mathsf{simpl}(\psi)$ with a new state w not appearing in $\mathsf{simpl}(\psi)$ with an arrow $w \xrightarrow{a} x$ (where x is the start state in $\mathsf{simpl}(\psi)$), and $\lambda(w) = \Sigma$. Consider any sentence ξ such that $\mathsf{simpl}(\langle a\rangle\psi) \models \xi$. Given the construction of $\mathsf{simpl}(\langle a\rangle\psi)$, ξ must be a conjunction of $\top$ and formulae of the form $\langle a \rangle \tau$. In the first case, $(\mathcal{L}', x')$ satisfies $\top$; in the second case, $(\mathcal{L}', x') \models \tau$ by the induction hypothesis and hence $(\mathcal{L}', w') \models \langle a \rangle \tau$.

Next, consider the case where $\phi = !A$, for some finite set $A \subset \Sigma$. From the definition of $\mathsf{simpl}()$, $\mathsf{simpl}(!A)$ is a model with one state s, no transitions, with $\lambda(s) = A$. Now the only formulae that are true in $\mathsf{simpl}(!A)$ are conjunctions of $\top$ and $!B$, for supersets $B \supseteq A$. If $\mathfrak{M} \models !A$ then by the semantic clause for $!$, $\lambda'(w') \subseteq A$, hence $\mathfrak{M}$ models all the formulae that are true in $\mathsf{simpl}(!A)$.

Finally, consider the case where $\phi = \psi_1 \wedge \psi_2$. Assume $\mathfrak{M} \models \psi_1$ and $\mathfrak{M} \models \psi_2$. We assume, by the induction hypothesis that $\mathsf{Th}(\mathsf{simpl}(\psi_1)) \subseteq \mathsf{Th}(\mathfrak{M})$ and $\mathsf{Th}(\mathsf{simpl}(\psi_2)) \subseteq \mathsf{Th}(\mathfrak{M})$. We need to show that $\mathsf{Th}(\mathsf{simpl}(\psi_1 \wedge \psi_2)) \subseteq \mathsf{Th}(\mathfrak{M})$. By the definition of $\mathsf{simpl}()$, $\mathsf{simpl}(\psi_1 \wedge \psi_2) = \mathsf{simpl}(\psi_1) \sqcap \mathsf{simpl}(\psi_2)$. If $\mathsf{simpl}(\psi_1)$ and $\mathsf{simpl}(\psi_2)$ are $\mathsf{inconsistent}$ (see the definition of $\mathsf{inconsistent}$ in Sect. 6.4) then $\mathfrak{M} = \bot$. In this case, $\mathsf{Th}(\mathsf{simpl}(\psi_1) \wedge \mathsf{simpl}(\psi_2)) \subseteq \mathsf{Th}(\bot)$. If, on the other hand, $\mathsf{simpl}(\psi_1)$ and $\mathsf{simpl}(\psi_2)$ are not $\mathsf{inconsistent}$, we shall show that $\mathsf{Th}(\mathsf{simpl}(\psi_1 \wedge \psi_2)) \subseteq \mathsf{Th}(\mathfrak{M})$ by reductio. Assume a formula ξ such that $\mathsf{simpl}(\psi_1 \wedge \psi_2) \models \xi$ but $\mathfrak{M} \not\models \xi$. Now $\xi \neq \top$ because all models satisfy $\top$. ξ cannot be of the form $\langle a \rangle \tau$ because, by the construction of merge (see Sect. 6.4), all transitions in $\mathsf{simpl}(\psi_1 \wedge \psi_2)$ are transitions from $\mathsf{simpl}(\psi_1)$ or $\mathsf{simpl}(\psi_2)$ and we know from the inductive hypothesis that $\mathsf{Th}(\mathsf{simpl}(\psi_1)) \subseteq \mathsf{Th}(\mathfrak{M})$ and $\mathsf{Th}(\mathsf{simpl}(\psi_2)) \subseteq \mathsf{Th}(\mathfrak{M})$. ξ cannot be $!A$ for some $A \subset \Sigma$, because, from the construction of merge, all state-labellings in $\mathsf{simpl}(\psi_1 \wedge \psi_2)$ are no more specific than the corresponding state-labellings in $\mathsf{simpl}(\psi_1)$ and $\mathsf{simpl}(\psi_2)$, and we know from the inductive hypothesis that $\mathsf{Th}(\mathsf{simpl}(\psi_1)) \subseteq \mathsf{Th}(\mathfrak{M})$ and $\mathsf{Th}(\mathsf{simpl}(\psi_2)) \subseteq \mathsf{Th}(\mathfrak{M})$. Finally, ξ cannot be $\xi_1 \wedge xi_2$ because the same argument applies to xi_1 and xi_2 individually. We have exhausted the possible forms of ξ, so conclude that there is no formula ξ such that $\mathsf{simpl}(\psi_1 \wedge \psi_2) \models \xi$ but $\mathfrak{M} \not\models \xi$. Hence $\mathsf{Th}(\mathsf{simpl}(\psi_1 \wedge \psi_2)) \subseteq \mathsf{Th}(\mathfrak{M})$. □

B.2 Proof of Lemma 6

If $\phi \models \psi$ then $\mathsf{simpl}(\phi) \preceq \mathsf{simpl}(\psi)$

Proof. By Theorem 1, $\mathsf{simpl}(\phi) \preceq \mathsf{simpl}(\psi)$ iff $\mathsf{Th}(\mathsf{simpl}(\psi)) \subseteq \mathsf{Th}(\mathsf{simpl}(\phi))$. Assume $\phi \models \psi$, and assume $\xi \in \mathsf{Th}(\mathsf{simpl}(\psi))$. We must show $\xi \in \mathsf{Th}(\mathsf{simpl}(\phi))$. Now $\mathsf{simpl}()$ is constructed so that:

$$\mathsf{simpl}(\psi) = \bigsqcup \{\mathfrak{M} \mid \mathfrak{M} \models \psi\}$$

So $\xi \in \mathsf{Th}(\mathsf{simpl}(\psi))$ iff for all models $\mathfrak{M}$, $\mathfrak{M} \models \psi$ implies $\mathfrak{M} \models \xi$. We must show that $\mathfrak{M} \models \phi$ implies $\mathfrak{M} \models \xi$ for all models $\mathfrak{M}$. Assume $\mathfrak{M} \models \phi$. Then since $\phi \models \psi$, $\mathfrak{M} \models \psi$. But since $\xi \in \mathsf{Th}(\mathsf{simpl}(\psi))$, $\mathfrak{M} \models \xi$ also. □

B.3 Proof of Lemma 7

If $\mathcal{I}(\psi) \subseteq \mathcal{I}(\phi)$ then $\mathcal{J}(\mathsf{simpl}(\psi)) \subseteq \mathcal{J}(\mathsf{simpl}(\phi))$

Proof. Assume $\mathcal{I}(\psi) \subseteq \mathcal{I}(\phi)$ and $\mathfrak{M} \sqcap \mathsf{simpl}(\psi) = \bot$. We need to show $\mathfrak{M} \sqcap \mathsf{simpl}(\phi) = \bot$. If $\mathcal{I}(\psi) \subseteq \mathcal{I}(\phi)$ then for all formulae ξ, if $\mathsf{simpl}(\xi) \sqcap \mathsf{simpl}(\psi) = \bot$ then $\mathsf{simpl}(\xi) \sqcap \mathsf{simpl}(\phi) = \bot$. Let ξ be $\mathsf{char}(\mathfrak{M})$. Given that $\mathfrak{M} \sqcap \mathsf{simpl}(\psi) = \bot$ and $\mathsf{simpl}(\mathsf{char}(\mathfrak{M})) \preceq \mathfrak{M}$, $\mathsf{simpl}(\mathsf{char}(\mathfrak{M})) \sqcap \mathsf{simpl}(\psi) = \bot$. Then as $\mathcal{I}(\psi) \subseteq \mathcal{I}(\phi)$, $\mathsf{simpl}(\mathsf{char}(\mathfrak{M})) \sqcap \mathsf{simpl}(\phi) = \bot$. Now as $\mathfrak{M} \preceq \mathsf{simpl}(\mathsf{char}(\mathfrak{M}))$, $\mathfrak{M} \sqcap \mathsf{simpl}(\phi) = \bot$. □

References

1. Haskell implementation of cathoristic logic. Submitted with the paper (2014)
2. Abramsky, S.: Computational interpretations of linear logic. TCS **111**, 3–57 (1993)
3. Allan, K. (ed.): Concise Encyclopedia of Semantics. Elsevier, Boston (2009)
4. Aronoff, M., Rees-Miller, J. (eds.): The Handbook of Linguistics. Wiley-Blackwell, Hoboken (2003)
5. Blackburn, P., de Rijke, M., Venema, Y.: Modal Logic. Cambridge University Press, Cambridge (2001)
6. Brachman, R., Levesque, H.: Knowledge Representation and Reasoning. Morgan Kaufmann, Burlington (2004)
7. Brandom, R.: Making It Explicit. Harvard University Press, Cambridge (1998)
8. Brandom, R.: Between Saying and Doing. Oxford University Press, Oxford (2008)
9. Davey, B.A., Priestley, H.A.: Introduction to Lattices and Order. Cambridge University Press, Cambridge (1990)
10. Davidson, D.: Essays on Actions and Events. Oxford University Press, Oxford (1980)
11. Enderton, H.B.: A Mathematical Introduction to Logic. Academic Press, Cambridge (2001)
12. Evans, R., Short, E.: Versu. http://www.versu.com. https://itunes.apple.com/us/app/blood-laurels/id882505676?mt=8
13. Evans, R., Short, E.: Versu - a simulationist storytelling system. IEEE Trans. Comput. Intell. AI Games **6**(2), 113–130 (2014)
14. Fikes, R., Nilsson, N.: Strips: a new approach to the application of theorem proving to problem solving. Artif. Intell. **2**, 189–208 (1971)
15. Girard, J.-Y.: Linear logic. TCS **50**, 1–101 (1987)
16. Hennessy, M.: Algebraic Theory of Processes. MIT Press Series in the Foundations of Computing. MIT Press, Cambridge (1988)
17. Hennessy, M., Milner, R.: Algebraic laws for non-determinism and concurrency. JACM **32**(1), 137–161 (1985)
18. Honda, K.: A Theory of Types for the π-Calculus, March 2001. http://www.dcs.qmul.ac.uk/~kohei/logics
19. Honda, K., Vasconcelos, V.T., Kubo, M.: Language primitives and type discipline for structured communication-based programming. In: Hankin, C. (ed.) ESOP 1998. LNCS, vol. 1381, pp. 122–138. Springer, Heidelberg (1998). https://doi.org/10.1007/BFb0053567
20. Honda, K., Yoshida, N.: A uniform type structure for secure information flow. SIGPLAN Not. **37**, 81–92 (2002)
21. O'Keeffe, A., McCarthy, M. (eds.): The Routledge Handbook of Corpus Linguistics. Routledge, Abingdon (2010)
22. Peregrin, J.: Logic as based on incompatibility (2010). http://philpapers.org/rec/PERLAB-2

23. Pitts, A.M.: Nominal Sets: Names and Symmetry in Computer Science. Cambridge Tracts in Theoretical Computer Science. Cambridge University Press, Cambridge (2013)
24. Russell, B.: An Inquiry into Meaning and Truth. Norton and Co, New York (1940)
25. Sangiorgi, D.: Introduction to Bisimulation and Coinduction. Cambridge University Press, Cambridge (2012)
26. Sassone, V., Nielsen, M., Winskel, G.: Models for concurrency: towards a classification. TCS **170**(1–2), 297–348 (1996)
27. Smith, D., Genesereth, M.: Ordering conjunctive queries. Artif. Intell. **26**, 171–215 (1985)
28. Sommers, F.: The Logic of Natural Language. Clarendon Press, Oxford (1982)
29. Takeuchi, K., Honda, K., Kubo, M.: An interaction-based language and its typing system. In: Halatsis, C., Maritsas, D., Philokyprou, G., Theodoridis, S. (eds.) PARLE 1994. LNCS, vol. 817, pp. 398–413. Springer, Heidelberg (1994). https://doi.org/10.1007/3-540-58184-7_118
30. Troelstra, A.S., Schwichtenberg, H.: Basic Proof Theory, 2nd edn. Cambridge University Press, Cambridge (2000)
31. Turbanti, G.: Modality in Brandom's incompatibility semantics. In: Proceedings of the Amsterdam Graduate Conference - Truth, Meaning, and Normativity (2011)
32. van Dalen, D.: Logic and Structure. Springer, Heidelberg (2004). https://doi.org/10.1007/978-3-540-85108-0
33. Wittgenstein, L.: Philosophische Bemerkungen. Suhrkamp Verlag, Frankfurt (1981). Edited by R. Rhees
34. Wittgenstein, L.: Tractatus Logico-Philosophicus: Logisch-Philosophische Abhandlung. Suhrkamp Verlag, Frankfurt (2003). Originally published: 1921

A Type Theory for Probabilistic λ–calculus

Alessandra Di Pierro(✉)

Dipartimento di Informatica, Università di Verona,
Strada le Grazie, 15, 37134 Verona, Italy
alessandra.dipierro@univr.it

Abstract. We present a theory of types where formulas may contain a choice constructor. This constructor allows for the selection of a particular type among a finite set of options, each corresponding to a given probabilistic term. We show that this theory induces a type assignment system for the probabilistic λ–calculus introduced in an earlier work by Chris Hankin, Herbert Wiklicky and the author, where probabilistic terms represent probability distributions on classical terms of the simply typed λ–calculus. We prove the soundness of the type assignment with respect to a probabilistic term reduction and a normalization property of the latter.

Keywords: Probabilistic reduction system · Type system · Probability distribution

1 Introduction

From a computer science viewpoint, the simply typed λ–calculus can be seen as an idealised language which takes into account some of the very basic concepts of programming languages, such as function application, scoping and evaluation strategy. In order to incorporate the features of the real-world programming languages, the simply typed λ–calculus has been extended in various ways by essentially including new types and related rules in the underlying type theory. These extensions range from the inclusions of data structure features, such as tuples, records, sums and variants (see e.g. [20] for a detailed account), to more function-oriented enrichments which are usually referred to as polymorphic extensions [7,8,14,20,24].

For probabilistic computation, a simply typed λ–calculus, i.e. a typed version of probabilistic λ–calculus, must include, together with the canonical type constructor ($\rightarrow$) that builds function types, also a type constructor for 'probabilistic' types, i.e. objects of a syntactic nature that are assigned to the probabilistic terms of the calculus. The definition of probabilistic terms we consider in this paper is the one introduced in [11] as a result of a collaborative intensive as much as enjoyable work of the author with Chris Hankin and Herbert Wiklicky on the semantics and analysis of probabilistic computation. The question of what

A. Di Pierro et al. (Eds.): Festschrift Hankin, LNCS 12065, pp. 86–102, 2020.
https://doi.org/10.1007/978-3-030-41103-9_3

all the above mentionned extensions to the classical theory of λ–calculus would mean in a probabilistic setting is not an easy one and gives a clue for further investigations. As a first step, we introduce here a novel theory of types and use it for assigning types to the probabilistic λ–calculus introduced in [11] so as to yield a *probabilistic simply-typed* λ*–calculus*.

A type in our theory must be thought of as a vector space of probability distributions, the latter representing the *probabilistic terms* inhabiting that type. In the model we have in mind, when a function is called on a probabilistic term M, the type of the actual argument selected probabilistically will be an appropriate subspace of the type of M, where the computation will take place with the probability associated to that argument.

We define a probabilistic reduction system with a strategy similar to the lazy Call-by-Value strategy (see e.g. [21]), and prove that it is correct with respect to an assignment system based on the new type theory.

In order to ensure type preservation under reduction, we equip our type theory with a suitable notion of subtyping. This is an essential feature of our calculus as it is used to define the selection of the correct components of a probabilistic term in the typing rule for the probabilistic terms.

The structure of the paper is as follows. In Sect. 2 we present the calculus, namely the type theory, the type assignment system. In Sect. 3 we define the reduction rules and establish their correctness with respect to the type system by means of a non standard subject reduction theorem. In particular, since we have no explicit subsumption rules, the subject reduction theorem establishes a kind of subtype preservation. Moreover, we prove a progress theorem ensuring that no well-typed closed terms are stuck during the reduction process. We also show an important property of our calculus, namely the normalization of the reduction system with respect to the adopted strategy. We then illustrate the main features of our calculus in Sect. 4 by means of a number of examples. In Sect. 5, we give an overview of other probabilistic calculi that have been proposed so far, and in Sect. 6 we discuss some future work.

2 A Probabilistic Typed λ-calculus

We introduce a type theory for the probabilistic (untyped) λ–calculus Λ_P defined in [11]. This calculus essentially consists of an extension of the classical untyped λ–calculus with *probabilistic terms*, i.e. terms of the form $\bigoplus_{i=1}^{n} p_i M_i$ representing probability distributions on classical deterministic terms M_i. A model of a type system for such a calculus must therefore consider mathematical structures for types that can suitably represent spaces of probability distributions. As these are subsets of vector spaces, a suitable interpretation of our types should be based on a vector space construction starting from the classical types; thus operations on terms will be the standard operations on vectors (e.g. vector sum, scalar product etc.), while operations on types will correspond to the standard ways of composing vector spaces (e.g. direct sum, tensor product, etc.). We won't take this discussion further in this paper and postpone the definition of a mathematical model for the semantics of our calculus to some future work, focusing here only on the formal theory of types, i.e. their syntax and the type system for Λ_P.

2.1 Probabilistic Types and Subtypes

Given a set of atomic classical types $\mathfrak{I} = \{\iota_1, \ldots, \iota_n\}$, the set of *concrete probabilistic types*, which we denote by pTypes, is given by the following syntax:

$$\tau ::= \iota_1 \mid \ldots \mid \iota_n \mid \tau \to \tau \mid \tau \oplus \tau\,.$$

Note that, although the syntax of the choice construct does not carry any label representing a probability, we call the types defined above *probabilistic* because of their intended use, namely as objects of a syntactic nature to be assigned to probabilistic lambda terms.

The choice type $\tau \oplus \tau$ will be used to type terms of the form $\{p_1 M_1^{\tau_1}, \ldots, p_n M_n^{\tau_n}\}$, and can therefore intutively be thought of as a form of 'union' of the types τ_i, corresponding e.g. to a categorical coproduct or, more concretely, to a direct sum of vector spaces (cf. e.g. [25]).

We will identify concrete types modulo associativity, symmetry and idempotency with respect to $\oplus$. We will also assume distributivity of the $\oplus$ operator over the $\to$ operator, as well as associativity of $\to$ on the right. More precisely, we define the relation $\equiv$ as the smallest congruence on pTypes which is compatible with the following laws:

$$\tau_1 \oplus (\tau_2 \oplus \tau_3) \equiv (\tau_1 \oplus \tau_2) \oplus \tau_3 \tag{1}$$

$$\tau_1 \oplus \tau_2 \equiv \tau_2 \oplus \tau_1 \tag{2}$$

$$\tau \oplus \tau \equiv \tau \tag{3}$$

$$(\tau_1 \oplus \tau_2) \to \rho \equiv (\tau_1 \to \rho) \oplus (\tau_2 \to \rho) \tag{4}$$

$$\rho \to (\tau_1 \oplus \tau_2) \equiv (\rho \to \tau_1) \oplus (\rho \to \tau_2) \tag{5}$$

Then *the set of probabilistic types* is the quotient set $\mathcal{T} = \mathsf{pTypes}_{/\equiv}$.

As usual, we will work with representatives of equivalence classes. With a little abuse of notation, the *syntactical equality between types* (namely equality between equivalence classes of concrete types) is also denoted by $\equiv$.

Equations (1)–(5), together with the assumption of the right associativity of function application, allow us to show the following result.

Lemma 1. *For all* $\tau = \tau_1 \to \tau_2$*, there exist* $\{\iota_1 \ldots, \iota_t\} \subseteq \mathfrak{I}$ *and* $\{\rho_1 \ldots, \rho_t\} \subseteq$ pTypes *such that* $\tau \equiv \bigoplus_{j=1}^{t} \iota_j \to \rho_j$.

Proof. By induction on the structure of type τ_1.

1. $\underline{\tau_1 = \iota \in \mathfrak{I}}$. This case is self-evident.

2. $\underline{\tau_1 = \sigma_1 \to \sigma_2}$. By the iductive hypothesis, there exist $\{\iota'_1 \ldots, \iota'_n\} \subseteq \mathfrak{I}$ and $\{\rho'_1 \ldots, \rho'_n\} \subseteq \mathsf{pTypes}$ such that $\tau_1 \equiv \bigoplus_{j=1}^n \iota'_j \to \rho'_j$. Therefore, we have that $\tau \equiv (\bigoplus_{j=1}^n \iota'_j \to \rho'_j) \to \tau_2$. By applying the equivalence law (4) and the right associativity of the arrow, we then obtain $\tau \equiv \bigoplus_{j=1}^n (\iota'_j \to \rho'_j \to \tau_2) \equiv \bigoplus_{j=1}^n \iota'_j \to (\rho'_j \to \tau_2)$. Finally, take $t = n$ and $\{\rho_i\}_{i=1}^t = \{\rho'_i \to \tau_2\}_{i=1}^t$.
3. $\underline{\tau_1 = \sigma_1 \oplus \sigma_2}$. By the equivalence law (4), $\tau \equiv (\sigma_1 \to \tau_2) \oplus (\sigma_2 \to \tau_2)$. Now apply the inductive hypothesis to σ_1 and σ_2, and the proof for case 2. □

Subtype Relation. In the classical theory of the λ–calculus, subtype polymorphism was introduced in order to make inferences more flexible: a term can be seen as belonging not only to its type but also to all types that are 'subsumed' according to some notion of type inclusion[1]. This means that, whenever τ is a subtype of ρ, any term of type τ can safely be used in a context where a term of type ρ is expected. In our probabilistic context a notion of subtype can be defined as follows.

Given a relation $\mathcal{R} \subseteq \mathfrak{I} \times \mathfrak{I}$, we define the *subtype relation* $<:$ on $\mathcal{T}$ as the smallest relation induced by the transitive and reflexive closure of the rules in Fig. 1.

$$\frac{\iota_1 \mathcal{R}\, \iota_2}{\iota_1 <: \iota_2}\ (1) \qquad \frac{\tau_1 <: \sigma_1 \quad \sigma_2 <: \tau_2}{\sigma_1 \to \sigma_2 <: \tau_1 \to \tau_2}\ (2)$$

$$\frac{\forall j \in [1,m]\ \exists i \in [1,n]\ \tau_i <: \sigma_j}{\bigoplus_{i=1}^n \tau_i <: \bigoplus_{j=1}^m \sigma_j}\ (3)$$

Fig. 1. Subtype relation

According to this definition, we may have, for example, *integer* $<:$ *double*, *integer* $<:$ *double* $\oplus$ *integer* and *double* $\oplus$ *integer* $<:$ *integer*, but *double* is not a subtype of *double* $\oplus$ *integer*.

We can show that any type in pTypes can be identified with a sum of arrows. To this purpose we extend the base types by introducing a type $\bot$ such that, for all $\tau \in \mathsf{pTypes}, \bot <: \tau$. We can then represent a base type ι by $\bot \to \iota$ without affecting the relation $\mathcal{R} \subseteq \mathfrak{I} \times \mathfrak{I}$. More formally, it is easy to show that

$$\text{If } \iota_1\ \mathcal{R}\ \iota_2 \text{ then } (\bot \to \iota_1) <: (\bot \to \iota_2).$$

In the following proposition we will informally denote by $\mathsf{pTypes} \setminus \{\tau \oplus \tau\}$ the set of all types that are either base types or constructed by means of the arrow constructor only, i.e. essentially the types of the simply typed λ-calculus without choice.

[1] Having in mind a model in which a type is a vector space, it is natural to consider its subspaces as its *subtypes*.

Proposition 1. *For any type $\tau \in$ pTypes there exist $\{\iota_i\}_{i=1}^n \subseteq \mathfrak{I} \cup \{\bot\}$ and $\{\rho_i\}_{i=1}^t \subseteq$ pTypes$\backslash\{\tau \oplus \tau\}$ such that $\tau \equiv \bigoplus_{i=1}^t \iota_i \to \rho_i$.*

Proof. By induction on the structure of the type τ.

1. $\underline{\tau = \iota \in \mathfrak{I}.}$ Take $t = 1$, $\iota_1 = \bot$ and $\rho_1 = \iota$.
2. $\underline{\tau = \tau_1 \to \tau_2.}$ By Lemma 1, there exist $\{\iota_1 \ldots, \iota_n\} \subseteq \mathfrak{I}$ and $\{\rho_1 \ldots, \rho_n\} \subseteq$ pTypes such that $\tau \equiv \bigoplus_{j=1}^n \iota_j \to \rho_j$. For each $k \in [1, n]$ such that $\rho_k = a \oplus b$, we can replace $\iota_k \to \rho_k$ by $(\iota_k \to a) \oplus (\iota_k \to b)$ by recursively applying the equivalence law (5) until a and b do not contain any occurrence of the $\oplus$ constructor.
3. $\underline{\tau = \tau_1 \oplus \tau_2.}$ This follows immediately from the inductive hypothesis applied to τ_1 and τ_2.

□

For a type τ we will take the form introduced in Proposition 1 as the *representative* of the equivalence class $[\tau]_\equiv$, and we will refer to it as $\bar{\tau}$.

Proposition 1 also allows us to define the 'complexity' of a type in terms of the actual range of the choice represented by the type, namely the number of choice components in the representative of the type.

Definition 1. *The size of a type $\tau \in \mathcal{T}$, with representative $\bar{\tau} = \bigoplus_{i=1}^n \iota_i \to \rho_i$, is defined as $\#\tau = n$.*

2.2 Terms and Type Assignment

We call Λ_P the calculus whose *raw terms* are defined by the following grammar:

$$\begin{array}{lll} x & ::= x_0 \mid x_1 \mid \ldots & \textit{variables} \\ M & ::= x^\tau \mid M_1 M_2 \mid \{p_1 M_1^{\tau_1}, \ldots, p_n M_n^{\tau_n}\} \mid & \\ & \quad \lambda x^\tau.M & \textit{terms} \end{array}$$

where $n \geq 1$, and $p_1, \ldots, p_n$ are rational numbers. This latter choice allows us to avoid problems with the computability of the numbers p_i. We adopt a Church typing discipline. In particular, in order to simplify the notation, we follow the approach of annotating directly into the terms each occurrence of a variable with its type. The difference with the set Λ of classical terms, is clearly in the presence of the syntactic construction $\{p_1 M_1^{\tau_1}, \ldots, p_n M_n^{\tau_n}\}$, representing probabilstic choice. This determines the structure of a term $M \in \Lambda_P$ as it may have nested choices.

For the properties of the probabilistic reduction relation we are going to define later, it is reasonable to define the set of values or normal forms exactly as in the classical lazy λ–calculus, i.e. formally as follows:

$$V, W ::= x \mid \lambda x^\tau M, M \in \Lambda.$$

We will use (possibly indexed) capital letters M, N and P to denote raw terms and V, W to denote values. We will also denote the set of terms by $Terms$ and the set of values by $\mathcal{V}$.

The notion of *free variables of a term*, $FV(M)$ is standard [15]; in particular, we have $FV(\{p_1 M_1^{\tau_1}, \ldots, p_n M_n^{\tau_n}\}) = FV(M_1) \cup \cdots \cup FV(M_n)$.

We work modulo *variable renaming* of terms and modulo permutation of the sub-terms $p_i M_i^{\tau_i}$ in $\{p_1 M_1^{\tau_1}, \ldots, p_n M_n^{\tau_n}\}$. Let us call $\equiv$ the corresponding equivalence relation between terms; the set of the actual terms is $\mathcal{M} = Terms/_{\equiv}$.

With $M_1 \equiv M_2$ we indicate that the (equivalence classes of) terms M_1 and M_2 are equivalent (equal). Since we are working modulo α-conversion, we can use the so called *Barendregt Convention on Variables* [4,6,15]: "In each mathematical context (a term, a definition, a proof...) the names chosen for bounded variables will always differ from those of the free ones".

We call *judgement* an expression $\vdash M : \tau$, where M is a term and τ is a type. The rules for assigning types to terms are given in Fig. 2.

In order to avoid possible misunderstanding, we point out that the fact that $\vdash M : \sigma$ is derivable does not imply that M is closed (note that we do not have judgements of the kind $\Gamma \vdash M : \sigma$).

The notation $\vdash M : \tau$ indicates that $M : \tau$ is *derivable* by means of the *typing rules* in Fig. 2; in this case we say that M *is well typed*.

$$\frac{}{x^\tau : \tau}\ (ax) \qquad \frac{M : \sigma}{\lambda x^\tau . M : \tau \to \sigma}\ (\lambda)$$

$$\frac{\{M_i : \tau_i\}_{i=1}^n \quad \forall i \in [1,n]\ p_i \in [0,1] \quad \sum_{i=1}^n p_i = 1}{\{p_i M_i^{\tau_i}\}_{i=1}^n : \bigoplus_{i=1}^n \tau_i}\ (\oplus)$$

$$\frac{M : \bigoplus_{i=1}^n (\tau_i \to \sigma_i) \quad N : \rho \qquad \forall j = i_1 \ldots i_m, 1 \le m \le n, \rho <: \tau_j}{MN : \bigoplus_{j=1}^m \sigma_j}\ (@)$$

Fig. 2. Typing rules

The rule (@) in Fig. 2 allows us to establish when an application term MN, where M is a choice of the form $\{p_i M_i^{\tau_i \to \sigma_i}\}_i$, can be considered well-formed. This is the case if there is at least one term M_j in the choice that can be applied to N, i.e. such that $\rho <: \tau_j$.

We avoid an explicit *subsumption rule*, and therefore the rules $\oplus$ and @ have explicit subtyping constraints. As a consequence of this choice we have the following property, easily provable by induction.

Proposition 2 (Uniqueness of type). *If* $\vdash M : \tau$ *and* $\vdash M : \sigma$ *then* $\tau \equiv \sigma$.

Substitution $M[N/x^\tau]$, with $N : \rho <: \tau$, up to α-equivalence is defined for well-typed terms in the usual way see [15], but for probability distributions of terms. For these terms substitution is defined as follows: if for each $i \in [1, n]$ we know that $M_i[N/x^\tau]$ is of type τ_i', then $\{p_1 M_1^{\tau_1}, \ldots, p_n M_n^{\tau_n}\}[N/x^\tau] \equiv \{p_1 M_1[N/x^\tau]^{\tau_1'}, \ldots, p_n M_n[N/x^\tau]^{\tau_n'}\}$.

The definition of substitution given above is supported by the following lemma.

Lemma 2 (Substitution Lemma). *If* x *is a variable of type* τ, $M : \sigma$ *and* $\vdash N : \tau'$ *with* $\tau' <: \tau$, *then there exists* $\sigma' <: \sigma$ *such that* $\vdash M[N/x^\tau] : \sigma'$.

Proof. By induction on the structure of M.

($M = y^\sigma$) If $y = x$ then $\sigma = \tau$ and we have that $M[N/x^\tau] \equiv N$. Thus, it suffices to take $\sigma' = \tau'$. If $y \neq x$ then we have that $M[N/x^\tau] \equiv y^\sigma$, and the claim is trivially shown.

($M = M_1 M_2$) Since M is well-typed, it must be $M_1 : \bigoplus_{i=1}^k \rho_i \to \gamma_i$ and $M_2 : \delta <: \rho_j$ for some $j = 1 \ldots m, 1 \leq m \leq k$ and $\sigma = \bigoplus_{j=1}^m \gamma_j$. By inductive hypothesis we have that $\vdash M_1[N/x^\tau] : \bigoplus_{i=1}^k \overline{\rho}_i \to \overline{\gamma}_i$, with $\bigoplus_{i=1}^k \overline{\rho}_i \to \overline{\gamma}_i <: \bigoplus_{i=1}^m \rho_i \to \gamma_i$, and $\vdash M_2[N/x^\tau] : \overline{\delta}$, with $\overline{\delta} <: \delta$. Therefore,

$$M[N/x^\tau] \equiv (M_1 M_2)[N/x^\tau] \equiv (M_1[N/x^\tau])(M_2[N/x^\tau])$$

and $(M_1[N/x^\tau])(M_2[N/x^\tau]) : \bigoplus_i^k \overline{\gamma}_i = \sigma'$ with $\sigma' <: \sigma$ by subtyping rule (2).

($M = \{p_1 M_1, \ldots, p_n M_n\}$) In this case, since M is well-typed, we have that $\sigma \equiv \bigoplus_{i=1}^k \sigma_i$, with $\sum_{i=1}^k p_i = 1$, $\approx_{i=1}^k \sigma_i$ and $\{M_i : \sigma_i\}_{i=1}^k$.
By i.h., we have that for all $i = 1 \ldots k$, $M_i[N/x^\tau] : \overline{\sigma}_i <: \sigma_i$. Now we can apply ($\oplus$) obtaining the well typed term $\{p_i M_i[N/x^\tau]\}_{i=1}^k : \bigoplus_{i=1}^k \overline{\sigma}_i$ which is exactly $M[N/x^\tau] : \bigoplus_{i=1}^k \overline{\sigma}_i$ and by subtyping rule (3), $\sigma' \equiv \bigoplus_{i=1}^k \overline{\sigma}_i <: \bigoplus_{i=1}^k \sigma_i = \sigma$.

($M = \lambda y^\delta.L$). Since M is well-typed, $\sigma = \delta \to \gamma$ must hold. By i.h., we know that $L[N/x^\tau] : \overline{\gamma} <: \gamma$. Then, $(\lambda y^\delta.L)[N/x^\tau] \equiv \lambda y.(L[N/x^\tau]) : \delta \to \overline{\gamma} \equiv \sigma'$. By subtyping rule (2), we can thus conclude that $\sigma' <: \sigma$, since $\delta <: \delta$ and $\overline{\gamma} <: \gamma$.

□

3 Probabilistic Reduction

Given the set of labels $\mathscr{L} = \{\beta, fc, pc, \pi\}$, for every $\alpha \in \mathscr{L}$ and for every $p \in \mathbb{Q}_{[0,1]}$, we define a relation $\to_\alpha^p \subseteq \mathcal{M} \times \mathcal{M}$ by the *reduction rules* in Fig. 3, where we assume that all the redexes are well-typed.

For a term $M \equiv \{p_i M_i^{(\oplus_{j_i} \tau_{j_i} \to \sigma_{j_i})}\}_{i=1}^n V$, with V a value of type ρ (values are defined in Sect. 2.2), we define the term $\mathbf{\Pi}[MV] \equiv \{\tilde{p}_k (M_k V)^{\alpha_k}\}_{TS_{\rho, M_k} \neq \emptyset}$, where $TS_{\rho,M} = \{\sigma_j \mid \rho <: \tau_j, 1 \leq j \leq n\}$, $\alpha_k \equiv \bigoplus_{\gamma \in TS_{\rho,M_k}} \gamma$, and $\tilde{p}_k$ are the probabilities p_i re-distributed over the terms M_k with $TS_{\rho,M_k} \neq \emptyset$, that is

$$\tilde{p}_k = \begin{cases} 0 & \text{if for all } i, TS_{\rho,M_k} = \emptyset, \\ \frac{p_k}{\sum_{TS_{\rho,M_i} \neq \emptyset} p_i} & \text{otherwise.} \end{cases}$$

We adopt a strategy similar to the Plotkin's *Call-by-Value Left Reduction* strategy [21], but for the fact that we allow reductions in the scope of a probabilistic λ–abstraction term. With respect to the standard Call-by-Value (lazy) λ–calculus [21], the new rules are (pc), (π) and (fc).

The reduction (π) selects all the functions M_i whose types are compatible with the type of V. The reduct is a new probability distribution containing all the terms M_iV for the selected M_i. It is important to note that the well–typedness of the redex, i.e. that rule (@) is applicable, guarantees that reduction effectively takes place.

The rule (pc) determines the probabilistic branching of the reduction tree, by selecting a term M_i with the associated probability. This is in contrast with reduction (π) which is not probabilistic, but rather selects all the enabled reductions in a function choice.

Finally, the rule (fc) reduces a choice inside a λ-abstraction to a term that is a choice on λ-abstractions. This rule is essential to obtain (classical) values as normal forms.

For any subset $\mathscr{S}$ of $\mathscr{L}$, we can construct a relation $\rightarrow^p_{\mathscr{S}}$ by just taking the union of $\rightarrow^p_\alpha$ over all $\alpha \in \mathscr{S}$. In particular, $\rightarrow^p$ will denote $\rightarrow^p_{\mathscr{L}}$, and $\rightarrow$ will denote $\rightarrow^1_{\mathscr{L}}$.

$$(\lambda x^\tau.M)V \rightarrow^1_\beta M[V/x^\tau], \qquad (\beta)$$

$$\lambda x^\tau.\{p_i M_i^{\tau_i}\}_{i=1}^n \rightarrow^1_{fc} \{p_i \lambda x^\tau.M_i^{\tau_i}\}_{i=1}^n \qquad (fc)$$

$$\{p_i M_i^{\tau_i}\}_{i=1}^n \rightarrow^{p_k}_{pc} M_k, \qquad (pc)$$

$$\{p_i M_i^{(\oplus_{j_i}\tau_{j_i} \rightarrow \sigma_{j_i})}\}_{i=1}^n V \rightarrow^1_\pi \mathbf{\Pi}[\{p_i M_i^{(\oplus_{j_i}\tau_{j_i} \rightarrow \sigma_{j_i})_i}\}_{i=1}^n V] \qquad (\pi)$$

$$\frac{P \rightarrow^p_\alpha N}{VP \rightarrow^p_\alpha VN} \qquad (ra)$$

$$\frac{M \rightarrow^p_\alpha N}{MP \rightarrow^p_\alpha NP} \qquad (la)$$

Fig. 3. Reduction rules

The transitive and reflexive closure of $\rightarrow^p$ is defined by means of the rules in Fig. 4.

$$\frac{}{M \hookrightarrow^1 M} \qquad \frac{M \rightarrow^p N}{M \hookrightarrow^p N} \qquad \frac{M \hookrightarrow^p P \quad P \hookrightarrow^q N}{M \hookrightarrow^{pq} N}$$

Fig. 4. Multistep reduction

A nice property of this transition system is the fact that at each reduction step the range of a type choice can only decrease, as shown by the following proposition.

Proposition 3. *If $\vdash M : \tau$, $\vdash N : \sigma$ and $M \rightarrow^p_\alpha N$ for some $\alpha \in \mathscr{L}$ and probability p, then $\#\sigma \leq \#\tau$.*

Proof. It is easy to see that for $\alpha \in \{\beta, fc, pc, ra, la\}$ the size remains unchanged. For transition (π), it suffices to observe that $\#(\bigoplus_{\gamma \in TS_{\rho,M}} \sigma) \leq \#(\bigoplus_{j=1}^n \sigma_j)$. □

Analogously, we can show that probabilistic reduction decreases the *depth* of a probabilistic term, i.e. the nesting level of the probabilistic choice in the term as defined below.

Definition 2. *Let $M \in \mathcal{M}$. We define the depth, $d(M)$, of M inductively as the the maximum nesting level in M:*

$$\begin{aligned} d(M) &= 0 \text{ if } M \in \Lambda \\ d(\{p_i M_i\}_{i \in [1,n]}) &= \max_{i \in [1,n]} d(M_i) + 1 \\ d(\lambda x.M) &= d(M) \\ d(M_1 M_2) &= d(M_1) + d(M_2) \end{aligned}$$

Note that the nesting level is defined for the choice term so that it is 1 when there is only one external choice, i.e. when the term is actually a choice on classical terms in Λ. The definition of the depth of an application term is justified by the fact that the substitution inside M_1 of the argument M_2 potentially increases the nesting level of M_1 of at least $d(M_2)$. In fact, there is no reasonable upper bound to the growth of such a term when M_2 is not a value, as one can see from the following example.

Example 1. Let $M_1 = \lambda z.z(z(x))$, $M_2 = \lambda y.\{\frac{1}{2}N_1, \frac{1}{2}N_1\}$, and assume that there is at least one occurrence of y in N_1. We have that $d(M_1) = 0$, $d(M_2) = 1$ but $d(M_1 M_2)$ grows indefinitely.

However, obviously if a term is applied to a value, the depth of the resulting term does not incerase, i.e. $d(MV) = d(M)$.

Proposition 4. *Let* $M = \{p_i M_i\}_{i=1}^n$ *and let* $M_i, N \in \mathcal{M}$, *for all* $i \in [1, n]$. *If* $M \hookrightarrow^p N$, *then* $d(N) < d(M)$.

Proof. As there at least one sub-term of M which is a probabilistic choice, rule (pc), must be applied at some point of the reduction, which will make $d(M)$ decrease by 1.

□

3.1 Soundness

In order to prove the correctness of the type system with respect to the reduction rules introduced in Sect. 2, we have to show that two properties of the computation defined by such rules hold, namely (1) *preservation* of the type during the computation, and (2) *progress* of a closed term reduction until a value is reached.

Theorem 1 (Subject Reduction). *If* $\vdash M : \sigma$ *and* $M \rightarrow_\alpha^p M'$, *then there exists* $\sigma' <: \sigma$ *such that* $\vdash M' : \sigma'$.

Proof. By induction on the length n of the derivation.

- If $n = 1$ then one of the rules (β), (fc), (pc) or (π) is applied. If $M \rightarrow_\beta^1 M'$ then the thesis follows by Lemma 2 (Substitution Lemma). If $M \rightarrow_{fc}^1 M'$, then $M : \tau \rightarrow \bigoplus_{i=1}^n \tau_i$, and $M' : \bigoplus_{i=1}^n \tau \rightarrow \tau_i$. Now observe that by the conguence on types defined in Sect. 2.1, law (5), we have that $\tau \rightarrow \bigoplus_{i=1}^n \tau_i \equiv \bigoplus_{i=1}^n \tau \rightarrow \tau_i$. If $M \rightarrow_{pc}^{p_j} M'$, $M = \{p_i M_i\}_{i=1}^k : \bigoplus_{i=1}^k \sigma_i$, $M' = M_i$, $\sigma' = \sigma_i$ $(i = 1 \dots k)$ and by subtyping rule (3) $\sigma_i <: \bigoplus_{i=1}^k \sigma_i \equiv \sigma$. In the case where the rule applied is (π), the claim is true by definition.
- If $n > 1$ a contextual rule is applied, that is either (ra) or (la).
  - (ra) In this case, $M = NL$, $M' = NL'$ with $L : \rho$, $L' : \overline{\rho}$ and $\overline{\rho} <: \rho$ by i.h. Since M is well typed, we must have that $N : \bigoplus_{i=1}^k \delta_i \rightarrow \gamma_i$ and $\sigma \equiv \bigoplus_{i=1}^k \gamma_i$. In this case $\sigma' \equiv \sigma$ and the thesis follows from the reflexivity of the subtyping relation.
  - (la) In this case, $M = NL$, $M' = N'L$ with $N : \bigoplus_{i=1}^k \delta_i \rightarrow \gamma_i \equiv \tau$, $N' : \bigoplus_{i=1}^k \overline{\delta_i} \rightarrow \overline{\gamma_i} \equiv \tau'$, $L : \rho <: \delta_i$ for all $i = 1 \dots k$. By i.h., $\tau' <: \tau$ and by subtyping rule (2) this means that $\delta_i <: \overline{\delta_i}$ and $\overline{\gamma_i} <: \gamma_i$. Since $\sigma' \equiv \bigoplus_{i=1}^k \overline{\gamma_i}$ and $\sigma \equiv \bigoplus_{i=1}^k \gamma_i$, by subtyping rule (3) we have the thesis.

□

Theorem 2 (Progress). *Let* $\vdash M : \tau$ *and let* M *be closed, then if* M *is not a value then there exist* N, p *such that* $M \rightarrow^p N$.

Proof. By induction on the structure of M.

First of all, we observe that M cannot be a variable since M is closed.

$(M \equiv \lambda x^\tau.\{p_1 M_1^{\tau_1}, \dots, p_n M_n^{\tau_n}\})$ Apply rule (fc).
$(M \equiv \{p_1 M_1^{\tau_1}, \dots, p_n M_n^{\tau_n}\})$ Apply rule (pc).

($M \equiv PQ$) If P is not a value, then apply the induction hypothesis and rule (la). If P is a value but Q is not, then apply the induction hypothesis and rule (ra). If both P and Q are values, then by the typing rule (@) we know (since P is a value) that $P \equiv \lambda x^\sigma.N$ and Q must have type $\rho <: \sigma$. Then we apply rule (β).

□

3.2 Normalization

We now prove that every well-typed term in our typed probabilistic λ-calculus is *normalizing*, i.e. it cannot reduce indefinitely according to the rules in Fig. 3.

We first introduce some useful definitions.

Definition 3. *Let $M \in \mathcal{M}$ be a probabilistic λ-term as defined in Sect. 2.2. A reduction sequence for M of length $n \leq \omega$ is a sequence $(M_i)_{0 \leq i \leq n}$, with $M_i \in \mathcal{M}$ for all $i = 1, \ldots, n$, such that $M_0 = M$ and $M_{i-1} \to_\alpha^{p_{i-1}} M_i$ for all $1 < i \leq n+1$, $\alpha \in \mathscr{L}$.*

Definition 4. *A term $M \in \mathcal{M}$ is:*

- *normalizing if all reduction sequences for M are finite;*
- *in normal form if for any p, α and N, $M \not\to_\alpha^p N$.*

In order to prove the normalization property for Λ_P, we will use the well known normalization result for the simply typed λ-calculus (see e.g. [5]), and the fact that the normal forms, i.e. the values $\mathcal{V}$, defined in Sect. 2.2, that we can obtain in our calculus are exactly the classical values of the simply typed λ-calculus[2]. More formally we show the following theorem.

Theorem 3. *If $\vdash M : \tau$ then M is normalizing.*

Proof. We will show the theorem by mutual induction on the structure of the terms and the size of the type τ.

1. Let $M = \lambda x^\rho.R^\sigma$ and $\tau = \rho \to \sigma$. If $R \in \Lambda$ then it is a value and the thesis is immediately true. Otherwise, R is of the form $R = \{p_i R_i\}_{i=1}^n$ for some $n \geq 1$ and $R_i \in \mathcal{M}$. By applying rule (fc) we obtain $M' = \{p_i \lambda x^\rho.R_i^{\sigma_i}\}_{i=1}^n$, with $\tau' = \rho \to \bigoplus_{i=1}^n \sigma_i$. Since for all $i \in [1, n]$, $\#(\sigma_i) < \#(\tau) = \#(\tau')$, we have by the inductive hypothesis that each R_i normalizes. Therefore, after a finite number of steps M reduces to a term M'' such that $d(M'') = 1$. This is because the depth of a term is always positive and by Proposition 4 it decreases during reduction. As each component of M'' is now in Λ, it normalizes and so M normalizes too.

[2] In fact, probabilities are used at the semantical level only to estimate the likelihood that a certain value is actually otained after the reduction process.

2. Let $M = \{p_i M_i^{\sigma_i}\}_{i=1,\ldots,n}$. Then $\tau = \bigoplus_{i=1}^{n} \sigma_i$. Since for all $i \in [1,n]$, $\#(\sigma_i) < \#(\tau) = \max_{i\in[1,n]} \#(\sigma_i)$, by inductive hypothesis M_i normalizes, i.e. for each M_i there exists a finite reduction sequence of length $l_i < \omega$ such that $M_i \hookrightarrow^{q_i} N_i$, with $N_i \in \mathcal{V}$ and probability q_i. Note that by effect of rule (π) the number of the final values may shrink to $m < n$ with probabilities normalized accordingly. At step $l = \max_{i=1,\ldots,n} l_i$ the resulting term is $\{q_j N_j\}_{j=1}^{m}$ to which only rule (pc) applies, so that we have

$$\{q_j N_j\}_{j=1}^{m} \rightarrow_{pc}^{q_j} N_j \not\rightarrow_{\alpha}^{p} .$$

3. If $M = PQ$, then from the (@) typing rule we have that $P : \bigoplus_{j=1}^{k} \gamma_j \rightarrow \rho_j$ and $Q : \sigma$, for some σ. By the inductive hypotesis, Q normalizes to a value $V^{\sigma'}$, with $\sigma' <: \sigma$ (by Theorem 1). Because of the adoption of a call-by-value strategy, we will be able to reduce the term PQ only after the finite reduction $Q \hookrightarrow^{q} V$ has been performed. We are then left with the term PV, to which the only applicable rule is the (π) reduction rule giving a term S with $\tau' = \bigoplus_{i=1}^{n} \rho_i$ for some $n \leq k$ and $\sigma' <: \gamma_i$ for all $i = 1, \ldots, n$. By Proposition 3, $\#(\tau') < \#(\tau)$. Thus, by the inductive hypothesis we deduce that S is normalizing and we can conclude that so is M.

□

4 Examples

In this section we illustrate the use of probabilistic terms and reductions in order to model probabilistic computation.

In the following example we assume that our calculus is enriched with constants for values of basic types and built-in functions.

Example 2. Assume $\mathfrak{I} = \{\mathsf{byte}, \mathsf{short}, \mathsf{char}, \mathsf{string}, \mathsf{int}, \mathsf{long}, \mathsf{float}, \mathsf{double}\}$, $\mathcal{R} \subseteq \mathfrak{I} \times \mathfrak{I}$ and $\mathcal{R} \subseteq \mathfrak{I} \times \mathfrak{I}$ defined as $\mathcal{R} = \{(\mathsf{byte}, \mathsf{short}), (\mathsf{short}, \mathsf{int}), (\mathsf{int}, \mathsf{long}), (\mathsf{long}, \mathsf{float}), (\mathsf{float}, \mathsf{double}), (\mathsf{char}, \mathsf{string}), (\mathsf{char}, \mathsf{int})\}$. Consider a further type string such that neither int <: string nor string <: int holds.

Let plus be (the encoding of) an overloaded function that behaves as an arithmetical sum on arguments of type double (and relative subtypes) and as a string concatenation on arguments of type string.

Consider now the terms

$$\begin{aligned}
M &= \{\tfrac{1}{3}\mathsf{plus}^{\mathsf{int}\rightarrow\mathsf{int}\rightarrow\mathsf{int}}, \tfrac{2}{3}\mathsf{plus}^{\mathsf{string}\rightarrow\mathsf{string}\rightarrow\mathsf{string}}\},\\
N_1 &= \overline{5} : \mathsf{int},\\
N_2 &= \overline{7} : \mathsf{int},\\
L_1 &= \{\tfrac{1}{4}\overline{5}^{\mathsf{int}}, \tfrac{3}{4}\mathtt{first}^{\mathsf{string}}\},\\
L_2 &= \mathtt{second} : \mathsf{string},
\end{aligned}$$

where $\overline{5}$ and $\overline{7}$ are the encoding of the numbers 5 and 7, respectively. We assume that all terms can be derived by means of well-typing rules. Note that $L_1 = \{\frac{1}{4}\overline{5}^{\mathsf{int}}, \frac{3}{4}\mathtt{first}^{\mathsf{string}}\}$ is a value, since $\mathsf{int} \neq \mathsf{string}$.

We depict in the following two possible probabilistic reductions based on the previous assumptions.

1. If we start with the term $((M)N_1)N_2 = ((\{\frac{1}{3}\mathsf{plus}^{\mathsf{int}\to\mathsf{int}\to\mathsf{int}}, \frac{2}{3}\mathsf{plus}^{\mathsf{string}\to\mathsf{string}\to\mathsf{string}}\})\overline{5})\overline{7}$, only the subterm $\mathsf{plus}^{\mathsf{int}\to\mathsf{int}\to\mathsf{int}}$ of M will be selected by the (π) rule and applied to N_1. Therefore, no probabilistic choice is made afterward.

$$((\{\tfrac{1}{3}\mathsf{plus}^{\mathsf{int}\to\mathsf{int}\to\mathsf{int}}, \tfrac{2}{3}\mathsf{plus}^{\mathsf{string}\to\mathsf{string}\to\mathsf{string}}\})\overline{5})\overline{7}$$

$$\downarrow 1\ \pi$$

$$(\{1\ (\mathsf{plus}\ \overline{5})^{\mathsf{int}\to\mathsf{int}}\})\overline{7}$$

$$\downarrow 1\ \pi$$

$$\{1\ (\mathsf{plus}\ \overline{5})\overline{7}\}$$

Example 3. Consider the situation where we have $\vdash M_1 : \tau_1 \to \alpha_1$, $\vdash M_2 : \tau_2 \to \alpha_2$, $\vdash V_1 : \rho$, $\vdash V_2 : \rho$, $\vdash W : \sigma$. Assume that $\tau_1 \not\equiv \tau_2, \tau_1 <: \tau_2, \rho <: \tau_1, \sigma <: \tau_2$ with σ not related to τ_1.

We depict in the following three possible probabilistic reductions based on the previous assumptions.

1. Suppose that we start with the term $\{p_1 M_1^{\tau_1\to\alpha_1}, p_2 M_2^{\tau_2\to\alpha_1}\}V_1$. Then it is possible to apply both M_1 and M_2 to V_1 and we obtain the reduction tree:

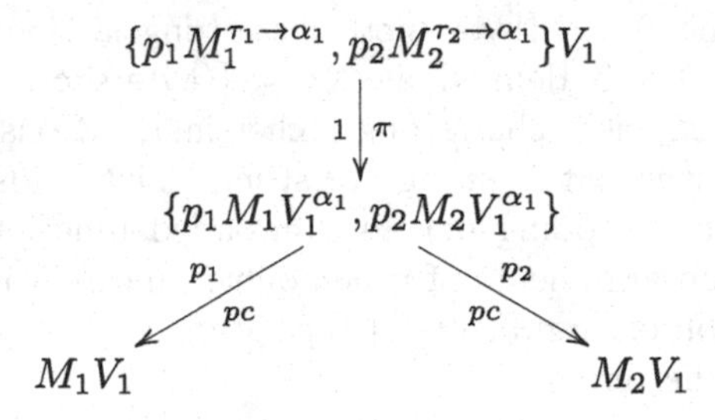

2. If we start with the term $\{p_1 M_1^{\tau_1\to\alpha_1}, p_2 M_2^{\tau_2\to\alpha_1}\}W$, then only M_2 can be selected so that no probabilistic choice is made afterwards:

$$\{p_1 M_1^{\tau_1\to\alpha_1}, p_2 M_2^{\tau_2\to\alpha_1}\}W \to_\pi^1 \{1 M_2 W^{\alpha_1}\} \to_\pi^1 M_2 W$$

3. Suppose now that the argument is a probabilistic term $\{q_1 V_1^\rho, q_2 V_2^\rho\}$. Following the Call-by-Value strategy, we must first reduce this argument to a value. We therefore have:

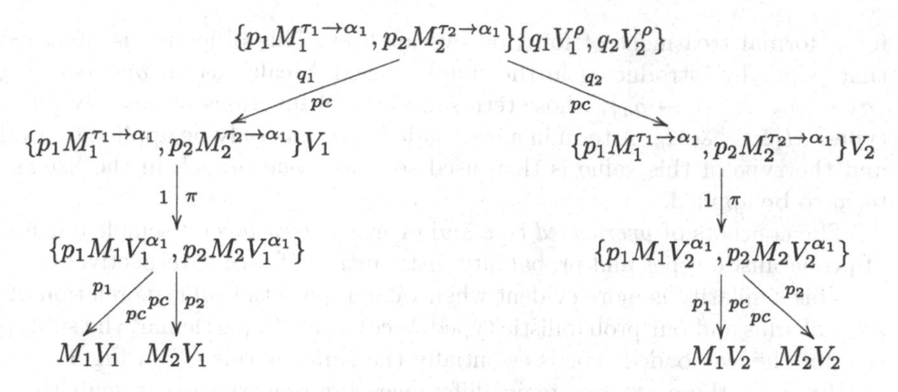

Example 4. Consider a situation where $\vdash M_1 : \alpha_1 \to (\beta_1 \to \gamma_1)$, $\vdash M_2 : \alpha_2 \to (\beta_2 \to \gamma_2)$, $\vdash M_2 : \alpha_3 \to (\beta_3 \to \gamma_2)$, $\vdash V : \alpha$, $\vdash W : \beta$. Suppose that $\alpha_1, \alpha_2, \alpha_3$ are pairwise different, $\alpha <: \alpha_1$, $\alpha <: \alpha_2$, $\alpha <: \alpha_3$. Moreover, let $\beta_1, \beta_2, \beta_3$ be pairwise different, and let $\beta <: \beta_2$, $\beta <: \beta_3$, with β, β_1 not related. Now consider the term $\{p_1 M_1^{\alpha_1 \to (\beta_1 \to \gamma_1)}, p_2 M_2^{\alpha_2 \to (\beta_2 \to \gamma_2)}, p_3 M_3^{\alpha_3 \to (\beta_3 \to \gamma_2)}\}$, and its application to two arguments V and W. We obtain the following reduction tree:

$$(\{p_1 M_1^{\alpha_1 \to (\beta_1 \to \gamma_1)}, p_2 M_2^{\alpha_2 \to (\beta_2 \to \gamma_2)}, p_3 M_3^{\alpha_3 \to (\beta_3 \to \gamma_2)}\} V) W$$

$$\downarrow 1 \; \pi$$

$$\{p_1 M_1 V^{\beta_1 \to \gamma_1}, p_2 M_2 V^{\beta_2 \to \gamma_2}, p_3 M_3 V^{\beta_3 \to \gamma_2}\} W$$

$$\downarrow 1 \; \pi$$

$$\{\tilde{p_2} M_2 V W^{\gamma_2}, \tilde{p_3} M_3 V W^{\gamma_2}\}$$

$$\swarrow \tilde{p_2} \; pc \qquad \tilde{p_3} \; pc \searrow$$

$$M_2 V W \qquad M_3 V W$$

where $\tilde{p_2}$ and $\tilde{p_3}$ are the the normalized probabilities $\frac{p_2}{p_2+p_3}$ and $\frac{p_3}{p_2+p_3}$, respectively.

5 Related Works

We have presented a simply typed probabilistic calculus. This can be seen as a typed version of the probabilistic calculus introduced in [11] with the objective of showing how probabilistic abstract interpretation [12] can be exploited in the context of static analysis of probabilistic programs even in presence of higher-order functions. The probabilistic calculus in [11] is an untyped λ-calculus whose syntax is itself an extension of pure λ-calculus with n-ary probabilistic choice. A theoretical study of a similar extension of untyped lambda calculus, equipped with a probabilistic operational semantics, can be found in [9].

The type theory we have proposed in this paper has a close relationship with the $\lambda\&$–calculus that was introduced in [8] with the aim of providing a basis for a formal treatment of function overloading. This objective is obtained in that paper by introducing in the simply typed λ–calculus an *overloaded type*, $\{\sigma_1 \to \rho_1, \ldots, \sigma_n \to \rho_n\}$, whose terms are lists of functions of possibly different types, $M_1 \& \ldots \& M_n$. A term in a overloaded type can only be applied to a value, and the type of this value is then used to select one branch in the list as the term to be applied.

The concepts of *overloaded type* and of *overloaded term* resemble our notion of probabilistic types and probability distribution of terms, respectively.

This similarity is more evident when we compare the subtype relation of the $\lambda\&$–calculus and our probabilistic typed λ–calculus. In particular, the subtyping rule for the overloaded type is essentially the same as rule (3) in Fig. 1.

However, there are two main differences between the two calculi that are related to the type formation and the computational rules, respectively.

In order to be well formed, an overloaded type $\{\sigma_1 \to \rho_1, \ldots, \sigma_n \to \rho_n\}$ must satisfy two strong conditions, namely:

- for each i, j in $[1, n]$ if $\sigma_i <: \sigma_j$ then $\rho_i <: \rho_j$ (ensuring the covariance of the family of the function codes);
- for each i, j in $[1, n]$ if there is σ such that $\sigma <: \sigma_i, \sigma_j$ then there must exist $\mathsf{inf}\{\sigma_i, \sigma_j\}$.

These conditions are necessary to guarantee the existence and the uniqueness of the type to be selected, and the correctness of the associated reduction, i.e. that types can only decrease during computation. In our calculus the reduction rule for the application of a probabilistic "polymorphic" function is defined so as to keep track of all the codes which can be applied to the argument; thus the restriction imposed by the two conditions above are not needed in our case. On the other hand, in our case the meaning of the probabilistic function choice is diffrent from a classical overloaded function: while for the latter the choice is made at runtime, in our case the probabilistic choice must be resolved at compile time by the static type checker.

Many proposals of probabilistic typed calculi are available in the literature which are more strongly oriented to programming languages and applications, as for example [17–19, 23].

More references and discussion on recent results related to the semantics of probabilistic typed lambda calculus can be found in the book [13]. An interesting difference with respect to our approach is the meaning assigned in that book to the probabilistic choice between two functions: this is the weighted average of the values of the two functions.

6 Future Work

The work that we have proposed in this paper can be variously carried on in several directions, some of which are discussed below.

Proof-Theoretical Analysis of Probabilistic Types. Our type system is reminiscent of the type system $\mathcal{C}$ defined in [10] (see also [3]), where the logical semantics of an extended version of lambda calculus with non-deterministic and parallel operators is studied. In particular, the type system $\mathcal{C}$ deals with both intersection types (able to characterize crucial subclasses of lambda terms, see for example [16]) and union types. More precisely, the following rule, where "+" represents non-deterministic choice and $\sigma \vee \rho$ is a union type,

$$\frac{\Gamma \vdash M : \sigma \quad \Gamma \vdash N : \rho}{\Gamma \vdash M + N : \sigma \vee \rho}$$

is admissible (note that the rule is defined for non-empty contexts, differently from our typing judgements). The full logical framework for union types opens up some problems, such as the loss of the subject reduction property (in accordance with the "problematic" treatment of or-elimination rule in natural deduction).

Even if at a first glance our probabilistic types have some similarities with union types, their interpretation as elements (vectors of coordinates) of a vectorial space makes their meaning quite different. We postpone to future investigations a deeper logical treatment of our probabilistic types, also including the study of the relationship with other type systems.

Semantics and Formal Analysis of Probabilistic Programs. The pure syntactical approach presented here is the base for a more semantical treatment of probabilistic λ-calculus, which can be obtained by means of suitable mathematical structures. As a first step we plan to extend the approaches originally developed for the simply typed λ–calculus so as to be able to handle the notion of probability distributions over terms. We believe that this will help us investigate suitable domain theoretic notions for the construction of a denotational semantics for the probabilistic (typed) λ-calculus. This will be the base for investigating methods and techniques of probabilistic static analysis extending those that have been developped in [1,2] for the non-probabilistic functional setting.

References

1. Abramsky, S., Hankin, C. (eds.): Abstract Interpretation of Declarative Languages. Halsted Press, Sydney (1987)
2. Abramsky, S., Jensen, T.P.: A relational approach to strictness analysis for higher-order polymorphic functions. In: Proceedings of ACM Symposium on Principles of Programming Languages, pp. 49–54 (1991)
3. Barbanera, F., Dezani-Ciancaglini, M., de'Liguoro, U.: Intersection and union types: syntax and semantics. Inf. Comput. **119**(2), 202–230 (1995). https://doi.org/10.1006/inco.1995.1086. https://dx.doi.org/10.1006/inco.1995.1086
4. Barendregt, H.P.: The Lambda Calculus, Studies in Logic and the Foundations of Mathematics, vol. 103, revised edn. North-Holland (1991)
5. Barendregt, H.P.: Lambda calculi with types. In: Abramsky, S., Gabbay, D.M., Maibaum, S.E. (eds.) Handbook of Logic in Computer Science, vol. 2, pp. 117–309. Oxford University Press Inc., New York (1992), http://dl.acm.org/citation.cfm?id=162552.162561

6. Barendregt, H.: The Lambda Calculus, Its Syntax and Semantics, Studies in Logic and the Foundations of Mathematics, vol. 103. Elsevier (1984)
7. Cardelli, L., Wegner, P.: On understanding types, data abstraction, and polymorphism. ACM Comput. Surv. **17**(4), 471–523 (1985). https://doi.org/10.1145/6041.6042
8. Castagna, G., Ghelli, G., Longo, G.: A calculus for overloaded functions with subtyping. Inform. Comput. **117**(1), 115–135 (1995). https://doi.org/10.1006/inco.1995.1033
9. Dal Lago, U., Zorzi, M.: Probabilistic operational semantics for the lambda calculus. RAIRO - Theor. Inform. Appl. **46**(3), 413–450 (2012)
10. Dezani, M.: Logical semantics for concurrent lambda calculus, ph.D. Thesis, Radboud University Nijmegen (1996)
11. Di Pierro, A., Hankin, C., Wiklicky, H.: Probabilistic lambda-calculus and quantitative program analysis. J. Logic Comput. **15**(2), 159–179 (2005)
12. Di Pierro, A., Wiklicky, H.: Measuring the precision of abstract interpretations. LOPSTR 2000. LNCS, vol. 2042, pp. 147–164. Springer, Heidelberg (2001). https://doi.org/10.1007/3-540-45142-0_9
13. Draheim, D.: Semantics of the Probabilistic Typed Lambda Calculus - Markov Chain Semantics, Termination Behavior, and Denotational Semantics. Springer, Heidelberg (2017). https://doi.org/10.1007/978-3-642-55198-7
14. Girard, J.Y.: Une extension de l'interprétation de Gödel à l'analyse, et son application à l'élimination des coupures dans l'analyse et la théorie des types. In: Proceedings of the 2nd Scandinavian Logic Symposium. Studies in Logic and the Foundations of Mathematics, vol. 63, pp. 63–92. North-Holland (1971)
15. Gunter, C.A.: Foundations of Computing. MIT Press, Cambridge (1992)
16. Coppo, M., Dezani-Ciancaglini, M., Venneri, B.: Functional characters of solvable terms. Math. Logic Q. **27**(2–6), 45–58 (1981)
17. Park, S.: A calculus for probabilistic languages. In: Proceedings of the ACM SIGPLAN International Workshop on Types in Languages Design and Implementation, pp. 38–49. ACM (2003)
18. Park, S., Pfenning, F., Thrun, S.: A monadic probabilistic language (2003). manuscript. http://www.cs.cmu.edu/~fp/papers/prob03.pdf
19. Park, S., Pfenning, F., Thrun, S.: A probabilistic language based on sampling functions. ACM Trans. Program. Lang. Syst. **31**(1), 4:1–4:46 (2008). https://doi.org/10.1145/1452044.1452048. https://doi.acm.org/10.1145/1452044.1452048
20. Pierce, B.C.: Types and Programming Languages. MIT Press, Cambridge (2002)
21. Plotkin, G.D.: Call-by-name, call-by-value and the λ-calculus. Theor. Comput. Sci. **1**(2), 125–159 (1975)
22. Plotkin, G.D.: LCF considered as a programming language. Theor. Comput. Sci. **5**(3), 223–255 (1977)
23. Ramsey, N., Pfeffer, A.: Stochastic lambda calculus and monads of probability distributions. In: Proceedings of ACM Symposium on Principles of Programming Languages, pp. 154–165. ACM Press (2002)
24. Reynolds, J.C.: Towards a theory of type structure. In: Robinet, B. (ed.) Programming Symposium. LNCS, vol. 19, pp. 408–425. Springer, Heidelberg (1974). https://doi.org/10.1007/3-540-06859-7_148
25. Roman, S.: Advanced Linear Algebra. Graduate Texts in Mathematics 135. Springer, New York (2008). https://doi.org/10.1007/978-0-387-72831-5

Program Analysis

Galois Connections for Recursive Types

Ahmad Salim Al-Sibahi[1,2(✉)], Thomas Jensen[1,3(✉)],
Rasmus Ejlers Møgelberg[4], and Andrzej Wąsowski[4]

[1] University of Copenhagen, Copenhagen, Denmark
ahmad@di.ku.dk
[2] Paperflow, Copenhagen, Denmark
[3] INRIA, Rennes, France
Thomas.Jensen@inria.fr
[4] IT University of Copenhagen, Copenhagen, Denmark

Abstract. Building a static analyser for a real language involves modeling of large domains capturing the many available data types. To scale domain design and support efficient development of project-specific analyzers, it is desirable to be able to build, extend, and change abstractions in a systematic and modular fashion. We present a framework for modular design of abstract domains for recursive types and higher-order functions, based on the theory of solving recursive domain equations. We show how to relate computable abstract domains to our framework, and illustrate the potential of the construction by modularizing a monolithic domain for regular tree grammars. A prototype implementation in the dependently typed functional language Agda shows how the theoretical solution can be used in practice to construct static analysers.

1 Introduction

It requires much work to construct a static analyzer using abstract interpretation. We must design suitable abstract domains for the properties we want to analyze, construct an abstract semantics that works with these domain, prove that this is sound with regards to the concrete semantics of our target programming language, and finally provide an implementation that reflects the developed theory and works efficiently in practice. A small change of the abstract domains to allow verifying new properties could have a large cascading effect w.r.t. the semantics, the proof of soundness and implementation. This is especially true for realistic programming language which have complex types, data structures and collections, and whose domains are often specified in a monolithic fashion.

There is therefore a need for compositional techniques for constructing abstract interpreters and proving their soundness, where multiple frameworks have been developed in recent years [6,14,16,17]. These frameworks are advantageous since they allow changing parts of a static analysis in an isolated

Authors ordered alphabetically.

A. Di Pierro et al. (Eds.): Festschrift Hankin, LNCS 12065, pp. 105–131, 2020.
https://doi.org/10.1007/978-3-030-41103-9_4

manner without completely changing the whole system. This makes it easier to continuously improve on static analyzers and exchange abstract domains for the target systems.

There is still a gap when it comes to compositionally analysing *recursive data structures*, which are present in many realistic programming languages. They are particularly challenging when combined with other domains, since recursive references can be present deeply inside data structures and nested inside complex domains like collections or functions. This paper proposes a systematic approach for designing and implementing abstract domains for complex types, based on the theory of solving recursive domain equations. We present a method for compositional construction of abstract domains that allows changing abstractions at the granularity of types. For instance changing the abstraction of shapes, without affecting the abstraction of integers or functions or vice versa.

The central theoretical contribution of this paper is to extend the construction of Galois connections for recursive data types. As far as we know, we are first to establish such a correspondence between concrete and abstract domains combinations that include recursive equations and higher-order functions. The solution is based on solving recursive domain equations over a category of complete lattices. Furthermore, we show how the construction naturally leads to a compositional implementation of concrete and abstract semantics for a recursive expression language (Sect. 6), in which the modularity of domains allows changing operations for select types without requiring rewriting the rest of semantics.

The general solution for abstracting recursive domains provides a modular precise mathematical formulation, but is not computable. We present two computable abstractions for recursive types: k-limited trees and typed regular tree grammars. k-limited trees allow us to show a property up to a limited depth k, while typed regular tree grammars allow us to infer a refined abstract data type. To see the latter, consider the negation normal form program in Fig. 1 (written in ML-like syntax). The program transforms abstract syntax of propositional formulas (e.g., $\neg(a \wedge (b \vee c))$) to equivalent ones where negations only appear in front of atoms (e.g., $\neg a \vee (\neg b \wedge \neg c)$). The typed regular tree grammar domain would *e.g.*, allow us to infer a grammar that precisely characterizes that the output of the program (nnf) only has negations in front of atoms as required:

$$NFml ::= \text{Atom}(\mathsf{int}) \mid \text{Neg}(\text{Atom}(\mathsf{int})) \mid \text{And}(NFml, NFml) \mid \text{Or}(NFml, NFml)$$

We show how to modularize the monolithic abstract domain containing these elements, dealing with the core challenges: representing syntactic recursion (e.g., over symbol NFml) while preserving typing, and compositionality of the lattice operations.

We show the correctness of the computable abstractions for recursive types by relating them to the general domain, thereby allowing reuse of proofs for most subcomponents (including base types, sum types and product types). The k-limited tree domain is fully modular and can be related using Galois connections

```
type Fml = Atom(int) | Neg(Fml) | And(Fml, Fml) | Or(Fml, Fml)

nnf (f: Fml) = case f of
| Atom(i) -> Atom(i) | Neg(Atom(i)) -> Neg(Atom(i))
| Neg(Neg(f)) -> nnf f
| Neg(And(f1, f2)) -> Or(nnf (Neg f1), nnf (Neg f2))
| Neg(Or(f1, f2)) -> And(nnf (Neg f1), nnf (Neg f2))
| Or(f1, f2) -> Or(nnf f1, nnf f2)
| And(f1, f2) -> And(nnf f1, nnf f2)
```

Fig. 1. Negation normal form

to the general abstract domain $[\![A]\!]^\sharp \xleftrightarrow[\alpha]{\gamma} [\![A]\!]^\sharp_{\mathrm{K}}$, and by composition to the concrete power-set domain $\mathscr{P}[\![A]\!] \xleftrightarrow[\alpha']{\gamma'} [\![A]\!]^\sharp_{\mathrm{K}}$. The more complex regular tree grammar domain is related to general abstract domain via concretization.

The paper is structured as follows. We define a simply-typed lambda calculus and its set-based collecting semantics in Sect. 2 to guide the presentation of our paper. Section 3 summarizes relevant background on abstract interpretation and Galois connections for unit type, sums, products and functions. Section 4 shows our solution for constructing Galois connections between concrete and abstract interpretation of recursive types, including the necessary proofs. Section 5 shows two computable compositional abstract domains for recursive types, and how to relate them to the general abstract domain. Section 7 presents related work and Sect. 8 concludes.

2 Expression Language and Its Collecting Semantics

We highlight the strengths of our technique using a typed λ-calculus with recursion.

The set of types include integers (int), the unit type (unit), sum types ($A_1 + A_2$), product types ($A_1 * A_2$), function types ($A_1 \Rightarrow A_2$), and recursive types ($\mu X.A$). The grammar of types is given by:

$$A ::= \mathsf{int} \mid \mathsf{unit} \mid \mu X.B \mid A_1 {*} A_2 \mid A_1 {+} A_2 \mid A_1 \Rightarrow A_2$$
$$B ::= \mathsf{int} \mid \mathsf{unit} \mid B_1 {*} B_2 \mid B_1 {+} B_2 \mid A \Rightarrow B \mid X$$

In order to accommodate recursive types, we introduce a type variable X which means that we now have closed types (metavariable A) and open types (metavariable B). For simplicity of presentation, we only have one type variable X which effectively disallows nested recursive types—we believe our results extend to them as well. The grammar for open types (metavariable B) restricts recursive types to be strictly positive, meaning that the type variable can not appear to the left of an arrow. This restriction is imposed in order to have a set theoretic interpretation of both higher order and recursive types.

$$\mathsf{bool} = \mathsf{unit}+\mathsf{unit} \quad x_0 : A_0, \ldots, x_n : A_n \vdash_{\mathsf{ex}} x_i : A_i \quad \Gamma \vdash_{\mathsf{ex}} n : \mathsf{int} \quad \Gamma \vdash_{\mathsf{ex}} \mathsf{tt} : \mathsf{unit}$$

$$\frac{\Gamma \vdash_{\mathsf{ex}} e_1 : \mathsf{int} \quad \Gamma \vdash_{\mathsf{ex}} e_2 : \mathsf{int}}{\Gamma \vdash_{\mathsf{ex}} e_1 \odot e_2 : \mathsf{int}} (\odot \in \{+, -, \times\}) \qquad \frac{\Gamma \vdash_{\mathsf{ex}} e_1 : \mathsf{int} \quad \Gamma \vdash_{\mathsf{ex}} e_2 : \mathsf{int}}{\Gamma \vdash_{\mathsf{ex}} e_1 \leq e_2 : \mathsf{bool}}$$

$$\frac{\Gamma \vdash_{\mathsf{ex}} e_1 : A_1 \quad \Gamma \vdash_{\mathsf{ex}} e_2 : A_2}{\Gamma \vdash_{\mathsf{ex}} (e_1, e_2) : A_1 * A_2} \qquad \frac{\Gamma \vdash_{\mathsf{ex}} e : A_1 * A_2}{\Gamma \vdash_{\mathsf{ex}} \mathsf{fst}\ e : A_1} \qquad \frac{\Gamma \vdash_{\mathsf{ex}} e : A_1 * A_2}{\Gamma \vdash_{\mathsf{ex}} \mathsf{snd}\ e : A_2}$$

$$\frac{\Gamma \vdash_{\mathsf{ex}} e : A_1}{\Gamma \vdash_{\mathsf{ex}} \mathsf{inl}\ e : A_1 + A_2} \qquad \frac{\Gamma \vdash_{\mathsf{ex}} e : A_2}{\Gamma \vdash_{\mathsf{ex}} \mathsf{inr}\ e : A_1 + A_2}$$

$$\frac{\Gamma \vdash_{\mathsf{ex}} e : A_1 + A_2 \quad \Gamma, x : A_1 \vdash_{\mathsf{ex}} e_1 : A \quad \Gamma, y : A_2 \vdash_{\mathsf{ex}} e_2 : A}{\Gamma \vdash_{\mathsf{ex}} \mathsf{case}\ e\ \mathsf{of}\ \mathsf{inl}\ x \to e_1 \parallel \mathsf{inr}\ y \to e_2 : A}$$

$$\frac{\Gamma \vdash_{\mathsf{ex}} e : B\,[\mu X.B/X]}{\Gamma \vdash_{\mathsf{ex}} \mathsf{fold}\ e : \mu X.B} \qquad \frac{\Gamma \vdash_{\mathsf{ex}} e : \mu X.B}{\Gamma \vdash_{\mathsf{ex}} \mathsf{unfold}\ e : B\,[\mu X.B/X]} \qquad \frac{\Gamma, x : A \vdash_{\mathsf{ex}} e : A}{\Gamma \vdash_{\mathsf{ex}} \mathsf{fix}\ x.e : A}$$

$$\frac{\Gamma, x : A_1 \vdash_{\mathsf{ex}} e : A_2}{\Gamma \vdash_{\mathsf{ex}} \lambda x.e : A_1 \Rightarrow A_2} \qquad \frac{\Gamma \vdash_{\mathsf{ex}} e_1 : A_2 \Rightarrow A_1 \quad \Gamma \vdash_{\mathsf{ex}} e_2 : A_2}{\Gamma \vdash_{\mathsf{ex}} e_1\ e_2 : A_1}$$

Fig. 2. Typing rules of the λ-calculus

The expression language is intrinsically typed and contains usual operations for each type, variables, and a fixed-point combinator ($\mathsf{fix}\ x.e$). The $\mathsf{unfold}\ e$ and $\mathsf{fold}\ e$ expressions respectively allow accessing/abstracting away the underlying representation of expressions of recursive types. The typing judgment for expressions has form $\Gamma \vdash_{\mathsf{ex}} e : A$, ensuring that expression e has type A under some typing context Γ. The associated rules are standard and given in Fig. 2.

The set-theoretic interpretation of types is the expected one, interpreting each syntactic construct into the corresponding semantic one. Because X is required to appear only strictly positively in B, the type can be interpreted as a functor on the category of sets $[\![B]\!] : \mathsf{Set} \to \mathsf{Set}$ defined by induction on B as follows

$$[\![\mathsf{int}]\!](X) = \mathbb{Z} \quad [\![\mathsf{unit}]\!](X) = \mathbb{1} \quad [\![X]\!](X) = X$$
$$[\![B_1 * B_2]\!](X) = [\![B_1]\!](X) \times [\![B_2]\!](X) \quad [\![B_1 + B_2]\!](X) = [\![B_1]\!](X) + [\![B_2]\!](X)$$
$$[\![A]\!](X) = [\![A]\!] \quad [\![A \Rightarrow B]\!](X) = [\![A]\!] \to ([\![B]\!](X))$$

where in the last case we assume that the closed type A has already been given an interpretation. For function types we can assume X does not appear in A (by strict positivity), so we can interpret it directly as a set. The only elaborate interpretation is the one for recursive types, which we will discuss in Sect. 4.

The collecting semantics for open terms (with context) is interpreted according to their types:

$$[\![\Gamma \vdash_{\mathsf{ex}} t : A]\!] : [\![\Gamma]\!] \to \mathscr{P}[\![A]\!]$$

where we interpret contexts as products of sets of values:

$$[\![x_1 : B_1, \ldots, x_n : B_n]\!] = \prod_i \mathscr{P}[\![B_i]\!]$$

There are alternative choices [11] for the collecting semantics domain depending on the target abstract domains, e.g. $\mathscr{P}(\prod_i [\![B_i]\!]) \to \mathscr{P}([\![A]\!])$ or $\mathscr{P}(\prod_i [\![B_i]\!] \to [\![A]\!])$. We chose ours because it is the simplest one that suffices for our compositional domain as proven in Sect. 4.

The collecting semantics provides a concrete interpretation of a term $\vdash_{\mathsf{ex}} e : A$ as the set of possible concrete values $v \in \mathscr{P}[\![A]\!]$ that the term can evaluate to. In our case, we first provide a compositional interpretation of program types into sets, so $[\![A]\!] : \mathsf{Set}$, and then present how to specify collecting semantics for terms in our language, including the challenging case of higher-order functions and fixed-points. The collecting semantics is particularly useful when relating to our abstract domains $[\![A]\!]^\sharp$ which are complete lattices, since power sets $\mathscr{P}[\![A]\!]$ form complete lattices themselves. This makes it possible to compositionally specify Galois connection and show soundness of the abstract interpretation framework, which we do in Sects. 3 and 4.

We define most cases straightforwardly, ensuring monotonicity of mapping w.r.t the context. We omit the explicit typing of syntax for readability of the semantics. We write $\mathscr{P}(f) : \mathscr{P}([\![A_1]\!] \times \cdots \times [\![A_n]\!]) \to \mathscr{P}[\![B]\!]$ to lift function $f : [\![A_1]\!] \times \cdots \times [\![A_n]\!] \to [\![B]\!]$ to the power set domain, mimicking notation from category theory.

$$[\![n]\!]\rho = \{\mathrm{n}\} \quad [\![x]\!]\rho = \rho(x) \quad [\![\mathsf{tt}]\!]\rho = \{\star\}$$
$$[\![e_1 \odot e_2]\!]\rho = \mathscr{P}(\odot)([\![e_1]\!]\rho, [\![e_2]\!]\rho)$$
$$[\![e_1 \leq e_2]\!]\rho = \mathscr{P}(\leq)([\![e_1]\!]\rho, [\![e_2]\!]\rho) \quad [\![(e_1, e_2)]\!]\rho = \mathscr{P}(,)([\![e_1]\!]\rho, [\![e_2]\!]\rho)$$
$$[\![\mathsf{fst}\ e]\!]\rho = \mathscr{P}(\pi_1)([\![e]\!]\rho) \quad [\![\mathsf{snd}\ e]\!]\rho = \mathscr{P}(\pi_2)([\![e]\!]\rho)$$
$$[\![\mathsf{inl}\ e]\!]\rho = \mathscr{P}(\imath_1)([\![e]\!]\rho) \quad [\![\mathsf{inr}\ e]\!]\rho = \mathscr{P}(\imath_2)([\![e]\!]\rho)$$
$$[\![e_1\ e_2]\!]\rho = \{f(a) | f \in [\![e_1]\!]\rho, a \in [\![e_2]\!]\rho\}$$
$$[\![\mathsf{case}\ e\ \mathsf{of}\ \mathsf{inl}\ x \to e_1 \parallel \mathsf{inr}\ y \to e_2]\!]\rho = \bigcup_{i \in \{1,2\}} [\![e_i]\!]\rho\,[x \mapsto \{v | \imath_i v \in [\![e]\!]\rho\}]$$

The interpretation for functions is more intricate. The outcome of analysing the body of a function for a given argument is a set of results, and this has to be lifted to a set of functions. To do this, we define the (left-invertible) mapping $\delta : ([\![A]\!] \to \mathscr{P}[\![B]\!]) \to \mathscr{P}([\![A]\!] \to [\![B]\!])$:

$$\delta(f) = \{g : [\![A]\!] \to [\![B]\!] | \forall a \in [\![A]\!].g(a) \in f(a)\}$$

Then we can interpret functions using δ into our target set of functions:

$$\begin{aligned}[\![\rho \vdash_{\mathsf{ex}} \lambda x.e : A \to B]\!]\rho &= \delta(\lambda a.[\![e]\!]\rho[x \mapsto \{a\}]) \\ &= \{f \in [\![A]\!] \to [\![B]\!] \mid \forall a \in [\![A]\!].f(a) \in [\![e]\!]\rho[x \mapsto \{a\}]\}\end{aligned}$$

Finally, we can then interpret fixed-points by relying on the power set domain $\mathscr{P}([\![A]\!])$ forming a complete lattice:

$$[\![\rho \vdash_{\mathsf{ex}} \mathsf{fix}\ x.e : A]\!]\rho = \mathsf{lfp}(\lambda X.[\![e]\!]\rho[x \mapsto X])$$

where the least fixed-point on complete lattices $\mathsf{lfp}(f) = \bigcap\{a \mid f(a) \subseteq a\}$ is given by the Knaster-Tarski Fixed-point Theorem [24] and Theorem 1. Now, that we have provided a compositional collecting semantics for all our language constructs, we can proceed with the abstract domains.

Theorem 1. *If* $\rho_1 \subseteq \rho_2$ *then* $[\![e]\!]\rho_1 \subseteq [\![e]\!]\rho_2$

3 Abstract Interpretation and Galois Connections

Abstract interpretation provides a theory for relating a concrete set of values for a type $\mathscr{P}[\![A]\!]$ to an abstract (usually complete) lattice of elements for the same type $[\![A]\!]^\sharp$. The relation is performed using Galois connections [10] $\mathscr{P}[\![A]\!] \underset{\alpha}{\overset{\gamma}{\leftrightarrows}} [\![A]\!]^\sharp$, which uses two monotone maps $\alpha : \mathscr{P}[\![A]\!] \to [\![A]\!]^\sharp$ and $\gamma : [\![A]\!]^\sharp \to \mathscr{P}[\![A]\!]$ to map between the two domains, so that the following relation is preserved $a \subseteq \gamma(a^\sharp) \iff \alpha(a) \sqsubseteq a^\sharp$. The existence of a Galois connection formalises the way in which an abstract property describes a set of concrete values, provides a notion of "best approximation" and even suggests a design methodology for deriving program analysers in a systematic way by composing the semantic function with an abstraction function α [9,18].

We are interested in abstract interpretation in which abstract domains are defined as compositional interpretations of types, where the semantics of a type is composed of the types it contains. E.g., the semantics of $A{*}B$ is composed of the semantics of A and B. For base types and first-order algebraic types there is a well-established theory for relating the concrete set-theoretic semantics $[\![A]\!]$ and the corresponding abstract interpretation to complete lattices $[\![A]\!]^\sharp$ in a compositional fashion, using Galois connections $\mathscr{P}[\![A]\!] \underset{\alpha}{\overset{\gamma}{\leftrightarrows}} [\![A]\!]^\sharp$. This design is what makes it possible to exchange abstractions at the granularity of types, while preserving the structure of other parts of the system.

The following theorem is useful for constructing Galois connections, by using a map $\alpha^1 : [\![A]\!] \to [\![A]\!]^\sharp$ that directly abstracts concrete values [20, p. 237].

Proposition 1. *Let E be a set and $E^\sharp$ a complete lattice. There is a bijective correspondence between maps $\alpha^1 : E \to E^\sharp$ and Galois connections $\mathscr{P}(E) \underset{\alpha}{\overset{\gamma}{\leftrightarrows}} E^\sharp$. The correspondence maps a Galois connection to α^1 defined as $\alpha^1(x) = \alpha(\{x\})$ and a map α^1 to the Galois connection:*

$$\alpha(X \subseteq E) = \bigsqcup_{x \in X} \alpha^1(x) \qquad\qquad \gamma(e^\sharp) = \{x \in E \mid \alpha^1(x) \sqsubseteq e^\sharp\}$$

The correspondence can be summarized in the following diagram.

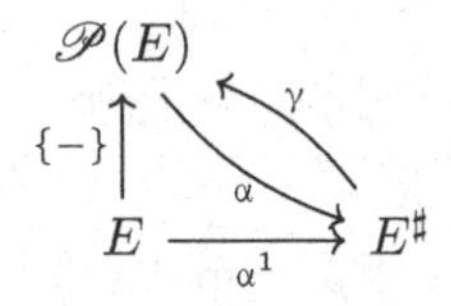

The systematic construction of Galois connections for base types, sums, and products is standard [20, sect. 4.4]—we recall them since they are needed for recursive types (Sect. 4).

Unit Type. We abstract the unit type as the 2-element lattice $[\![\mathsf{unit}]\!]^\sharp = \Sigma = \{\bot, \top\}$ where $\bot \leq \top$. The induced Galois connection is the isomorphism $\mathscr{P}(\mathbb{1}) \cong \Sigma$ mapping $\emptyset$ to $\bot$ and $\{\star\}$ to $\top$. Similarly, we abstract tt as $\top$, $[\![\mathsf{tt}]\!]^\sharp\rho^\sharp = \top$.

Sums. The abstract interpretation of a sum type is the product of lattices (with the pointwise ordering):$[\![A_1 + A_2]\!]^\sharp = [\![A_1]\!]^\sharp \times [\![A_2]\!]^\sharp$. The intuition is that each component of a pair (a_1, a_2) in the product $[\![A_1]\!]^\sharp \times [\![A_2]\!]^\sharp$ describes a property of the sum value *if* the value belongs to the corresponding summand; for concrete values $\imath_1(v)$ (corr. $\imath_2(v)$) we will have that v satisfies the property a_1 (corr. a_2). The element abstraction map is defined as:

$$\alpha^1_{A_1+A_2}(\imath_1(x)) = (\alpha^1_{A_1}(x), \bot) \qquad \alpha^1_{A_1+A_2}(\imath_2(y)) = (\bot, \alpha^1_{A_2}(y))$$

where $\imath_1$ and $\imath_2$ are the injections into the disjoint sum.

Products. The simplest useful abstraction of pairs is by using the smash product of abstract domains ($A^\sharp \otimes B^\sharp$). The smash product is a product that disallows pairs where only one component is the bottom element:

$$[\![A_1 * A_2]\!]^\sharp = A^\sharp \otimes B^\sharp = \{(a, b) | a = \bot \iff b = \bot\}$$

We will implicitly convert a pair which has a single $\bot$ component, to one where both components are $\bot$. Using this, the element abstraction map is defined as

$$\alpha^1_{A_1 * A_2}(x, y) = (\alpha^1_{A_1}(x), \alpha^1_{A_2}(x))$$

Other abstract domain constructions for products are based on ordinary Cartesian products or tensor products [19] which keep relational information between the two components. We conjecture that the theory extends to these other types of products but we leave the formal verification of this for further work.

Function Types. We abstractly interpret function types as the set of monotone functions:

$$[\![A_1 \Rightarrow A_2]\!]^\sharp = \{f : [\![A_1]\!]^\sharp \to [\![A_2]\!]^\sharp \mid \forall x, y. x \leq y \Longrightarrow f(x) \leq f(y)\}$$

By induction, we can assume the Galois connection is given for the types A_1 and A_2 and define,

$$\alpha^1_{A_1 \Rightarrow A_2} : [\![A_1 \Rightarrow A_2]\!] \to [\![A_1 \Rightarrow A_2]\!]^\sharp$$

to map $f : [\![A_1]\!] \to [\![A_2]\!]$ to the composition

$$\alpha_{A_2} \circ \mathscr{P}(f) \circ \gamma_{A_1}$$

Note that $\mathscr{P}(f)$ is monotone, and so the composition is monotone, but since γ_{A_1} is not necessarily continuous, neither is the composition. This is one reason $[\![A_1 \Rightarrow A_2]\!]^\sharp$ needs to be the set of monotone, rather than continuous maps. Monotone functions over complete lattices form themselves a complete lattice.

4 Recursive Types

We now extend the theory of Sect. 3 with recursive types. We shall define abstract domains for recursive types as solutions to recursive domain equations. Categorically, we rely on *initial algebras* to describe inductive types. An initial algebra for some functor $F : \mathsf{Set} \to \mathsf{Set}$ consists of a set X (the carrier) and a map $f : FX \to X$ such that for any other map $g : FY \to Y$, there exists a unique map $h : X \to Y$ making the following diagram commute:

$$\begin{array}{ccc} FX & \xrightarrow{Fh} & FY \\ {\scriptstyle f}\downarrow & & \downarrow{\scriptstyle g} \\ X & \xrightarrow{h} & Y \end{array}$$

For example, the initial algebra for the functor $F(X) = 1 + \mathbb{Z} \times X$ is exactly the set of lists of integers, and the initiality property described above captures exactly the induction principle for lists. Initial algebras generalise directly from Set to other categories, see e.g., Awodey [4, Chp 10.5] for more details.

We rely on strict positivity to interpret $\mu X.B$, since the corresponding functor $[\![B]\!](-)$ is known to have an initial algebra [1]. Formally, $\mu X.B$ is interpreted as the carrier of the initial algebra for $[\![B]\!](-)$.

4.1 Recursively Defined Lattices

To define the abstract interpretation of recursive types, open types must be interpreted as functors on a category of complete lattices. For this, let cLat denote the category of complete lattices and continuous maps, i.e., maps preserving all least upper bounds. Recall that this implies that maps are monotone and preserve the bottom element.

Given two complete lattices $X^\sharp, Y^\sharp$, the set of maps (hom-set) between them $\mathsf{cLat}(X^\sharp, Y^\sharp)$ is itself a complete lattice under the pointwise ordering, i.e., $f \sqsubseteq g$ if $f(x) \sqsubseteq g(x)$ for all x. A functor $F : \mathsf{cLat} \to \mathsf{cLat}$ is *locally continuous* if the lifting of maps $f \in \mathsf{cLat}(X^\sharp, Y^\sharp)$ to work at the functor level $Ff \in \mathsf{cLat}(F(X^\sharp), F(Y^\sharp))$ is itself continuous, i.e., $F(\bigsqcup fs) = \bigsqcup_{f \in fs} F(f)$.

Theorem 2. *Any locally continuous functor $F : \mathsf{cLat} \to \mathsf{cLat}$ has a fixed point $\mathsf{Fix}\, F \cong F(\mathsf{Fix}\, F)$.*

Proof. The proof is completely standard [2,23], but we recall the construction of the fixed point for future reference. Let $\mathbb{1} = \{\bot\}$ be the singleton lattice. There are continuous maps $e_0 : \mathbb{1} \to F\mathbb{1}$ (mapping $\bot$ to the bottom element of $F(\mathbb{1})$) and $p_0 : F\mathbb{1} \to \mathbb{1}$ forming an embedding-projection pair, i.e., satisfying $p_0 \circ e_0 = \mathsf{id}$ and $e_0 \circ p_0 \sqsubseteq \mathsf{id}$. Defining $p_n = F^n p_0$ and $e_n = F^n e_0$ gives an embedding-projection pair from $F^n(\mathbb{1})$ to $F^{n+1}(\mathbb{1})$. The fixed point is defined as the limit of the chain of p_n maps, i.e., $\mathsf{Fix}\, F = \{(x_1, x_2, \dots) \mid \forall n. x_n \in F^n(\mathbb{1}), x_n = p_n(x_{n+1})\}$.

To define the lattice interpretation of types, it thus suffices to construct an interpretation of each open type expression as a locally continuous functor $\mathsf{cLat} \to \mathsf{cLat}$. The interpretation is defined using the constructions of Sect. 3:

$$[\![X]\!]^\sharp(X^\sharp) = X^\sharp \qquad [\![B_1 * B_2]\!]^\sharp(X^\sharp) = [\![B_1]\!]^\sharp(X^\sharp) \otimes [\![B_2]\!]^\sharp(X^\sharp)$$
$$[\![B_1 + B_2]\!]^\sharp(X^\sharp) = [\![B_1]\!]^\sharp(X^\sharp) \times [\![B_2]\!]^\sharp(X^\sharp)$$
$$[\![A \Rightarrow B]\!]^\sharp(X^\sharp) = [\![A]\!]^\sharp \underset{\text{mono}}{\to} ([\![B]\!]^\sharp(X^\sharp))$$

Here, the notation $\underset{\text{mono}}{\to}$ refers to the lattice of monotone functions.

4.2 Defining the Galois Connection

Having defined the interpretation of each type A as a set $[\![A]\!]$ and a lattice $[\![A]\!]^\sharp$ respectively, we must now construct a Galois connection from $\mathscr{P}[\![A]\!]$ to $[\![A]\!]^\sharp$. This will be defined by induction on the structure of A and so must also take open types into account. For this, we define for each open type B and each complete lattice $X^\sharp$ a map

$$\alpha^1_B(X^\sharp) : [\![B]\!](X^\sharp) \to [\![B]\!]^\sharp(X^\sharp)$$

where the codomain is the underlying set of the complete lattice $[\![A]\!]^\sharp(X^\sharp)$ (essentially forgetting the lattice structure). For most type constructors we just show the case of the open type, since the closed case is essentially the same. The basic cases are as follows:

$$\alpha^1_X(X^\sharp)(x) = x \qquad \alpha^1_{B_1 * B_2}(X^\sharp)(x,y) = (\alpha^1_{B_1}(X^\sharp)(x), \alpha^1_{B_2}(X^\sharp)(y))$$
$$\alpha^1_{B_1+B_2}(X^\sharp)(\imath_1(x)) = (\alpha^1_{B_1}(X^\sharp)(x), \bot) \qquad \alpha^1_{B_1+B_2}(X^\sharp)(\imath_2(x)) = (\bot, \alpha^1_{B_2}(X^\sharp)(x))$$

For function type $A \Rightarrow B$, by induction we have

$$\alpha^1_A : [\![A]\!] \to [\![A]\!]^\sharp \qquad \alpha^1_B(X^\sharp) : [\![B]\!](X^\sharp) \to [\![B]\!]^\sharp(X^\sharp)$$

(since A is a closed type). These induce

$$\gamma_A : [\![A]\!]^\sharp \to \mathscr{P}([\![A]\!]) \qquad \alpha_B(X^\sharp) : \mathscr{P}([\![B]\!](X^\sharp)) \to [\![B]\!]^\sharp(X^\sharp)$$

by Proposition 1. Thus, we can define

$$\alpha^1_{A \Rightarrow B}(X^\sharp) : ([\![A]\!] \to [\![B]\!](X^\sharp)) \to ([\![A]\!]^\sharp \underset{\text{mono}}{\to} [\![B]\!]^\sharp(X^\sharp))$$

to map g to the composition $\alpha_B(X^\sharp) \circ \mathscr{P}(g) \circ \gamma_A$.

In the case of recursive types, recall that $[\![\mu X.B]\!]$ is defined as the initial algebra for the functor $[\![B]\!] : \mathsf{Set} \to \mathsf{Set}$. To define $\alpha^1_{\mu X.B}$ we therefore define an algebra structure

$$[\![B]\!]([\![\mu X.B]\!]^\sharp) \to [\![\mu X.B]\!]^\sharp$$

for $[\![\mu X.B]\!]^\sharp$ as the composite of

$$\alpha^1_B([\![\mu X.B]\!]^\sharp) : [\![B]\!]([\![\mu X.B]\!]^\sharp) \to [\![B]\!]^\sharp([\![\mu X.B]\!]^\sharp)$$

and the isomorphism

$$\mathsf{fold}^\sharp : [\![B]\!]^\sharp([\![\mu X.B]\!]^\sharp) \to [\![\mu X.B]\!]^\sharp$$

Thus, we define $\alpha^1_{\mu X.B}$ to be the unique map that makes the diagram commute:

$$\begin{array}{ccc} [\![B]\!]([\![\mu X.B]\!]) & \xrightarrow{[\![B]\!](\alpha^1_{\mu X.B})} & [\![B]\!]([\![\mu X.B]\!]^\sharp) \\ \downarrow & & \downarrow \\ [\![\mu X.B]\!] & \xrightarrow{\alpha^1_{\mu X.B}} & [\![\mu X.B]\!]^\sharp \end{array}$$

The vertical map on the left is the algebra structure for $[\![\mu X.B]\!]$ and the map $\alpha^1_{\mu X.B}$ exists uniquely by the initial algebra property.

Note that this construction crucially uses that the set-interpretation of recursive types is as initial algebras[1]. On the other hand, we do not use anything specific about the lattice interpretation, and in fact, the construction of the Galois connection simply requires a map

$$[\![B]\!]^\sharp([\![\mu X.B]\!]^\sharp) \to [\![\mu X.B]\!]^\sharp$$

not necessarily an isomorphism. We will use this fact in Sect. 5, and for this reason we state the main theorem in a more general setting.

Theorem 3. *Suppose the abstract interpretation of recursive types*

$$\mathsf{fold}^\sharp : [\![B\,[\mu X.B/X]]\!]^\sharp \to [\![\mu X.B]\!]^\sharp \qquad \mathsf{unfold}^\sharp : [\![\mu X.B]\!]^\sharp \to [\![B\,[\mu X.B/X]]\!]^\sharp$$

satisfies the equations

$$\mathsf{fold}^\sharp \circ \mathsf{unfold}^\sharp = \mathsf{id} \qquad \mathsf{unfold}^\sharp \circ \mathsf{fold}^\sharp \sqsupseteq \mathsf{id}$$

If $\vdash_{\mathsf{ex}} t : A$ *then* $[\![t]\!] \subseteq \gamma([\![t]\!]^\sharp)$*, or equivalently* $\alpha([\![t]\!]) \sqsubseteq [\![t]\!]^\sharp$

Theorem 3 is proved by induction on typing judgements, and must therefore be extended to open terms. This is done in the following lemma, which uses an extension of the maps γ and α to contexts defined pointwise:

Lemma 1. *Suppose* $\Gamma \vdash_{\mathsf{ex}} e : A$*, and that* $\rho \in [\![\Gamma]\!]$ *and* $\rho^\sharp \in [\![\Gamma]\!]^\sharp$*, are such that* $\rho \subseteq \gamma(\rho^\sharp)$*. If the conditions on the abstract interpretation of recursive types of Theorem 3 are satisfied, then* $[\![t]\!]\rho \subseteq \gamma([\![t]\!]^\sharp\rho^\sharp)$ *or, equivalently,* $\alpha([\![t]\!]\rho) \subseteq [\![t]\!]^\sharp\rho^\sharp$

Theorem 3 obviously follows from this as a special case. Before proving Lemma 1 we need a few lemmas.

[1] The construction would not work if we used final coalgebras.

Lemma 2. *If B is an open type and A a closed one, then $[\![B]\!]([\![A]\!]) = [\![B\,[A/X]]\!]$ and $[\![B]\!]^\sharp([\![A]\!]^\sharp) = [\![B\,[A/X]]\!]^\sharp$*

Proof. Easy induction on B, which we omit.

Lemma 3. *For any open type B and closed type A the following equality holds*

$$\alpha^1_B([\![A]\!]^\sharp) \circ [\![B]\!](\alpha^1_A) = \alpha^1_{B[A/X]}$$

In diagram style, this is

$$\begin{array}{ccccc}
[\![B\,[A/X]]\!] & \xrightarrow{\mathsf{id}} & [\![B]\!]([\![A]\!]) & \xrightarrow{[\![B]\!](\alpha^1_A)} & [\![B]\!]([\![A]\!]^\sharp) \\
\downarrow{\scriptstyle \alpha^1_{B[A/X]}} & & & & \downarrow{\scriptstyle \alpha^1_B([\![A]\!]^\sharp)} \\
[\![B\,[A/X]]\!]^\sharp & & \xrightarrow{\mathsf{id}} & & [\![B]\!]^\sharp([\![A]\!]^\sharp)
\end{array}$$

Proof. Induction on B (Appendix A).

Lemma 4. *If B is an open type, then the following diagram commutes*

$$\begin{array}{ccc}
\mathscr{P}([\![B\,[\mu X.B/X]]\!]) & \xrightarrow{\alpha_{B[\mu X.B/X]}} & [\![B\,[\mu X.B/X]]\!]^\sharp \\
\downarrow{\scriptstyle \mathscr{P}(\mathsf{fold})} & & \downarrow{\scriptstyle \mathsf{fold}^\sharp} \\
\mathscr{P}([\![\mu X.B]\!]) & \xrightarrow{\alpha_{\mu X.B}} & [\![\mu X.B]\!]^\sharp
\end{array}$$

Proof. It suffices to show that $\alpha^1_{\mu X.B} \circ \mathsf{fold} = \mathsf{fold}^\sharp \circ \alpha^1_{B[\mu X.B/X]}$ since the two compositions in the diagram are the unique extensions of the two sides of this equation to continuous maps. By definition of $\alpha^1_{\mu X.B}$ we get

$$\alpha^1_{\mu X.B} \circ \mathsf{fold} = \mathsf{fold}^\sharp \circ \alpha^1_B([\![\mu X.B]\!]^\sharp) \circ [\![B]\!](\alpha^1_{\mu X.B})$$

By Lemma 3 the right hand side of this is equal to $\mathsf{fold}^\sharp \circ \alpha^1_{B[\mu X.B/X]}$ which concludes the proof.

We can now prove Lemma 1.

Proof (Lemma 1). By induction on typing derivation $\Gamma \vdash_{\mathsf{ex}} e : A$. We show the interesting cases for recursive types and the rest are in Appendix A. Case $t = \mathsf{fold}\ u : \mu X.A$,

$$\alpha([\![\mathsf{fold}\ u]\!]\rho) = \alpha(\mathscr{P}(\mathsf{fold})([\![u]\!]\rho)) = \mathsf{fold}^\sharp(\alpha([\![u]\!]\rho)) \sqsubseteq \mathsf{fold}^\sharp([\![u]\!]^\sharp\rho^\sharp) = [\![\mathsf{fold}\ u]\!]^\sharp\rho^\sharp$$

- Case $t = \mathsf{unfold}\ u : A\,[\mu X.A/X]$
  First note that

$$\alpha \circ \mathscr{P}(\mathsf{unfold}) \sqsubseteq \mathsf{unfold}^\sharp \circ \alpha$$

since

$$\alpha \circ \mathscr{P}(\mathsf{unfold}) \sqsubseteq \mathsf{unfold}^\sharp \circ \mathsf{fold}^\sharp \circ \alpha \circ \mathscr{P}(\mathsf{unfold})$$
$$= \mathsf{unfold}^\sharp \circ \alpha \circ \mathscr{P}(\mathsf{fold} \circ \mathsf{unfold}) = \mathsf{unfold}^\sharp \circ \alpha$$

and so

$$\alpha([\![\mathsf{unfold}\ u]\!]\rho) = \alpha(\mathscr{P}(\mathsf{unfold})([\![u]\!]\rho))$$
$$\sqsubseteq \mathsf{unfold}^\sharp(\alpha([\![u]\!]\rho)) \overset{\text{by IH}}{\sqsubseteq} \mathsf{unfold}^\sharp([\![u]\!]^\sharp\rho^\sharp) = [\![\mathsf{unfold}\ u]\!]^\sharp\rho^\sharp$$

We can finally provide a compositional interpretation for program expressions manipulating recursive types, based on the corresponding concrete and abstract semantic operations for recursive types:

$$[\![\mathsf{fold}\ e]\!]\rho = \mathscr{P}(\mathsf{fold})([\![e]\!]\rho) \qquad [\![\mathsf{unfold}\ e]\!]\rho = \mathscr{P}(\mathsf{unfold})([\![e]\!]\rho)$$
$$[\![\mathsf{fold}\ e]\!]^\sharp\rho^\sharp = \mathsf{fold}^\sharp([\![e]\!]^\sharp\rho^\sharp) \qquad [\![\mathsf{unfold}\ e]\!]^\sharp\rho^\sharp = \mathsf{unfold}^\sharp([\![e]\!]^\sharp\rho^\sharp)$$

We present an elaborate example illustrating how the constructs can be explicitly instantiated for lists of integers in Sect. 4.3. This is primarily to provide a more concrete formal intuition about the solutions to the fixed-points, since our theory works with all inductive types in general.

4.3 Example: Integer Lists

We illustrate the theoretical construction of Galois connections for recursive types on the simplest case of such types: lists of integers. Recall that the recursive type of an integer list is $\mu X.\mathsf{unit}{+}\mathsf{int}{*}X$, where the first component of the sum represents the empty list and the second component represents a pair of an integer element—representing the head element—and the rest of the list (tail).

In the standard semantics, the interpretation of this type is the initial algebra of the functor $[\![\mathsf{unit}{+}\mathsf{int}{*}X]\!](X) = 1 + \mathbb{Z} \times X$, which is simply the set of lists of integers. For the abstract semantics we interpret integers using the standard Sign abstraction, which abstracts a set of integers by the sign of elements (+, − or 0) it contains (if all elements have the same sign), or otherwise the bottom $\bot$ (representing the empty set) or $\top$ (abstracting of sets with elements of mixed signs).

Then, $[\![\mu X.\mathsf{unit}{+}\mathsf{int}{*}X]\!]^\sharp$ is the fixed point of the functor

$$[\![\mathsf{unit}{+}\mathsf{int}{*}X]\!]^\sharp(X) = \Sigma \times (\mathsf{Sign} \otimes X)$$

The fixed point is constructed as in the proof of Theorem 2, i.e., as a limit of a chain obtained by applying F countably many times to the singleton lattice $\mathbb{1}$. We start by computing a few iterations.

$$F(\mathbb{1}) = \Sigma \times (\mathsf{Sign} \otimes \mathbb{1}) \cong \Sigma$$
$$F^2(\mathbb{1}) \cong \Sigma \times (\mathsf{Sign} \otimes \Sigma) \cong \{(b, m_0, b_0) \in \Sigma \times \mathsf{Sign} \times \Sigma \mid m_0 = \bot \iff b_0 = \bot\}$$
$$F^3(\mathbb{1}) \cong \Sigma \times (\mathsf{Sign} \otimes (\Sigma \times (\mathsf{Sign} \otimes \Sigma)))$$
$$= \left\{ (b, m_0, b_0, m_1, b_1) \in \Sigma \times (\mathsf{Sign} \times \Sigma)^2 \;\middle|\; \begin{array}{c} (m_0 = \bot \iff (b_0 = \bot \land m_1 = \bot))\land \\ (m_1 = \bot \iff b_1 = \bot) \end{array} \right\}$$

In all cases, the map $F^{n+1}(\mathbb{1}) \to F^n(\mathbb{1})$ forgets the last two elements. In general

$$F^n(\mathbb{1}) \cong \{(b,(m_i,b_i)_i) \in \Sigma \times (\mathsf{Sign} \times \Sigma)^{n-1} \mid m_{n-1} = \bot \iff b_{n-1} = \bot$$
$$\forall i < n.(m_i = \bot \iff \forall k \geq 0.(b_{i+k} = \bot \wedge m_{i+k+1} = \bot))\}$$

and the limit becomes

$$[\![\mu X.\mathsf{unit}{+}\mathsf{int}{*}X]\!]^\sharp \cong \{(b,(m_i,b_i)_i) \in \Sigma \times (\mathsf{Sign} \times \Sigma)^{\mathbb{N}} \mid$$
$$\forall i.(m_i = \bot \iff \forall k.(b_{i+k} = \bot \wedge m_{i+k+1} = \bot))\}$$

The algebra structure on the underlying set of this lattice has type

$$1 + \mathbb{Z} \times ([\![\mu X.\mathsf{unit}{+}\mathsf{int}{*}X]\!]^\sharp) \to [\![\mu X.\mathsf{unit}{+}\mathsf{int}{*}X]\!]^\sharp$$

and maps $\imath_1(\star)$ (where $\star$ is the unique element in 1) to

$$(\top,(\bot,\bot)_n)$$

and an element $\imath_2(m,(b,(m_i,b_i)_i))$ to $(\bot,(m'_i,b'_i)_i)$ where

$$(m'_i,b'_i) = \begin{cases} (m,b) & i = 0 \\ (m_{i-1},b_{i-1}) & i > 0 \end{cases}$$

The map

$$\alpha^1_{\mu X.\mathsf{unit}{+}\mathsf{int}{*}X} : [\![\mu X.\mathsf{unit}{+}\mathsf{int}{*}X]\!] \to [\![\mu X.\mathsf{unit}{+}\mathsf{int}{*}X]\!]^\sharp$$

which is defined using the initial algebra property of the set of lists thus acts as follows

$$\alpha^1_{\mu X.\mathsf{unit}{+}\mathsf{int}{*}X}([]) = (\top,(\bot,\bot)_n)$$
$$\alpha^1_{\mu X.\mathsf{unit}{+}\mathsf{int}{*}X}([x_0,\ldots,x_n]) = (\bot,(\mathsf{sign}(x_0),\bot),\ldots,(\mathsf{sign}(x_n),\top),(\bot,\bot),\ldots)$$

The abstraction map

$$\alpha_{\mu X.\mathsf{unit}{+}\mathsf{int}{*}X} : \mathscr{P}[\![\mu X.\mathsf{unit}{+}\mathsf{int}{*}X]\!] \to [\![\mu X.\mathsf{unit}{+}\mathsf{int}{*}X]\!]^\sharp$$

maps a set X to the least upper bound of $\alpha^1_{\mu X.\mathsf{unit}{+}\mathsf{int}{*}X}$ applied to the elements of the set. For example, if

$$A = \left\{ xs \in [\![\mu X.\mathsf{unit}{+}\mathsf{int}{*}X]\!] \;\middle|\; \begin{array}{l} \exists n.\mathsf{length}(xs) = 2n. \\ \forall i < n.\mathsf{sign}(xs[2i]) = +, \mathsf{sign}(xs[2i+1]) = - \end{array} \right\}$$

is the set whose elements are lists of even length with alternating sign (starting with positive), then

$$\alpha_{\mu X.\mathsf{unit}{+}\mathsf{int}{*}X}(A) = (\top,(+,\bot),(-,\top),(+,\bot),(-,\top),\ldots)$$

is the corresponding abstraction. Concretely, we have $\bot$ every second time (so $b_{2i} = \bot$ for all i) to ensure that the length of the list is even.

The concretization map

$$\gamma_{\mu X.\mathsf{unit+int*}X} : [\![\mu X.\mathsf{unit+int*}X]\!]^\sharp \to \mathscr{P}[\![\mu X.\mathsf{unit+int*}X]\!]$$

maps an abstract element to the set of all lists whose abstraction (by $\alpha_{\mu X.\mathsf{unit+int*}X}$) are below it. For example, it maps $\alpha_{\mu X.\mathsf{unit+int*}X}(A)$ back to A.

Another example is the top element of the abstract domain which is:

$$\top_{\mu X.\mathsf{unit+int*}X} = (\top_\Sigma, (\top_{\mathsf{Sign}}, \top_\Sigma) \ldots)$$

It basically, uses the top element of each constituting abstract domain, while having the infinite form abstracting over all sizes of concrete lists, as required by the limit.

Limitation of the Compositional Solution. A limitation of the compositional abstract solution for recursive types is that it cannot precisely capture constraints across the recursive structure. E.g., there is no precise abstract domain that characterizes lists where the elements are sorted. Providing a solution that can capture richer constraints is future work.

5 Computable Abstractions

The general abstraction of recursive types (Theorem 2) contains elements with a non-finitary structure, which makes it hard to use directly for terminating analyses. In this section, we show how to construct two computable abstract domains for recursive types: k-limited trees and typed regular tree expressions. We show how these can be constructed modularly and then easily related to the concrete powerset domain through the general abstract domain.

5.1 k-Limited Trees

We can get a computable analysis from our general abstract domain, by only considering the subset of prefixes of recursive structures up to some fixed depth k. This subset of finitary elements is called k-limited trees. Formally, we define the k-limited trees abstract domain as follows

$$[\![\mu X.B]\!]^\sharp_{\mathrm{K}} = ([\![B]\!]^\sharp)^k(\Sigma)$$

where $[\![B]\!]^\sharp : \mathsf{cLat} \to \mathsf{cLat}$ is the abstract interpretation of B from Sect. 4.2.

We define $\mathsf{fold}^\sharp_k = ([\![B]\!]^\sharp)^k(!)$ where $! : [\![B]\!]^\sharp(X^\sharp) \to \Sigma$ is a morphism that forgets additional structure in $X^\sharp$ by mapping non-bottom elements to the top element and $\mathsf{unfold}^\sharp_k = ([\![B]\!]^\sharp)^k(\mathsf{inj})$ where $\mathsf{inj} : \Sigma \to [\![B]\!]^\sharp(\Sigma)$ is a morphism that maps bottom/top to bottom/top respectively. This interpretation of $\mathsf{fold}^\sharp_k$ and $\mathsf{unfold}^\sharp_k$ is suitable: only $\mathsf{fold}^\sharp_k$ introduces over-approximation, which is consistent with the conditions in Theorem 3. It also makes it clear why k is limiting: we can only unfold k times before losing information. The Galois connection of Sect. 4.2 extends to give a Galois connection from $\mathscr{P}([\![\mu X.A]\!])$ and $[\![\mu X.A]\!]^\sharp_{\mathrm{K}}$ by composition through $[\![\mu X.A]\!]^\sharp$.

Example: 3-Limited Binary Trees. Consider the recursive equation

$$\mathsf{Tree} = \mathsf{unit}{+}\mathsf{int}{*}\mathsf{Tree}{*}\mathsf{Tree}$$

which represents binary trees of integers. Examples include the empty leaf $(\imath_1\star)$, and the balanced tree with 5, −3 and −6 as elements $(\imath_2(5, \imath_2(-3, \imath_1\star, \imath_1\star), \imath_2(-6, \imath_1\star, \imath_1\star)))$. The abstract lattice interpretation of 3-limited binary trees (with signs representing integers) is

$$([\![\mathsf{Tree}]\!]^\sharp)^3(\Sigma) = \Sigma \times (\mathsf{Sign} \otimes ([\![\mathsf{Tree}]\!]^\sharp)^2(\Sigma) \otimes ([\![\mathsf{Tree}]\!]^\sharp)^2(\Sigma))$$

which defines an inductive abstract representation of binary trees that is three levels deep, and ends with $([\![\mathsf{Tree}]\!]^\sharp)^0(\Sigma) = \Sigma$.

Concrete examples of elements of the abstract 3-limited binary tree domains include binary trees that have either depth of 0 or 2 and alternate between positive and negative elements:

$$\begin{aligned}(\top, (+,&\\ &(\bot, (-, (\top, \bot), (\top, \bot))),\\ &(\bot, (-, (\top, \bot), (\top, \bot)))))\end{aligned}$$

This abstraction captures our concrete balanced tree presented above.

Another example is the top element of the domain:

$$\begin{aligned}(\top, (\top,&\\ &(\top, (\top, (\top, (\top, \top, \top)), (\top, (\top, \top, \top)))),\\ &(\top, (\top, (\top, (\top, \top, \top)), (\top, (\top, \top, \top))))))\end{aligned}$$

Here all the elements of Σ and Sign are represented by their top elements, and the tree does so recursively until it reaches the limit.

5.2 Modular Typed Regular Tree Expressions

For practical program analyses, we often would like to go beyond prefixes and also capture inductive invariants. A classical way to capture these inductive invariants, is by relying on regular tree expressions [12] which capture the structure of inductive types using grammars. We will show how we can build an extended strongly typed version of regular tree expressions (RTEs) in our framework. This will provide an idea on how to convert otherwise monolithic domains to be compositional. We further show how this domain can be directly related by concretization to the general solution for abstracting recursive types provided in Sect. 4; this intermediate relation is useful for compositionality of analyses and allows reusing soundness proofs based on the general solution.

In RTEs we capture inductive invariants by constructing a syntactic grammar over possible values. The RTE domain contains these syntactic grammars as elements, and is ordered by language inclusion.

Example: Positive-Negative Binary Trees. Reconsider the binary tree example from the previous section. We can represent the 3-limited alternating positive-negative tree abstraction as the following non-recursive grammar in the RTE domain:

$$P_0 ::= \text{Leaf} \qquad N_1 ::= \text{Node}(-, P_0, P_0) \qquad P_2 ::= \text{Node}(+, N_1, N_1)$$

We can further generalize the grammar above in the RTE domain, to capture the invariant inductively at any depth. We do this using the following recursive grammar:

$$P ::= \text{Leaf} \mid \text{Node}(+, N, N) \qquad N ::= \text{Node}(-, P, P)$$

Here, P and N are symbols that can be referenced recursively on the right-hand side, thus inductively describing the required invariants. Notice how the grammar contains syntactic references (P and N) deeply nested inside the constructors on the right-hand side. We would like these references to only be valid given the grammar context, so we do not refer to undefined symbols or symbols of the wrong type. Because of this, the domain is usually implemented in a monolithic fashion. Our goal is to make this domain more modular, while preserving strong type safety.

The representation of the positive-negative tree grammar in our strongly typed framework is as follows:

$$[P \mapsto (\mathsf{unit}+\mathsf{int}*X*X, (\top, (+, N, N))), \quad N \mapsto (\mathsf{unit}+\mathsf{int}*X*X, (\bot, (-, P, P)))]$$

In particular, we represent the grammar as an environment mapping variables to a pair where the first component is the target program type B (excluding function types $A \Rightarrow B$) and the second component represents the interpretation of that program type $[\![B]\!]^\sharp_{\mathrm{E}}$: $\mathsf{Type} \to \mathsf{Set}$ into our modular abstract domains. Formally, given a countable set of symbols $a, b \in \mathsf{Sym}$ we define an RTE pre-environment

$$\Gamma^{-\sharp} : \mathsf{RTEnv}^{-\sharp} = \mathsf{Sym} \to \{(B, t^\sharp) \mid B : \mathsf{Type}, t^\sharp \in [\![B]\!]^\sharp_{\mathrm{E}}(\mu X.B)\}$$

where the first argument to the abstract interpretation is the type of the recursive references. We define the interpretation of types as follows:

$$\begin{aligned}
[\![X]\!]^\sharp_{\mathrm{E}}(B) &= \{(a, B) \mid a \in \mathsf{Sym}\} & [\![\mathsf{int}]\!]^\sharp_{\mathrm{E}}(B) &= \mathsf{Sign} \\
[\![\mathsf{unit}]\!]^\sharp_{\mathrm{E}}(B) &= \Sigma & [\![\mu X.B_1]\!]^\sharp_{\mathrm{E}}(B) &= \{(b, \mu X.B_1) \mid b \in \mathsf{Sym}\} \\
[\![B_1+B_2]\!]^\sharp_{\mathrm{E}}(B) &= [\![B_1]\!]^\sharp_{\mathrm{E}}(B) \times [\![B_2]\!]^\sharp_{\mathrm{E}}(B) & [\![B_1*B_2]\!]^\sharp_{\mathrm{E}}(B) &= [\![B_1]\!]^\sharp_{\mathrm{E}}(B) \otimes [\![B_2]\!]^\sharp_{\mathrm{E}}(B)
\end{aligned}$$

Our interpretation is similar to the general interpretation given in Sect. 4, except that type variables and recursive types are replaced by typed grammar symbols. Pre-environments are not well-formed since they allow arbitrary symbols, which might have the wrong type or not be defined. We need to add a well-formedness constraint $\mathsf{WFEnv}(\Gamma^{-\sharp})$ on pre-environments $\Gamma^{-\sharp}$ to get proper environments:

$$\mathsf{WFEnv}(\Gamma^{-\sharp}) = \forall a \in \mathsf{dom}\ \Gamma^{-\sharp}.\mathsf{WFType}(\Gamma^{-\sharp}(a), \Gamma^{-\sharp})$$

Here, $\mathsf{WFType}(B, t^\sharp, \Gamma^{-\sharp})$ is a predicate[2] on type B, its abstract interpretation $t^\sharp \in [\![B]\!]^\sharp_{\mathrm{E}}(\mu X.B)$ and pre-environment $\Gamma^{-\sharp}$, that checks whether the mapped values in the pre-environment map to existing symbols of the correct type:

[2] We have flattened the tuple in the first argument, to improve presentation.

$$\mathsf{WFType}(\mathsf{int}, t^\sharp, \Gamma^{-\sharp}) \quad \mathsf{WFType}(\mathsf{unit}, \star, \Gamma^{-\sharp})$$
$$\mathsf{WFType}(X, (a, B), \Gamma^{-\sharp}) \iff \exists t^\sharp. \Gamma^{-\sharp}(a) = (B, t^\sharp)$$
$$\mathsf{WFType}(B_1 + B_2, (t_1^\sharp, t_2^\sharp), \Gamma^{-\sharp}) \iff \mathsf{WFType}(B_1, t_1^\sharp, \Gamma^{-\sharp}) \wedge \mathsf{WFType}(B_2, t_2^\sharp, \Gamma^{-\sharp})$$
$$\mathsf{WFType}(B_1 * B_2, (t_1^\sharp, t_2^\sharp), \Gamma^{-\sharp}) \iff \mathsf{WFType}(B_1, t_1^\sharp, \Gamma^{-\sharp}) \wedge \mathsf{WFType}(B_2, t_2^\sharp, \Gamma^{-\sharp})$$
$$\mathsf{WFType}(\mu X.B, (b, \mu X.B), \Gamma^{-\sharp}) \iff B \neq X \wedge \exists t^\sharp. \Gamma^{-\sharp}(b) = (\mu X.B, t^\sharp)$$

Abstract elements of type int and unit are always well-formed, abstract sums $B_1 + B_2$ and product $B_1 * B_2$ elements must check well-formedness recursively, and type variables X and recursive types $\mu X.B$ must check that the referenced symbols map to values of the correct type in the pre-environment. We disallow direct recursion $\mu X.X$, since they would make our concretization non-productive.

We can now define environments as pre-environments which are well-formed:

$$\Gamma^\sharp \in \mathsf{RTEnv}^\sharp = \{\Gamma^{-\sharp} \mid \Gamma^{-\sharp} \in \mathsf{RTEnv}^{-\sharp} \wedge \mathsf{WFEnv}(\Gamma^{-\sharp})\}$$

Similarly, we can define our top-level interpretation of types as abstract elements closed under an environment:

$$[\![A]\!]^\sharp_{\mathrm{RTE}} = \{(\Gamma^\sharp, t^\sharp) \mid \Gamma^\sharp \in \mathsf{RTEnv}^\sharp \wedge t^\sharp \in [\![A]\!]^\sharp_{\mathrm{E}} \wedge \mathsf{WFType}(A, t^\sharp, \Gamma^\sharp)\}$$

Our top-level types A are closed, but have a similar interpretation to open types B without the Type argument needed for recursion.

Semantic Operations. Most semantic operations stay the same as in the previous section, the only exception is that we must define $\mathsf{unfold}^\sharp$ and $\mathsf{fold}^\sharp$ for our RTE interpretation of recursive types. We define $\mathsf{unfold}^\sharp : [\![\mu X.B]\!]^\sharp_{\mathrm{RTE}} \rightarrow [\![B\,[\mu X.B/X]]\!]^\sharp_{\mathrm{RTE}}$ as follows:

$$\mathsf{unfold}^\sharp(\Gamma^\sharp, a) = (\Gamma^\sharp, t^\sharp) \quad \mathsf{where} \quad (B, t^\sharp) = \Gamma^\sharp(a)$$

The definition is intuitively simple: we look up the target symbol in the environment to expose the underlying structure of the recursive type.

Formally, we need to check a few conditions to ensure it is correct. The well-formedness condition for the environment ensures that the target symbol a is in the environment and that its mapped value $\Gamma^\sharp a$ is well-formed as well. To ensure that the result $(B, t^\sharp)$ is in the target domain $[\![B\,[\mu X.B/X]]\!]^\sharp_{\mathrm{RTE}}$, we only need to check $t^\sharp$ since $\Gamma^\sharp$ stays the same. Recall that $t^\sharp$ must be in $[\![B]\!]^\sharp_{\mathrm{E}}(\mu X.B)$ by definition of pre-environments, which means that all direct type variable references X are required to be in the set $\{(a, \mu X.B) \mid a \in \mathsf{Sym}\}$. The second component of the result domain $[\![B\,[\mu X.B/X]]\!]^\sharp_{\mathrm{RTE}}$ is in $[\![B\,[\mu X.B/X]]\!]^\sharp_{\mathrm{E}}$, which had all its direct type variable references X syntactically replaced with $\mu X.B$, and whose interpretation is exactly the same set we got from the lookup $\{(a, \mu X.B) \mid a \in \mathsf{Sym}\}$. Our interpretation of $\mathsf{unfold}^\sharp$ is therefore correct.

We now define $\mathsf{fold}^\sharp : [\![B\,[\mu X.B/X]]\!]^\sharp_{\mathrm{RTE}} \rightarrow [\![\mu X.B]\!]^\sharp_{\mathrm{RTE}}$ as follows:

$$\mathsf{fold}^\sharp(\Gamma^\sharp, t^\sharp) = (\Gamma^\sharp[a \mapsto (B, t^\sharp)], a) \quad \mathsf{where} \quad a \text{ fresh}$$

Essentially, we create a new definition in our environment pointed to by a fresh symbol a that maps to the given input structure $t^\sharp$. Since a is fresh, the well-formedness of other values in the environment is unchanged and the well-formedness of the result is immediate from the extension. It is also immediate that our result lies in the required set $\{(a, \mu X.B) \mid a \in \mathsf{Sym}\}$. Our equations from Theorem 3 are satisfied provided we quotient by equivalent grammars, since folding/unfolding grammars does not lose information.

Lattice Operations. The RTE forms a bounded but not complete lattice, since some sets of RTEs have non-regular trees as least upper bound. This means that there can be programs which do not have the best available abstract interpretation in this domain (their least fixed-point does not exist), and so we must settle for some over-approximating fixed-point instead.

The bottom element is $(\Gamma^\sharp, \bot)$ and the top element is $(\Gamma^\sharp, \top)$ for any environment $\Gamma^\sharp$. To allow defining the other operations [3], it is necessary to redefine the parameterized composite lattices (like products) to pass down information to their parameters. This makes definitions slightly less modular, but more general. We define $\sqsubseteq_e$ on RTEs as passing down a map $e : \mathsf{Sym} \to \mathscr{P}\mathsf{Sym}$ that dynamically maps a symbol to the set of symbols that over-approximates it in the current context. Then we can define inclusion on symbols as follows:

$$(\Gamma^\sharp, (B, a)) \sqsubseteq_e (\Gamma^\sharp, (B, b)) \iff \begin{cases} b \in e(a) \\ b \notin e(a) \wedge (\Gamma^\sharp, \Gamma^\sharp(a)) \sqsubseteq_{e[a \mapsto e(a) \cup \{b\}]} (\Gamma^\sharp, \Gamma^\sharp(b)) \end{cases}$$

The first case states that a is included in b if it is assumed so in e. In the second case the inclusion is delegated to the relevant abstract lattice as before, with the minor addition that $e[a \mapsto e(a) \cup \{b\}]$ should be passed down recursively to when new symbols are met. This shows inclusion by bisimulation, since if the right-hand side of the symbol a in $\Gamma^\sharp$ is included in the right-hand side of the symbol b, then we can safely assume that b covers at least the same cases as a.

Similarly, we pass down a partial symmetric map $u : \mathsf{Sym} \times \mathsf{Sym} \rightharpoonup \mathsf{Sym}$ to the least upper bound operation to keep track of already merged symbols:

$$(\Gamma^\sharp, (A, a)) \sqcup_u (\Gamma^\sharp, (A, b)) = (\Gamma^\sharp, c) \quad \text{if } u(a, b) = c$$

$$(\Gamma^\sharp, (A, a)) \sqcup_u (\Gamma^\sharp, (A, b)) = (\Gamma^{\sharp\prime}[c \mapsto t^\sharp], t^\sharp) \quad \text{if } \begin{cases} (a, b) \notin \mathsf{dom}\ u \\ c \text{ fresh} \\ u' = u[(a, b) \mapsto c] \\ (\Gamma^{\sharp\prime}, t^\sharp) = \\ \qquad (\Gamma^\sharp, \Gamma^\sharp(a)) \sqcup_{u'} (\Gamma^\sharp, \Gamma^\sharp(b)) \end{cases}$$

The greatest lower bound case is analogous. If the two environments of the input to the operations are different, we can simply merge them by renaming all the symbols in one of the environment and then extending the other with it.

Concretization. We can define a concretization to sets of concrete values from RTEs $\gamma : [\![A]\!]^\sharp_{\mathrm{RTE}} \to \mathscr{P}[\![A]\!]$ by composition of a concretization to the intermediate abstract domain presented in Sect. 4 ($\gamma'_A : [\![A]\!]^\sharp_{\mathrm{RTE}} \to [\![A]\!]^\sharp$) and its concretization to sets of concrete types ($\gamma'' : [\![A]\!]^\sharp \to \mathscr{P}[\![A]\!]$). It is easier to define γ' than γ directly, and it allows changing abstraction of a type without requiring to redo the complete soundness proof that was required in Sect. 4.

We define $\gamma'_A : [\![A]\!]^\sharp_{\mathrm{RTE}} \to [\![A]\!]^\sharp$ co-inductively as follows:

$$\gamma'_X(\Gamma^\sharp, (a, A)) = \gamma'_A(\Gamma^\sharp, t^\sharp) \ \ \text{where} \ \ (A, t^\sharp) = \Gamma^\sharp(a)$$

$$\gamma'_{\mathsf{int}}(\Gamma^\sharp, n^\sharp) = n^\sharp \qquad \gamma'_{\mathsf{unit}}(\Gamma^\sharp, x^\sharp) = x^\sharp$$

$$\gamma'_{\mu X.A}(\Gamma^\sharp, (a, \mu X.B)) = \gamma'_A(\Gamma^\sharp, t^\sharp) \ \ \text{where} \ \ (B, t^\sharp) = \Gamma^\sharp(a)$$

$$\gamma'_{A_1 + A_2}(\Gamma^\sharp, (t_1^\sharp, t_2^\sharp)) = (\gamma'_{A_1}(\Gamma^\sharp, t_1^\sharp), \gamma'_{A_2}(\Gamma^\sharp, t_2^\sharp))$$

$$\gamma'_{A_1 * A_2}(\Gamma^\sharp, (t_1^\sharp, t_2^\sharp)) = (\gamma'_{A_1}(\Gamma^\sharp, t_1^\sharp), \gamma'_{A_2}(\Gamma^\sharp, t_2^\sharp))$$

The concretization simply recursively unfolds the definitions for symbols, with interpretation of most types being direct. The definition is productive since we disallow direct recursive definitions $\mu X.X$ in the WFType predicate.

6 Implementation

We have implemented a first-order version of the expression language with its concrete and abstract semantics for k-limited trees. The implementation is in Agda [21], and shows how one could implement the ideas modularly in practice[3].

We first define the syntax of the types in Agda. The opn and cls, stratify the syntax to match the open (B) and closed (A) types from Sect. 1.

```
data State : Set where opn cls : State
data Type' : State -> Set where
  Int Bool : Type' cls
  Const : Type' cls -> Type' opn
  Var : Type' opn
  _*t_ _+t_ : forall {s} -> Type' s -> Type' s -> Type' s
  Rec : Type' opn -> Type' cls

Type : Set
Type = Type' cls

Ctx : Nat -> Set
Ctx n = Vec Type n
```

The core of our implementation depends on a modular interface of semantic operations (SemanticOps), which our language interpretation[4] relies on. It relies on a field [[_]]t to provide an interpretation of our syntax of types into concrete and abstract domains[5]. The rest of the semantic operations are specified in a

[3] https://github.com/ahmadsalim/agda-moddom.

[4] In Language.Rec and Language.Semantics modules.

[5] In Domains.Concrete and Domains.Abstract modules respectively.

strongly typed fashion with regards to that implementation, which ensures that we can modularly exchange the domains for a particular type t. We parameterize over m to allow possible general monadic computations in the computation rule for sums, E-case, since it contains function arguments. Adding new types is easy as well: we update the interface with the semantic operations for that type and implement them in the corresponding concrete and abstract interpretations; this can be done in an isolated manner—without affecting other operations—because of the strong typing of the interface.

```
record SemanticOps (m : Set -> Set) : Set1 where
  field
    [[_]]t : Type -> Set

    P-pair : forall {t s} -> [[ t ]]t -> [[ s ]]t -> [[ t *t s ]]t
    P-fst : forall {t s} -> [[ t *t s ]]t -> [[ t ]]t
    P-snd : forall {t s} -> [[ t *t s ]]t -> [[ s ]]t
    E-left : forall {t s} -> [[ t ]]t -> [[ t +t s ]]t
    E-right : forall {t s} -> [[ s ]]t -> [[ t +t s ]]t
    E-case : forall {t s w} -> [[ t +t s ]]t -> ([[ t ]]t -> m [[ w ]]t)
                                  -> ([[ s ]]t -> m [[ w ]]t) -> m [[ w ]]t
    S-abs : forall {t} -> [[ t < Rec t > ]]t -> [[ Rec t ]]t
    S-rep : forall {t} -> [[ Rec t ]]t -> [[ t < Rec t > ]]t
```

We need to use a dependently typed language like Agda to get a strongly typed interpretation, because recursive types require substitution which is a type-level operation (_<_>). An inherent challenge that arises is that it is sometimes necessary to make explicit proofs to allow type checking to succeed. For example, we had to show the fundamental lemma of substitution for both the concrete and abstract interpretation of types, in order for Agda to accept our definitions of S-abs and S-rep:

```
    SynSemSub : forall t t' -> Equiv [[ t < t' > ]]t ([[ t ]]t' [[ t' ]]t)
```

This proof obligation is a reasonable price to pay for achieving modularity, and often implies less work than implementing ad-hoc, monolithic domains.

7 Related Work

Cousot and Cousot [13] present a framework for compositional analysis of programs that relies on symbolic relational analysis for sharing information. Existing work on specifying inductive properties for relational domains [7, 8] requires fixed shape for inductive structures. Extending our modular construction to support relational information is future work. Rival et al. [22, 25] discuss how to provide a way to modularize symbolic memories used by pointer-manipulating by decompositing them into distinct sub-memories which share information. This is suitable if different parts of the program need to be analyzed with different abstractions. Our work instead focuses on combining domains together in a modular fashion, that allows easily changing parts of the abstractions.

Benton [5] presents a systematic mathematically sound way of deriving abstractions for strictness properties for various algebraic data types (sums, products, inductive types). Our approach is more general since it allows modular construction of a large set of abstractions by combining those for specific types.

Jensen [15] presents a generic framework to derive abstract domains for various analyses using a program logic. The framework is able to reason precisely about disjunctive program properties, also inside inductive values, but does not systematically define a Galois connection between concrete and abstract domains like we do here. Combining these techniques could be interesting future work.

Darais et al. [14] present ways to reuse program implementations for various modes of concrete and abstract interpretation. Their work focuses on using monad transformers as a mean to provide interpretation of semantic operations that can be given both concrete and abstract semantics. Similarly, Keidel et al. [17] present a compositional semantics for concrete and abstract interpretation based on category theory, including a general technique for proving soundness and implementation based on Haskell's *arrows* library. These ideas can be used well with our modular domain construction, which allows further modularity at the type level and easily replacing core operations for complex structures. Both works contribute toward systematically writing abstract interpreters: Keidel et al.'s work focuses on making it easier to prove correctness of concrete abstract interpreters, while our work focuses on providing a general mathematical framework that works modularly with inductive types and higher-order functions.

8 Conclusion

We have shown that the theory for solving recursive domain equations can be applied to abstract interpretation for systematically constructing abstract domains for recursive data types, as well as the accompanying Galois connection with the concrete domains. This extends and completes the existing theory of compositionally combining abstract domains for base types, products, sums and functions. The abstract domain provided by the solution can be further abstracted to yield computable abstractions. We demonstrate this by providing a Galois connection for k-limited trees and regular tree expressions. The framework does not capture all abstract domains that have been proposed for specific data types. In particular, relational properties between sub-structures of recursive types (*e.g.* "the elements of a list are sorted") are not captured by our compositional construction. Nevertheless, we see a great potential for the use of the framework with language specification and engineering tools for building domain-specific languages in particular.

A Proofs

Theorem 1. *If $\rho_1 \subseteq \rho_2$ then $[\![e]\!]\rho_1 \subseteq [\![e]\!]\rho_2$*

Proof. Most cases follow straightforwardly by induction over the syntax of terms e and functoriality of the powerset lifting $\mathscr{P}$. We will consider the interesting cases below:

- Case $e = x$: follows by monotonicity of premise.
- Case $e = \mathsf{case}\ e_0\ \mathsf{of}\ \mathsf{inl}\ x \to e_1 \parallel \mathsf{inr}\ y \to e_2$: follows by monotonicity of extension and induction hypothesis.
- Case $e = x \Rightarrow e_0$: follows by monotonicity of extension and induction hypothesis.
- Case $e = \mathsf{fix}\ x.e_0$: follows by monotonicity of extension, induction hypothesis and the fact that the intersection of a set of monotone functions is itself a set of monotone functions.

Proposition 1. *Let E be a set and $E^\sharp$ a complete lattice. There is a bijective correspondence between maps $\alpha^1 : E \to E^\sharp$ and Galois connections $\mathscr{P}(E) \underset{\alpha}{\overset{\gamma}{\leftrightarrows}} E^\sharp$. The correspondence maps a Galois connection to α^1 defined as $\alpha^1(x) = \alpha(\{x\})$ and a map α^1 to the Galois connection:*

$$\alpha(X \subseteq E) = \bigsqcup_{x \in X} \alpha^1(x) \qquad \gamma(e^\sharp) = \{x \in E \mid \alpha^1(x) \sqsubseteq e^\sharp\}$$

The correspondence can be summarized in the following diagram.

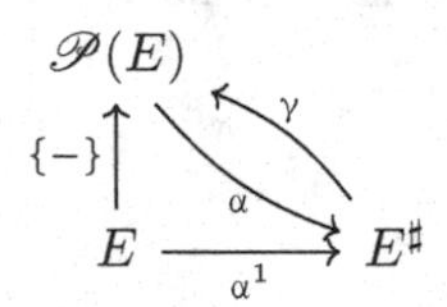

Proof.

$$\begin{aligned}\alpha(X) \sqsubseteq e^\sharp &\iff \left(\bigsqcup_{x \in X} \alpha^1(x)\right) \sqsubseteq e^\sharp \\ &\iff \forall x \in X.\alpha^1(x) \sqsubseteq e^\sharp \\ &\iff X \subseteq \gamma(e^\sharp)\end{aligned}$$

Theorem 3. *Suppose the abstract interpretation of recursive types*

$$\mathsf{fold}^\sharp : [\![B\,[\mu X.B/X]]\!]^\sharp \to [\![\mu X.B]\!]^\sharp \qquad \mathsf{unfold}^\sharp : [\![\mu X.B]\!]^\sharp \to [\![B\,[\mu X.B/X]]\!]^\sharp$$

satisfies the equations

$$\mathsf{fold}^\sharp \circ \mathsf{unfold}^\sharp = \mathsf{id} \qquad \mathsf{unfold}^\sharp \circ \mathsf{fold}^\sharp \sqsupseteq \mathsf{id}$$

If $\vdash_{\mathsf{ex}} t : A$ then $[\![t]\!] \subseteq \gamma([\![t]\!]^\sharp)$, or equivalently $\alpha([\![t]\!]) \sqsubseteq [\![t]\!]^\sharp$

Theorem 3 is proved by induction on typing judgements, and must therefore be extended to open terms. This is done in the following lemma, which uses an extension of the maps γ and α to contexts defined in the obvious (pointwise) way.

Lemma 1. *Suppose $\Gamma \vdash_{ex} e : A$, and that $\rho \in [\![\Gamma]\!]$ and $\rho^\sharp \in [\![\Gamma]\!]^\sharp$, are such that $\rho \subseteq \gamma(\rho^\sharp)$. If the conditions on the abstract interpretation of recursive types of Theorem 3 are satisfied, then $[\![t]\!]\rho \subseteq \gamma([\![t]\!]^\sharp\rho^\sharp)$ or, equivalently, $\alpha([\![t]\!]\rho) \subseteq [\![t]\!]^\sharp\rho^\sharp$*

Theorem 3 obviously follows from this as a special case. Before proving Lemma 1 we need a few lemmas.

Lemma 2. *If B is an open type and A a closed one, then $[\![B]\!]([\![A]\!]) = [\![B\,[A/X]]\!]$ and $[\![B]\!]^\sharp([\![A]\!]^\sharp) = [\![B\,[A/X]]\!]^\sharp$*

Proof. Easy induction on B, which we omit.

Lemma 3. *For any open type B and closed type A the following equality holds*

$$\alpha^1_B([\![A]\!]^\sharp) \circ [\![B]\!](\alpha^1_A) = \alpha^1_{B[A/X]}$$

In diagram style, this is

$$\begin{array}{ccccc}
[\![B\,[A/X]]\!] & \xrightarrow{\mathsf{id}} & [\![B]\!]([\![A]\!]) & \xrightarrow{[\![B]\!](\alpha^1_A)} & [\![B]\!]([\![A]\!]^\sharp) \\
\downarrow{\scriptstyle \alpha^1_{B[A/X]}} & & & & \downarrow{\scriptstyle \alpha^1_B([\![A]\!]^\sharp)} \\
[\![B\,[A/X]]\!]^\sharp & & \xrightarrow{\qquad\mathsf{id}\qquad} & & [\![B]\!]^\sharp([\![A]\!]^\sharp)
\end{array}$$

Proof. Induction on B. The cases of integers and unit type are trivial. In the case of a type variable X, we get

$$\alpha^1_X([\![A]\!]^\sharp) \circ [\![X]\!](\alpha^1_A) = \mathsf{id}_{[\![A]\!]^\sharp} \circ \alpha^1_A = \alpha^1_A = \alpha^1_{X[A/X]}$$

In the case of a product type we get

$$\begin{aligned}
\alpha^1_{B_1 * B_2}([\![A]\!]^\sharp) \circ [\![B_1 * B_2]\!](\alpha^1_A)(x,y) &= \alpha^1_{B_1 * B_2}([\![A]\!]^\sharp)([\![B_1]\!](\alpha^1_A)(x), [\![B_2]\!](\alpha^1_A)(y)) \\
&= (\alpha^1_{B_1}([\![A]\!]^\sharp)([\![B_1]\!](\alpha^1_A)(x)), \alpha^1_{B_2}([\![A]\!]^\sharp)([\![B_2]\!](\alpha^1_A)(y))) \\
&= (\alpha^1_{B_1[A/X]}(x), \alpha^1_{B_2[A/X]}(y)) \\
&= \alpha^1_{B_1[A/X] * B_2[A/X]}(x,y) \\
&= \alpha^1_{(B_1 * B_2)[A/X]}(x,y)
\end{aligned}$$

using the induction hypothesis in the third equality. The case for sum types is somewhat similar, except here we must branch over the input being an injection on the left or right. In the first case we get

$$\begin{aligned}
\alpha^1_{B_1 + B_2}([\![A]\!]^\sharp) \circ [\![B_1 + B_2]\!](\alpha^1_A)(\imath_1(x)) &= \alpha^1_{B_1 + B_2}([\![A]\!]^\sharp) \circ (\imath_1([\![B_1]\!](\alpha^1_A)(x))) \\
&= (\alpha^1_{B_1}([\![A]\!]^\sharp)([\![B_1]\!](\alpha^1_A)(x)), \bot) \\
&= (\alpha^1_{B_1[A/X]}(x), \bot) \\
&= \alpha^1_{B_1[A/X] + B_2[A/X]}(\imath_1(x)) \\
&= \alpha^1_{(B_1 + B_2)[A/X]}(\imath_1(x))
\end{aligned}$$

again using the induction hypothesis in the third equality. The case of the input being an injection on the right is similar.

The final case is that of functions: $B = (A_1 \Rightarrow B_1)$. If $f : [\![A_1]\!] \to [\![B_1]\!]([\![A]\!]^\sharp)$ then

$$\begin{aligned}\alpha^1_{A_1 \Rightarrow B_1}([\![A]\!]^\sharp) \circ [\![A_1 \Rightarrow B_1]\!](\alpha^1_A)(f) &= \alpha^1_{A_1 \Rightarrow B_1}([\![A]\!]^\sharp)([\![B_1]\!](\alpha^1_A) \circ f)\\ &= \alpha_{B_1}([\![A]\!]^\sharp) \circ \mathscr{P}([\![B_1]\!](\alpha^1_A) \circ f) \circ \gamma_{A_1}\\ &= \alpha_{B_1}([\![A]\!]^\sharp) \circ \mathscr{P}([\![B_1]\!](\alpha^1_A)) \circ \mathscr{P}(f) \circ \gamma_{A_1}\end{aligned}$$

By the induction hypothesis

$$\alpha^1_{B_1}([\![A]\!]^\sharp) \circ [\![B_1]\!](\alpha^1_A) = \alpha^1_{B_1[A/X]}$$

which implies

$$\alpha_{B_1}([\![A]\!]^\sharp) \circ \mathscr{P}([\![B_1]\!](\alpha^1_A)) = \alpha_{B_1[A/X]}$$

since either side of this equation is the unique extension of the corresponding sides of the equation above to continuous maps. Thus, we get

$$\begin{aligned}\alpha^1_{A_1 \Rightarrow B_1}([\![A]\!]^\sharp) \circ [\![A_1 \Rightarrow B_1]\!](\alpha^1_A)(f) &= \alpha_{B_1[A/X]} \circ \mathscr{P}(f) \circ \gamma_{A_1}\\ &= \alpha^1_{A_1 \Rightarrow (B_1[A/X])}(f)\\ &= \alpha^1_{(A_1 \Rightarrow B_1)[A/X]}(f)\end{aligned}$$

proving the case and concluding the proof.

Lemma 4. *If B is an open type, then the following diagram commutes*

$$\begin{array}{ccc}\mathscr{P}([\![B\,[\mu X.B/X]]\!]) & \xrightarrow{\alpha_{B[\mu X.B/X]}} & [\![B\,[\mu X.B/X]]\!]^\sharp\\ \downarrow{\scriptstyle \mathscr{P}(\mathsf{fold})} & & \downarrow{\scriptstyle \mathsf{fold}^\sharp}\\ \mathscr{P}([\![\mu X.B]\!]) & \xrightarrow{\alpha_{\mu X.B}} & [\![\mu X.B]\!]^\sharp\end{array}$$

Proof. It suffices to show that $\alpha^1_{\mu X.B} \circ \mathsf{fold} = \mathsf{fold}^\sharp \circ \alpha^1_{B[\mu X.B/X]}$ since the two compositions in the diagram are the unique extensions of the two sides of this equation to continuous maps. By definition of $\alpha^1_{\mu X.B}$ we get

$$\alpha^1_{\mu X.B} \circ \mathsf{fold} = \mathsf{fold}^\sharp \circ \alpha^1_B([\![\mu X.B]\!]^\sharp) \circ [\![B]\!](\alpha^1_{\mu X.B})$$

By Lemma 3 the right hand side of this is equal to $\mathsf{fold}^\sharp \circ \alpha^1_{B[\mu X.B/X]}$ which concludes the proof.

We can now prove Lemma 1.

Proof (Lemma 1). By induction on typing derivation:

- Case $t = x$ trivial

- Case $t = \mathsf{fix}\ \lambda x.s.$:
  By IH we get:

$$\begin{aligned}\llbracket s\rrbracket\rho[x \mapsto \gamma(\llbracket t\rrbracket^\sharp\rho^\sharp)] &\subseteq \gamma(\llbracket s\rrbracket^\sharp\rho^\sharp[x \mapsto \llbracket t\rrbracket^\sharp\rho^\sharp]) \\ &= \gamma(\llbracket t\rrbracket^\sharp\rho^\sharp)\end{aligned}$$

  So $\gamma(\llbracket t\rrbracket^\sharp\rho^\sharp)$ is a post-fixpoint of $\lambda a.\llbracket s\rrbracket\rho[x \mapsto a]$ and so greater than the greatest lower bound of these ($\llbracket t\rrbracket\rho = \bigcap\{a \mid \llbracket s\rrbracket\rho[x \mapsto a] \subseteq a\}$).
- Case $t = s\ u$:
  Recall that $\alpha_{A\to B} : \mathscr{P}(\llbracket A\rrbracket \to \llbracket B\rrbracket) \to \llbracket A\rrbracket^\sharp \to \llbracket B\rrbracket^\sharp$ is defined as:

$$\alpha(F) = \bigsqcup_{f\in F} \alpha_B \circ \mathscr{P}(f) \circ \gamma_A$$

  Then

$$\begin{aligned}\alpha_B(\llbracket s\ u\rrbracket\rho) &= \alpha_B(\{f(a) \mid f \in \llbracket s\rrbracket\rho, a \in \llbracket u\rrbracket\rho\}) \\ &= \alpha_B(\bigcup_{f\in\llbracket s\rrbracket\rho} \mathscr{P}(f)(\llbracket u\rrbracket\rho)) \\ &= \bigsqcup_{f\in\llbracket s\rrbracket\rho} \alpha_B(\mathscr{P}(f)(\llbracket u\rrbracket\rho)) \\ &\sqsubseteq \bigsqcup_{f\in\llbracket s\rrbracket\rho} \alpha_B(\mathscr{P}(f)(\gamma_A(\alpha_A(\llbracket u\rrbracket\rho)))) \\ &= \alpha_{A\to B}(\llbracket s\rrbracket\rho)(\alpha(\llbracket u\rrbracket\rho)) \\ &\sqsubseteq \llbracket s\rrbracket^\sharp\rho^\sharp(\llbracket u\rrbracket^\sharp\rho^\sharp) \text{ By IH}\end{aligned}$$

- Case $t = \lambda x.u : A \to B$
  We must show that for all $f \in \llbracket\lambda x.u\rrbracket\rho$ it holds that

$$\alpha \circ \mathscr{P}(f) \circ \gamma \sqsubseteq \llbracket\lambda x.u\rrbracket^\sharp\rho^\sharp$$

  or equivalently

$$\mathscr{P}(f) \circ \gamma \sqsubseteq \gamma \circ \llbracket\lambda x.u\rrbracket^\sharp\rho^\sharp \tag{1}$$

  We know by IH that for all $a^\sharp$ it holds

$$\llbracket u\rrbracket\rho[x \mapsto \gamma(a^\sharp)] \subseteq \gamma(\llbracket u\rrbracket^\sharp\rho^\sharp[x \mapsto \alpha(\gamma(a^\sharp))]) \tag{2}$$

$$\subseteq \gamma(\llbracket u\rrbracket^\sharp\rho^\sharp[x \mapsto a^\sharp]) \tag{3}$$

  To show Eq. (1) we must show that for all $a \in \gamma(a^\sharp)$ it holds:

$$\begin{aligned}f(a) &\in \gamma((\llbracket\lambda x.u\rrbracket^\sharp\rho^\sharp)a^\sharp) \\ &= \gamma(\llbracket u\rrbracket^\sharp\rho^\sharp[x \mapsto a^\sharp])\end{aligned}$$

  The assumption $f \in \llbracket\lambda x.u\rrbracket\rho$ means precisely that $f(a) \in \llbracket u\rrbracket\rho[x \mapsto \{a\}] \subseteq \llbracket u\rrbracket\rho[x \mapsto \gamma(a^\sharp)]$ (by monotonicity), so by Eq. (3) we conclude.

- Case $t = \mathsf{fold}\ u : \mu X.A$,
  The case is proved as follows

$$\begin{aligned}\alpha([\![\mathsf{fold}\ u]\!]\rho) &= \alpha(\mathscr{P}(\mathsf{fold})([\![u]\!]\rho))\\ &= \mathsf{fold}^\sharp(\alpha([\![u]\!]\rho))\\ &\sqsubseteq \mathsf{fold}^\sharp([\![u]\!]^\sharp\rho^\sharp)\\ &= [\![\mathsf{fold}\ u]\!]^\sharp\rho^\sharp\end{aligned}$$

  using Lemma 4 for the second equality.
- Case $t = \mathsf{unfold}\ u : A\,[\mu X.A/X]$
  First note that

$$\alpha \circ \mathscr{P}(\mathsf{unfold}) \sqsubseteq \mathsf{unfold}^\sharp \circ \alpha$$

  since

$$\begin{aligned}\alpha \circ \mathscr{P}(\mathsf{unfold}) &\sqsubseteq \mathsf{unfold}^\sharp \circ \mathsf{fold}^\sharp \circ \alpha \circ \mathscr{P}(\mathsf{unfold})\\ &= \mathsf{unfold}^\sharp \circ \alpha \circ \mathscr{P}(\mathsf{fold} \circ \mathsf{unfold})\\ &= \mathsf{unfold}^\sharp \circ \alpha\end{aligned}$$

  and so

$$\begin{aligned}\alpha([\![\mathsf{unfold}\ u]\!]\rho) &= \alpha(\mathscr{P}(\mathsf{unfold})([\![u]\!]\rho))\\ &\sqsubseteq \mathsf{unfold}^\sharp(\alpha([\![u]\!]\rho))\\ &\sqsubseteq \mathsf{unfold}^\sharp([\![u]\!]^\sharp\rho^\sharp) \text{ by IH}\\ &= [\![\mathsf{unfold}\ u]\!]^\sharp\rho^\sharp\end{aligned}$$

References

1. Abbott, M.G., Altenkirch, T., Ghani, N.: Containers: constructing strictly positive types. Theoret. Comput. Sci. **342**(1), 3–27 (2005)
2. Abramsky, S., Jung, A.: Domain theory. In: Abramsky, S., Gabbay, D.M., Maibaum, T.S.E. (eds.) Handbook of Logic in Computer Science, vol. 3, pp. 1–168. Clarendon Press, Oxford (1994)
3. Aiken, A., Murphy, B.R.: Implementing regular tree expressions. In: Hughes, J. (ed.) FPCA 1991. LNCS, vol. 523, pp. 427–447. Springer, Heidelberg (1991). https://doi.org/10.1007/3540543961_21
4. Awodey, S.: Category Theory. Oxford University Press, Oxford (2011)
5. Benton, P.N.: Strictness properties of lazy algebraic datatypes. In: Cousot, P., Falaschi, M., Filé, G., Rauzy, A. (eds.) WSA 1993. LNCS, vol. 724, pp. 206–217. Springer, Heidelberg (1993). https://doi.org/10.1007/3-540-57264-3_42
6. Bodin, M., Gardner, P., Jensen, T., Schmitt, A.: Skeletal semantics and their interpretations. PACMPL **3**(POPL), 44:1–44:31 (2019)
7. Chang, B.E., Rival, X.: Relational inductive shape analysis. In: Necula, G.C., Wadler, P. (eds.) POPL 2008, pp. 247–260. ACM (2008)

8. Chang, B.-Y.E., Rival, X., Necula, G.C.: Shape analysis with structural invariant checkers. In: Nielson, H.R., Filé, G. (eds.) SAS 2007. LNCS, vol. 4634, pp. 384–401. Springer, Heidelberg (2007). https://doi.org/10.1007/978-3-540-74061-2_24
9. Cousot, P.: The calculational design of a generic abstract interpreter. In: Broy, M., Steinbrüggen, R. (eds.) Calculational System Design. NATO ASI Series F. IOS Press, Amsterdam (1999)
10. Cousot, P., Cousot, R.: Abstract interpretation: a unified lattice model for static analysis of programs by construction or approximation of fixpoints. In: Graham, R.M., Harrison, M.A., Sethi, R. (eds.) POPL 1977, pp. 238–252. ACM (1977)
11. Cousot, P., Cousot, R.: Invited talk: higher order abstract interpretation (and application to comportment analysis generalizing strictness, termination, projection, and PER analysis. In: Bal, H.E. (ed.) Proceedings of the IEEE Computer Society 1994 International Conference on Computer Languages, Toulouse, France, 16–19 May 1994, pp. 95–112. IEEE Computer Society (1994)
12. Cousot, P., Cousot, R.: Formal language, grammar and set-constraint-based program analysis by abstract interpretation. In: FPCA 1995, pp. 170–181. ACM (1995)
13. Cousot, P., Cousot, R.: Modular static program analysis. In: Horspool, R.N. (ed.) CC 2002. LNCS, vol. 2304, pp. 159–179. Springer, Heidelberg (2002). https://doi.org/10.1007/3-540-45937-5_13
14. Darais, D., Labich, N., Nguyen, P.C., Horn, D.V.: Abstracting definitional interpreters (functional pearl). PACMPL **1**(ICFP), 12:1–12:25 (2017)
15. Jensen, T.P.: Disjunctive program analysis for algebraic data types. ACM Trans. Program. Lang. Syst. **19**(5), 751–803 (1997)
16. Journault, M., Miné, A., Ouadjaout, A.: Modular static analysis of string manipulations in C programs. In: Podelski, A. (ed.) SAS 2018. LNCS, vol. 11002, pp. 243–262. Springer, Cham (2018). https://doi.org/10.1007/978-3-319-99725-4_16
17. Keidel, S., Poulsen, C.B., Erdweg, S.: Compositional soundness proofs of abstract interpreters. PACMPL **2**(ICFP), 72:1–72:26 (2018)
18. Midtgaard, J., Jensen, T.: A calculational approach to control-flow analysis by abstract interpretation. In: Alpuente, M., Vidal, G. (eds.) SAS 2008. LNCS, vol. 5079, pp. 347–362. Springer, Heidelberg (2008). https://doi.org/10.1007/978-3-540-69166-2_23
19. Nielson, F., Nielson, H.R.: The tensor product in wadler's analysis of lists. Sci. Comput. Program. **22**(3), 327–354 (1994)
20. Nielson, F., Nielson, H.R., Hankin, C.: Principles of Program Analysis. Springer, Heidelberg (1999). https://doi.org/10.1007/978-3-662-03811-6
21. Norell, U.: Dependently typed programming in agda. In: Koopman, P., Plasmeijer, R., Swierstra, D. (eds.) AFP 2008. LNCS, vol. 5832, pp. 230–266. Springer, Heidelberg (2009). https://doi.org/10.1007/978-3-642-04652-0_5
22. Rival, X., Toubhans, A., Chang, B.-Y.E.: Construction of abstract domains for heterogeneous properties (position paper). In: Margaria, T., Steffen, B. (eds.) ISoLA 2014. LNCS, vol. 8803, pp. 489–492. Springer, Heidelberg (2014). https://doi.org/10.1007/978-3-662-45231-8_40
23. Streicher, T.: Domain-Theoretic Foundations of Functional Programming. World Scientific Publishing Company, Singapore (2006)
24. Tarski, A.: A lattice-theoretical fixpoint theorem and its applications. Pac. J. Math. **5**(2), 285–309 (1955)
25. Toubhans, A., Chang, B.-Y.E., Rival, X.: Reduced product combination of abstract domains for shapes. In: Giacobazzi, R., Berdine, J., Mastroeni, I. (eds.) VMCAI 2013. LNCS, vol. 7737, pp. 375–395. Springer, Heidelberg (2013). https://doi.org/10.1007/978-3-642-35873-9_23

Incremental Abstract Interpretation

Helmut Seidl, Julian Erhard(✉), and Ralf Vogler

Technische Universität München, Garching, Germany
seidl@in.tum.de, {julian.erhard,ralf.vogler}@tum.de

Abstract. Non-incremental static analysis by abstract interpretation has to be rerun every time the code to be analyzed changes. For large code bases, this incurs a significant overhead, in particular, if the individual changes to the code are small. In order to accelerate the analysis on changing code bases, incremental static analysis reuses analysis results computed for earlier versions of the source code where possible. We show that this behavior can seamlessly be achieved for the analysis of C programs if a local generic solver such as the top-down solver is used as the fixed-point engine. This solver maintains a set of stable unknowns for which fixpoint iteration has already stabilized and it recursively destabilizes dependent unknowns on change. We indicate how this machinery can be applied to selectively invalidate results for those unknowns that may be directly or indirectly affected by program changes. We also explain the technical difficulties faced when realizing this basic idea within an analysis infra-structure such as GOBLINT. We also report the results of a preliminary experimental evaluation concerning the impact of incrementalization on analysis performance.

1 Introduction

Static analysis allows to verify safety properties of software and hint developers at potential program faults. Excessive analysis times, however, limit its usefulness within a software development process where a given code base evolves by means of series of rather small successive modifications. Full re-analysis after each modification thus is no realistic option. This has been repeatedly observed [5,9,18]. Quite some research effort therefore has been invested to design *incremental* static analyzers. A first summary over work in that area occurred as a separate section in [20]. Ideally, such an analyzer only analyzes that part of the code that has changed. That kind of minimal recomputation is supported if the overall analysis setting is *modular* [8,18]. Other approaches are restricted on particular *distributive* analyses as expressed, e.g., in the IDE- or IFDS frameworks [4] where variants of *semi-naive* fixpoint iteration can be applied to further restrict recomputation only to invalidated *parts* of the analysis results. Many useful analyses, though, cannot be expressed within these analysis frameworks. This is already the case for *interval* analysis in C programs which determines (hopefully tight) upper and lower bounds for each numerical variable of the program.

A. Di Pierro et al. (Eds.): Festschrift Hankin, LNCS 12065, pp. 132–148, 2020.
https://doi.org/10.1007/978-3-030-41103-9_5

Here, we take another point of view. We would not like to restrict the analysis engineer in the design or implementation of the desired analysis. Instead, incrementalization should be provided by the analysis infrastructure itself. We claim that this can be achieved by means of *generic local solvers.*

Generic local solvers are the algorithmic core of our static analyzer GOBLINT of multi-thread C code [1,10,25]. Several new generic solvers have recently been additionally experimented with [3,23]. These new solvers enhance earlier proposals for top-down analysis of PROLOG [14,16], e.g., by dedicated support for *widening* and *narrowing* in order to handle abstract domains also for more involved program properties.

In general, a local generic solver for a given system of equations is called with some initial query to some unknown x_0 and then explores only those unknowns of the system whose values may be necessary for answering the initial query, i.e., for determining the value of x_0. Since generic solvers cannot preprocess the system of equations, they have to determine dependencies between unknowns on-the-fly. Our claim is that, due to demand-driven query evaluation and automatic detection of dependencies, such solvers can be applied to construct an incremental analysis infrastructure with little effort. In order to substantiate this claim, we report on the construction of such an incremental extension of the analysis infrastructure GOBLINT based on the top-down solvers in [23]. The most substantial change was to add functionality for detecting which functions of the C source code have been semantically altered, and to provide a naming scheme for program constituents such as variables, types and functions across versions, so that results obtained for earlier versions can be made available for later versions as well.

The rest of the paper is organized as follows. In the next section, we briefly discuss related work. In Sect. 3, we recall background on local generic solving and present, as one such solver, a version of the top-down solver from [23]. In Sect. 4, we present the contract satisfied by our solver and argue how incremental reanalysis can be obtained. In Sect. 5, practical issues are discussed to incorporate the principal approach into the static analyzer GOBLINT. In Sect. 6, we report the results of preliminary experiments, while Sect. 7 concludes.

2 Related Work

Due to potential performance benefits, incremental analysis is an important feature provided by some commercial analysis tools. According to [9], incremental analysis has already been supported in 2008, e.g., by the (unsound) commercial bug-hunting tool COVERITY. Later, it has been adopted by KLOCWORK [5], to mention another example. O'Hearn reports that incremental analysis is of crucial importance for providing fast feedback to software developers at FACEBOOK [18]. Their tool INFER is based on a modular approach to analysis combined with abduction, i.e., optimistic assumptions on the behavior of the environment. A more fundamental study on modular analysis is [8]. Therein, Cousot and Cousot propose to split programs into parts, which then are analyzed separately. Subsequently, the analysis results for the individual program parts are used to compute the analysis result for the whole program. In the best case, on program change, only the analysis of the affected modules and the fixed-point iteration for the program, defined in terms of solutions of its parts, has to be redone, while the re-analysis of unaffected modules can be avoided.

Already in the nineties, Ramalingam and Reps present an incremental solution for context-free graph reachability [21], which can be applied to incrementalize simple interprocedural data flow analyses of imperative languages [22]. Recently, that idea has been taken up by Arzt and Bodden [4]. For incremental re-analysis, their tool REVISOR starts by computing a superset of changed nodes using a graph-diff algorithm. Abstract values of nodes which are reachable from these are possibly affected by the program change and have to be updated. Thus, the set of all unaffected nodes that directly precede any affected node is used as a starting point for a forward propagation.

Interesting contributions have also been made for incrementalizing the analysis of PROLOG or, more generally, logic programs. An incremental algorithm for the analysis of constraint logic programs has been proposed by Hermenegildo et al. [12, 19]. For incremental runs, the algorithm makes use of dependencies that have been tracked previously. This incremental algorithm is specifically tailored towards logic programs, such that a special handling for the addition and deletion of clauses is given. Further, the algorithm differs from ours in that it uses a priority queue to determine the iteration order. Though certain analysis problems for imperative languages have been translated to analysis problems for constraint logic programs, such as in [17], the applicability and performance of incremental algorithms for logic programs when applied to analyzing imperative programs has, to the best of our knowledge, not been tested. Particular issues, clearly, are the flexibility of the framework when it comes to arbitrary abstract domains and the peculiarities of the C semantics.

Quite recently, Garcia-Contreras, Morales and Hermenegildo have reconsidered incremental analysis for constrained Horn clause programs [11]. They consider changes to programs on the module level and explicitly maintain dependencies between modules. They describe how existing analyzers, as long as they satisfy certain conditions, are used to realize incremental re-analysis of affected modules only. For C programs, computing dependencies needs not necessarily be syntactic and therefore requires an analysis run of its own. It is for that reason that our approach solely relies on the dependencies detected by the solver itself during the analysis.

3 The Top-Down Solver

In the following, we present an enhanced variant of the *top-down* solver **TD**. The vanilla version of this solver was introduced in [14, 16] with the analysis of PROLOG in mind. It supported a demand driven exploration of a query over a complete lattice and provided Kleene fixpoint iteration in case that the query turns out to be recursive. Thus, termination of this solver can only be guaranteed if the given complete lattice has no infinite strictly ascending chains. Non-trivial analyses of imperative languages such as C on the other hand, resort to domains violating this assumption or, at least have impractically long strictly ascending chains. This is already the case for *interval* analysis which tries to determine tight upper and lower bounds on the run-time values of integer or floating point variables.

In [3], the top-down solver was therefore enhanced with *widening* and *narrowing* which are intertwined. The latter solver, though, could only be proven to terminate in case that only finitely many unknowns are queried, and the system of equations has *monotonic* right-hand sides only. This was the starting-point of a series of further enhancements to **TD** as presented in [23]. Here, we recall the simplest version from there, which beyond the solver from [3], is guaranteed to terminate for systems of equations with or without monotonic right-hand sides whenever only the number of unknowns encountered during solving is *finite*. For simplicity, we henceforth refer to this enhanced version of the top-down solver as **TD**. In the following subsection, we make our notions and concepts precise where we closely follow the exposition in [23]. After recalling the solver **TD** in Subsect. 3.2, we indicate in Sect. 5, how that solver naturally gives rise to incremental analysis.

3.1 Abstract Semantics, Systems of Equations and Their Lower Monotonization

Static analysis by means of abstract interpretation is based on an *abstract semantics* of the program. Here, we assume that the abstract semantics is based on a set $\mathcal{X}$ of *unknowns* and a complete lattice $\mathbb{D}$ whose elements represent pieces of information over the program. For imperative programs, e.g., $\mathcal{X}$ could be the set of *program points* or, in case of interprocedural analysis in the style of [2,6,13,24], the set of pairs $\langle u, d\rangle$ of program points and abstract calling contexts $d \in \mathbb{D}$. The *abstract* semantics of the program then is obtained from a system of equations

$$x = f_x \quad (x \in \mathcal{X}) \tag{1}$$

where the right-hand sides $f_x : (\mathcal{X} \to \mathbb{D}) \to \mathbb{D}$ formalize how the abstract value of x is related to the abstract values of all other unknowns in the system. Given that all right-hand side functions f_x are monotonic, the abstract semantics then is defined as the *least* solution of the given system of equations. When $\mathbb{D}$ has infinite ascending chains, however, this least solution need not be easily computable. In that case, any *post-solution* of the system may provide less precise but still *sound* information about the program in question. Here, a post-solution is any function $\rho : \mathcal{X} \to \mathbb{D}$ where $\rho\, x \sqsupseteq f_x\, \rho$ holds for all $x \in \mathcal{X}$.

Example 1. Consider a minimalistic program with functions main and g only, where main calls g twice:

```
   void main() {
0:     g();
1:     g();
2: }
```

Assume that the abstract state before the call to main is d. Then the definition of main gives rise to the following equations for the program points $0, 1, 2$ of main in context d:

$$\langle 0, d\rangle = f_0 \quad \langle 1, d\rangle = f_1 \quad \langle 2, d\rangle = f_2$$

where the right-hand side functions f_i are given by:

$$\begin{aligned} f_0\,\rho &= s \\ f_1\,\rho &= \rho\,\langle r_g, \rho\,\langle 0, d\rangle\rangle \\ f_2\,\rho &= \rho\,\langle r_g, \rho\,\langle 1, d\rangle\rangle \end{aligned}$$

where r_g is the end point of the function g. Here, the function g must be analyzed for the abstract state d as well as for the state $\rho\,\langle r_g, d\rangle$ attained at the end point of g in calling context d. □

We remark that already for this minimalistic program, right-hand side functions are not necessarily *monotonic*. This is due to the fact that the set of unknowns whose values are queried, itself depends on the values of unknowns. For non-monotonic right-hand sides, on the other hand, systems of equations need no longer have uniquely defined least (post) solutions. In [10], however, it has been observed that (at least in presence of a Galois connection between the concrete and the abstract domains and reasonable assumptions on the relationship between concrete and abstract semantics) non-monotonic right-hand sides can be replaced by their *lower monotonization*. For a function $f : (\mathcal{X} \to \mathbb{D}) \to \mathbb{D}$, the lower monotonization $\underline{f} : (\mathcal{X} \to \mathbb{D}) \to \mathbb{D}$ is defined as

$$\underline{f}\,\rho = \bigsqcap\{f\,\rho' \mid \rho \sqsubseteq \rho'\} \tag{2}$$

By construction, $\underline{f}$ is monotonic with $\underline{f} \sqsubseteq f$. The lower monotonization of the abstract system of Eq. (1) then is given by

$$x = \underline{f}_x \quad (x \in \mathcal{X}) \tag{3}$$

This system now has a unique least solution. Accordingly, the goal of static analysis can be refined to compute a post solution of the system (3).

3.2 Generic Local Solving, Variable Dependencies and Computation Trees

Static analyzers generally, are complex software systems. From a software engineering point of view, therefore, it is desirable to separate engineering of the analysis itself, i.e., crafting appropriate complete lattices and systems of equations, from the algorithmic part, i.e., the actual computation of a reasonably precise post solution. That insight triggered the development of *generic* solvers: the term "generic" here means that the solver makes as little assumptions as possible on the system of equations. One obvious such solver is *round-robin* iteration, which, however, has the drawback of repeatedly having to reevaluate the right-hand sides of *all* unknowns in a row—no matter of whether their values require re-computation or not. More refined solvers, on the other hand, gain efficiency by explicitly tracking *dependencies* between unknowns. As generic solvers essentially treat right-hand sides of equations as black box functions, no precomputations on the given system can be performed to determine or at least over-approximate these dependencies.

For that reason, the generic solvers in [3,10,23] resort to *self-monitoring* in order to detect dependencies between unknowns on-the-fly. Technically, this means that the right-hand side f_x for some unknown x is not evaluated on the current assignment ρ itself, but on an instrumented lookup function which, when queried for the value of some unknown z, records the dependence of x on z, before returning the current value of z. A sufficient prerequisite for such an implementation is that each right-hand side is *pure* [15]. A terminating function f is pure iff it can be represented as a *computation tree*, i.e., an element of the datatype

$$\mathcal{T} \quad ::= \quad A\,\mathbb{D} \mid Q\,(\mathcal{X}, \mathbb{D} \to \mathcal{T})$$

The computation tree $A\,d$ immediately returns the answer d, while the computation tree $Q(x, f)$ queries the value of the unknown x in order to apply the continuation f to the obtained value.

Example 2. Consider the right-hand side f_1 in Example 1 defined by

$$f_1\,\rho = \rho\,\langle r_g, \rho\,\langle 0, d\rangle\rangle$$

This function is represented by the computation tree

$$\begin{aligned} t \; = Q\,(\langle 0, d\rangle, &\,\textbf{fun}\, d' \;\to \\ &Q(\langle r_g, d'\rangle, \textbf{fun}\, d'' \to \\ &\quad A\,d''))\end{aligned}$$

The semantics of computation trees is best described by means of state transformer monads. Assume that S is a set of states. Then the state transformer monad $\mathcal{M}_S(A)$ for values of type A consists of all functions $S \to S \times A$. In particular, $\mathcal{M}_S(A)$ provides the functions $\mathsf{return} : A \to \mathcal{M}_S(A)$ and $\mathsf{bind} : \mathcal{M}_S(A) \to (A \to \mathcal{M}_S(B)) \to \mathcal{M}_S(B)$ by:

$$\begin{aligned} \mathsf{return}\, a \;&= \mathsf{fun}\, s \to (s, a) \\ \mathsf{bind}\, m\, f &= \mathsf{fun}\, s \to \mathsf{let}\,(s', a) = m\, s \\ &\qquad\qquad\quad \mathsf{in}\, f\, a\, s' \end{aligned}$$

Accordingly, the semantics of trees is a function $[\![t]\!] : (\mathcal{X} \to \mathcal{M}_S(\mathbb{D})) \to \mathcal{M}_S(\mathbb{D})$.

$$\begin{aligned} [\![\mathsf{A}\, d]\!]\, \mathsf{get}\, s \quad\;\; &= (s, d) \\ [\![\mathsf{Q}\,(x, f)]\!]\, \mathsf{get}\, s &= \textbf{let}\;(s', d) = \mathsf{get}\, x\, s \\ &\quad\;\; \textbf{in}\; [\![f\, d]\!]\, \mathsf{get}\, s' \end{aligned}$$

The function $[\![t]\!]$ is polymorphic in the set of states used by the state transformer monad. In fact, it is even parametric w.r.t. monads in general, as it can be defined by means of the monad operations return and bind alone. We have:

$$\begin{array}{ll} [\![\mathsf{A}\,d]\!]\,\mathsf{get} & = \mathsf{return}\,d \\ [\![\mathsf{Q}\,(x,f)]\!]\,\mathsf{get} & = \mathsf{bind}\,(\mathsf{get}\,x)\,(\mathsf{fun}\,d \to [\![f\,d]\!]\,\mathsf{get}) \end{array}$$

One particular instance of state transformer monads for our purpose is a monad which tracks the unknowns accessed during the evaluation. For that, we consider the set of states $S = (\mathcal{X} \to \mathbb{D}) \times 2^{\mathcal{X}}$ together with the function

$$\begin{array}{l} \mathsf{get}\,x\,(\rho, X) = \mathbf{let}\ d = \rho\,x \\ \qquad\qquad\qquad\quad \mathbf{in}\ ((\rho, X \cup \{x\}), d) \end{array}$$

The following proposition formalizes that evaluation of a computation tree only depends on the values of variables accessed during the evaluation.

Proposition 1. *For a mapping* $\rho : \mathcal{X} \to \mathbb{D}$ *and* $s = (\rho, \emptyset)$, *assume that* $[\![t]\!]\,\mathsf{get}\,s = (s_1, d)$. *Then for* $s_1 = (\rho_1, X)$ *the following holds:*

1. $\rho = \rho_1$;
2. *Assume that* $\rho' : \mathcal{X} \to \mathbb{D}$ *is another mapping and* $s' = (\rho', \emptyset)$. *Let* $[\![t]\!]\,\mathsf{get}\,s' = ((\rho', X'), d')$. *If* ρ' *agrees with* ρ *on* X, *i.e.,* $\rho|_X = \rho'|_X$, *then* $X' = X$ *and* $d' = d$ *holds.*

In light of this proposition, we define for a pure function $f = [\![t]\!]$,

$$\begin{array}{l} \mathsf{dep}(f, \rho) = \mathbf{let}\ ((\_, X'), \_) = f\ \mathsf{get}\ (\rho, \emptyset) \\ \qquad\qquad\quad\ \mathbf{in}\ X' \end{array} \tag{4}$$

Another obstacle for solving abstract systems of equations is that the number of unknowns mentioned in the system may be unfeasibly large, if not infinite. This is already the case for interprocedural constant propagation when contexts comprise also the values of program variables of some numerical type. It turns out, though, that in many cases only a rather small set of calling contexts for each function is of interest, i.e., is required to describe all concrete calling contexts that may occur during a concrete execution of the program. *Local* solvers therefore try to restrict the evaluation of right-hand sides to a set of unknowns that is as small as possible. A local solver is started with a single unknown x_0 as initial *query* and then recursively proceeds to solve those unknowns whose values may contribute to the evaluation of the right-hand side f_{x_0} of x_0. For interprocedural analysis of C programs, the initial query is given by $\langle r_{\mathsf{main}}, d_0 \rangle$ where r_{main} is the end point of the initial function main and d_0 is the initial abstract calling context of main. The local solver, when started with $\langle r_{\mathsf{main}}, d_0 \rangle$, recursively traverses the program and analyzes function definitions for those abstract calling contexts which may contribute to the execution of the program.

As a result, a local solver, when applied to some initial query $x_0 \in \mathcal{X}$, should return a triple $(\rho, \mathsf{stable}, \mathsf{infl})$ consisting of an assignment $\rho : \mathcal{X} \to \mathbb{D}$, a subset stable of unknowns together with an assignment $\mathsf{infl} : \mathcal{X} \to 2^{\mathcal{X}}$ such that for each $x \in \mathsf{stable}$, the following three properties hold:

- $\rho\, x \sqsupseteq \underline{f}_x\, \rho$;
- $\mathsf{dep}(x, \rho) \subseteq \mathsf{stable}$;
- If $y \in \mathsf{dep}(x, \rho)$, then $x \in \mathsf{infl}\; y$.

In this case, we call the triple *consistent*. Moreover, we demand that after termination of the local solver, $x_0 \in \mathsf{stable}$.

As one specific example of a local generic solver, Listing 1.1 displays the base version of the top-down solver TD from [23]. In order to enforce termination, that solver employs a *widening* operator ∇ together with a narrowing operator Δ [7]. These operators should satisfy the following properties:

$$a \sqcup b \sqsubseteq a \nabla b$$
$$a \sqcap b \sqsubseteq a \,\Delta\, b \sqsubseteq a$$

where for all initial values $a_0, a_0' \in \mathbb{D}$ and any sequence $b_i \in \mathbb{D}, i \geq 1$, the sequences

$$a_i = a_{i-1} \nabla b_i \qquad (i \geq 1)$$
$$a_i' = a_{i-1}' \,\Delta\, b_i \qquad (i \geq 1)$$

are ultimately stable.

For better readability, we refrained from threading the program state explicitly through the code, but have adopted a (call-by-value) functional notation in the spirit of OCAML where the state is manipulated by means of side effects.

```
(* functions on Map.t and Set.t *)
val find    : 'a -> ('a, 'b) Map.t -> 'b
val replace : 'a -> 'b -> ('a, 'b) Map.t -> unit
val mem     : 'a -> 'a Set.t -> bool
val add     : 'a -> 'a Set.t -> unit
val rem     : 'a -> 'a Set.t -> unit

(* mutable data structures *)
let rho    = Map.create () : (X, D) Map.t in
let infl   = Map.create () : (X, X Set.t) Map.t in
let called = Set.create () : X Set.t in
let stable = Set.create () : X Set.t in
let point  = Set.create () : X Set.t in

let destabilize x =
    let w = find x infl in
    replace x (Set.create ()) infl;
    iter (fun y ->
        rem y stable;
        if not (mem y called) then
            destabilize y
        ) w
in
let rec eval x y =
```

```
    if mem y called then
        add y point
    else
        solve ∇ y;
    add x (find y infl);
    find y rho
and solve p x =
    if not (mem x stable || mem x called) then (
        add x stable;
        add x called;
        let tmp = (f_x (eval x))_1 in
        rem x called;
        let tmp =
            if mem x point then p (find x rho) tmp
            else tmp
        in
        if find x rho = tmp then
            if p = ∇ && mem x point then (
                rem x stable;
                solve Δ x
            )
        else (
            replace x tmp rho;
            destabilize x;
            solve p x
        )
    )
```

Listing 1.1. A basic version of the top-down solver. Code is taken from [23] with minor adaptations.

The solver is started by calling `solve p x`, which triggers the evaluation of the unknown `x`. The parameter `p` determines whether widening (∇) or narrowing (Δ) should be performed on the unknown—given that it is a widening point. Initially, we call `solve ∇ x` on all unknowns `x` we are interested in, e.g., the return node of the main function of the analyzed program. After termination, the result of the solver provides global data structures `rho`, `infl` and `stable`, respectively, for:

- a mapping $\rho : \mathcal{X} \to \mathbb{D}$ from unknowns to abstract values;
- a mapping $\mathsf{infl} : \mathcal{X} \to 2^{\mathcal{X}}$ from unknowns to sets of unknowns recording the encountered dependencies between unknowns; and
- a set `stable` of unknowns signifying those unknowns z for which $\rho\, z \sqsupseteq \underline{f}_z\, \rho$ holds.

During evaluation, the solver additionally maintains sets `called` and `point`. Initially both sets are empty. The set `called` is meant to keep track of those unknowns z for which the evaluation of the right-hand side f_z has been initiated, but not yet completed. The set `point` is meant to collect a sufficient subset of unknowns where widening/narrowing should be applied to enforce termination of the iteration.

The call `solve p x` first checks whether the unknown `x` is already contained either in `stable` or in `called`. In both cases, the call immediately returns. Otherwise, it inserts `x` into both sets. Then it triggers the evaluation of the right-hand side $f_{\mathtt{x}}$ of `x`. That function, however, is not evaluated on the current assignment `rho` directly, but on the wrapper function `eval`, partially applied to `x`.

Each call `eval x y` performs four tasks.

- First, it checks whether `y` is already contained in the set `called`. If so, a cyclic dependence is detected, implying that `y` is made a widening/narrowing point, i.e., is added to the set `point`.
- Second, `solve` ∇ `y` is called in order to determine the currently best possible value for the unknown `y`.
- Third, the unknown `x` is added to the set `find y infl` in order to record the dependence between `x` and `y`.
- Finally, the value of `y` is looked up and returned.

Returning to the implementation of the function `solve` for arguments `p` and `x`, we find that the value provided by evaluating the right-hand side of `x`, is stored in the temporary variable `tmp`. Having completed the evaluation of $f_{\mathtt{x}}$, the unknown `x` is removed from the set `called`. If `x` is contained in `point`, `tmp` is updated to receive the combination of the old value for `x` in `rho` with the new value provided by the right-hand side by means of the operator `p`. The value stored in `rho` for `x` is now compared with the given value `tmp`. Two cases may occur.

Either the new value equals `find x rho`. In this case, the current iteration on `x` has terminated. If it was a widening iteration (signified by $\mathtt{p} = \Delta$), `x` can not yet be considered as having received the best possible value. Therefore, `x` is removed from `stable` and `solve` is called for `x`, but now with the operator ∇. If the current iteration on `x` was already a narrowing iteration (signified by $\mathtt{p} = \nabla$), the current call returns.

It remains to consider the case when the new value `tmp` differs from `find x rho`. Then all unknowns which directly or indirectly may depend on `x`, are removed from the set `stable`. This task is delegated to the function `destabilize`. When called for some unknown z, it removes all unknowns in the set `find z infl` from the set `stable`. Before recursively calling `destabilize` y for all y in that set which are not in `called`, the set `find z infl` is reset to the empty set. Subsequently, `solve` is again called for the actual parameters `p` and `x`.

4 Invariants

Let $\rho : \mathcal{X} \to \mathbb{D}$ and $\mathsf{infl} : \mathcal{X} \to 2^{\mathcal{X}}$ denote the assignments

$$\begin{aligned} \rho\, x &= \mathtt{find}\ \ x\ \mathtt{rho} \\ \mathsf{infl}\, x &= \mathtt{find}\ \ x\ \mathtt{infl} \end{aligned}$$

The core function `solve` of the generic local solver **TD** satisfies the following contract.

Assume that before the call of `solve p x` for some unknown `x`, the triple $(\rho, \texttt{stable}, \mathsf{infl})$ is consistent. Then the attained triple after termination of the call is again consistent where additionally, $x \in \texttt{stable}$. Recall that in order to perform interprocedural abstract interpretation of some program, the function `solve` is called with ∇ and the initial query $x_0 = \langle r_{\mathsf{main}}, d_0 \rangle$ where r_{main} is the end point of the entry function of the program and d_0 is the initial calling context. After modification of the program, our goal is to recompute a consistent triple with $x_0 \in \texttt{stable}$ with as few re-evaluations of right-hand sides as possible. Here, we consider modifications of programs at the function level only. Every new version of a function f is represented by a control-flow graph whose set of program points is disjoint from all program points introduced so far—with the only exception of the start and end points s_f, r_f of f, since these serve as interfaces to the surrounding program.

In order to trigger the necessary re-evaluations, we rely on the function `destabilize` as provided by the solver **TD**. Let X' denote the subset of unknowns in `stable` which, by modification of the program, have received a new right-hand side. Since the right-hand sides of the unknowns in X' have not yet been evaluated, these may violate consistency. Therefore, X' is removed from `stable`, and `destabilize` z is called for each unknown $z \in X'$. By that, consistency is re-established.

For a modified function f, this means that all unknowns of the form $\langle r_f, d \rangle$ are removed from `stable` for any $d \in \mathbb{D}$, and all other unknowns which directly or indirectly depend on any such $\langle r_f, d \rangle$ are removed from `stable` as well.

Two cases can occur. *Either* after destabilization, the initial context d_0 has not changed and $\langle r_{\mathsf{main}}, d_0 \rangle$ is still contained in the set `stable`. Then no re-evaluation is required since the modifications only affect irrelevant parts of the abstract system of equations. *Alternatively*, we either have a new initial calling context d_0, or $\langle r_{\mathsf{main}}, d_0 \rangle$ is removed from the set `stable`. Then, the function `solve` must again be called for the unknown $\langle r_{\mathsf{main}}, d_0' \rangle$ where d_0' is the possibly new calling context of the initial function.

In the course of this reevaluation, however, not the whole program is re-analyzed, since the abstract values of all unknowns still contained in `stable` will not be recomputed, but looked up.

5 Practical Incremental Analysis

In order to set the re-evaluation of modified code in a Git repository to work, we adapt the pipeline of our static analyzer GOBLINT in the following way. First, after each run of the analyzer on some version of the program, we persistently store the analysis results, i.e., in particular the data structures `rho`, `stable`, and `infl`. Additionally, we store a map of function names to their current (preprocessed) implementations. Second, we add *change detection* for the source code in order to identify those parts of the program which have been altered. Finally, we modify the solver to reuse old results and selectively invalidate them, based upon the detected program changes.

We note that the analyzer proceeds differently, depending on whether or not a previous version of the program has already been analyzed. In case the code base has not been analyzed yet, the program is analyzed from scratch. In a run on a changed version of the repository, the stored data structures for the most recent older commit which was analyzed are recovered. The analysis does not need to be run for every commit—changes between an analyzed commit and any other commit can be accumulated and analyzed together. The analyzer is searching older commits for results by default. In principle, the direction might also be changed, if one wants to analyze going backwards from the current commit, e.g., to find out when a problem was introduced in the program. One could also imagine to search in both directions and use the closest results (in terms of commits or even set of changes) to speed up a search using `git bisect`. After preprocessing the new version, we compare it to the preprocessed version of the previous version. We do this comparison on the function level. For each function in the current version, we check whether it has been changed. In order to enable the reuse of data computed in the former run of GOBLINT, further technical adaptions had to be performed. Most importantly, the identity of program entities, such as functions, global and local variables, must be preserved across multiple versions in order to allow for reuse of analysis results computed in previous runs of GOBLINT.

Granularity. Change can be considered at various levels of granularity. These include file level, function level or the level of single nodes in the control flow graph. The finer the granularity is, at which changes are detected, the less results may have to be invalidated. At a finer granularity, on the other hand, change detection becomes more subtle to realize. For a programming language such as C, change detection at the level of functions seems a decent compromise. Changes may affect the function's type, its formal or local variables, or its body. All information and values computed for a modified function g will no longer be considered as valid. In particular, this refers to all unknowns where program points of that function (excluding start and end points) are involved. Conceptually, fresh equations are introduced for the new body in all possible contexts. In particular, the end points $\langle r_g, d \rangle$ of g, queried in some abstract calling context d, receive fresh right-hand sides. By removing these unknowns from the restored set `stable` and subsequent destabilization, a consistent state is reestablished.

Intermediate Language Stage. We check for program changes after the code is preprocessed into the intermediate language of the compiler framework CIL, as opposed to the source code itself. This has the advantage that some source code changes that have no impact on the program semantics are already eliminated. CIL applies simple constant folding and normalizes several syntactic constructs.

Syntactic and Semantic Changes. It is important to note that some rather local changes to the code may still have global semantic effects. In particular, this is true for changes in type definitions which may affect all functions where that type is used. Other changes affecting the syntax of a function may not alter the semantics of a function. When, e.g., the type declaration of a local variable is changed to use a type alias, the semantics is not affected: type synonyms are interchangeable and should denote the same type. Addition or removal of white space (as long as it does not affect the abstract syntax tree), can be ignored as well.

We always compare the version of the function to analyze with the function—if any—with the same name in the previously analyzed version. A refactoring with mere renaming of functions, thus has the unfortunate effect that the renamed functions, as well as its callers are considered as changed, causing their complete reanalysis. In a more refined implementation, it therefore would be desirable to provide dedicated support at least for basic refactoring transformations (such as renaming of functions or variables) applied to the source code, by providing corresponding transformations of the analysis results (such as updating the corresponding names of variables and functions).

Another issue is the treatment of globals. In principle, changes to global variables may affect the *whole* program—thus causing a full re-analysis. One immediate solution is provided by the analyzer GOBLINT. In order to increase scalability, this analyzer offers the option to analyze global data structures *context-* and *control*-insensitively by means of a single *global invariant*. When this option is selected, the values of globals are no longer included into the calling contexts of functions. This means that now the impact of changes to a global x is restricted to those functions which explicitly access x for reading or writing. Technically, the required combination of analyses which differ in their sensitivities in GOBLINT is based on an enhancement of the underlying framework and solver with side-effects [2,23,25].

6 Experimental Evaluation

We evaluate the performance of the incremental algorithm in a whole-program analysis setting. The analyzer is meant to perform an context-sensitive base analysis of constant values and pointers together with a interval analysis for `int` variables. For the experiment, we use the repository of *figlet*[1], which we chose because it is written in ANSI-C and a version history is available[2]. Notably, the first revision of *figlet* tracked by the version control system already contains the majority of the source code while the following commits contain bug fixes as well as modifications not relating to program logic. We combined source files using the `Makefile` provided by the project. In total, the first commit contains 5107 lines of code (including comments) in `.c` and `.h` files. Two of these files, namely `chkfont.c` and `getopt.c`, initially totalling 565 lines, were not included in the

[1] http://www.figlet.org/.
[2] https://github.com/cmatsuoka/figlet.

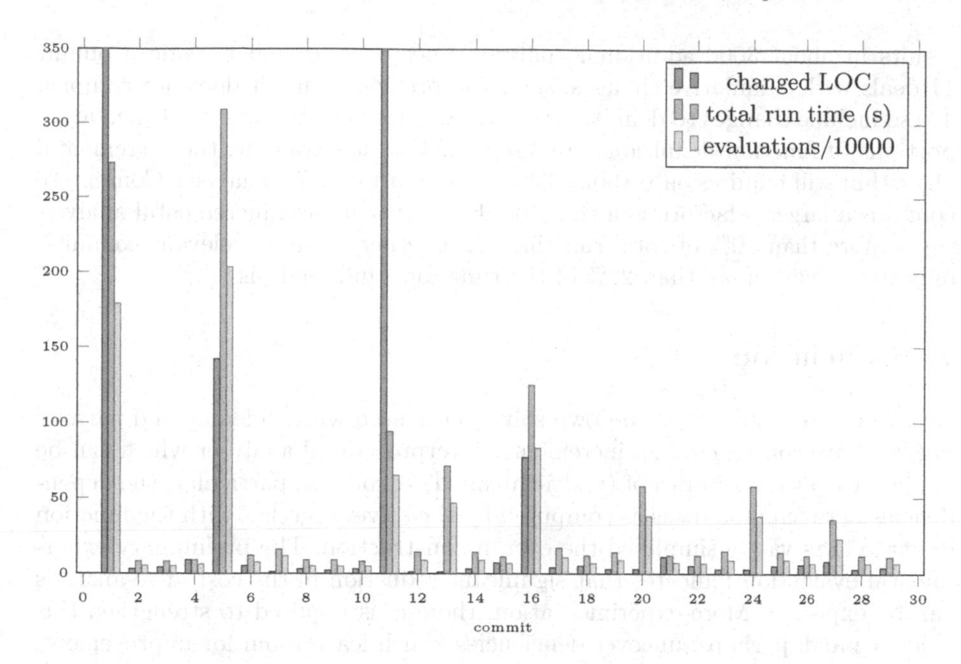

Fig. 1. Run-time, number of evaluations and changed lines of code for the *figlet* repository. The first commit inserts 4542 lines into the analyzed source files; commit 11 adds 507 lines compared to commit 10 (graph is cropped).

main program, and thus have not been part of the analysis. Starting from the first commit, we subsequently analyze the following ones with the incremental algorithm.

Out of the 105 commits, only 41 had source file changes (including the initial commit). One of these commits introduced function calls with too few arguments. While the commit could be compiled with `gcc`, it produced an error with GOBLINT's front-end. The affected function definition, however, was fixed to expect less arguments in the next commit. The bogus commit therefore was excluded from the experiment. Out of the remaining 40 commits, only 29 had *semantic* changes and resulted in partial re-analysis. Only these are reported in Fig. 1. As an indicator of the size of the program change, the number of lines of code (`.c` and `.h` files) is given in which the analyzed commit differs from the last analyzed commit, including changes made to comments. Technically, we take the maximum of the number of removed and added lines, respectively. Efficiency of the incremental analysis is measured via the number of evaluations of right-hand sides as well as the total run-times of GOBLINT per analyzed commit. For a comparison, the total run-time of the *non-incremental* version of GOBLINT on any commit does not differ significantly from the total run-time on the initial commit, which is 237 s.

Changes to functions behave mostly benevolent—with some exceptions that are more expensive to analyze (cf. Fig. 1): Commit 5 changes the Makefile which results in about 3000 additional analyzed lines of intermediate code. Commit 11 deals with cumulative changes, since the previous commit does not compile. These include changes to definitions of global variables. At the same time, larger portions of function definitions are adapted. For that commit, the incremental algorithm still requires only about 35% of the effort of a full analysis. Commit 16 contains a larger refactoring of the code. Even in that case, incremental analysis saves more than 50% of total run time. Altogether, from 29 relevant commits, only five require more than 22% of the time for a full analysis.

7 Conclusion

Based on a version of the top-down solver, enhanced with widening and narrowing, we have constructed an incremental interprocedural analyzer which can be applied to Git repositories of (multi-threaded) C code. In particular, the dependencies between unknowns as computed by the solver together with the function `destabilize` vastly simplified the overall construction. The preliminary experimental evaluation indicates that significant reduction in the cost of re-analysis can be expected. More experimentation, though, is required to strengthen this evidence and, perhaps, uncover deficiencies which leave room for improvement. One apparent deficiency is that modifications to global data-structures in the C code may cause re-analysis of major parts of the program. Currently, we tried to confine the impact of such changes by tracking context - as well as flow-insensitive information for globals only. While this may be the method of choice for the analysis of larger programs, still alternative techniques are sought for, which allow to analyze functions within trimmed contexts where all *irrelevant* globals have been omitted.

For larger programs, not only analysis time becomes an issue—but also the excessive space consumption for storing analysis results. For that reason, another version of the top-down solver has been proposed which maintains abstract values only for a bare minimum of unknowns while recovering the values for the remaining unknowns on demand [23]. It remains for future experimentation in how far incremental analysis may benefit from that further optimization.

References

1. Amato, G., Scozzari, F., Seidl, H., Apinis, K., Vojdani, V.: Efficiently intertwining widening and narrowing. Sci. Comput. Program. **120**, 1–24 (2016). https://doi.org/10.1016/j.scico.2015.12.005
2. Apinis, K., Seidl, H., Vojdani, V.: Side-effecting constraint systems: a swiss army knife for program analysis. In: Jhala, R., Igarashi, A. (eds.) APLAS 2012. LNCS, vol. 7705, pp. 157–172. Springer, Heidelberg (2012). https://doi.org/10.1007/978-3-642-35182-2_12
3. Apinis, K., Seidl, H., Vojdani, V.: Enhancing top-down solving with widening and narrowing. In: Probst, C.W., Hankin, C., Hansen, R.R. (eds.) Semantics, Logics, and Calculi. LNCS, vol. 9560, pp. 272–288. Springer, Cham (2016). https://doi.org/10.1007/978-3-319-27810-0_14

4. Arzt, S., Bodden, E.: Reviser: efficiently updating IDE-/IFDS-based data-flow analyses in response to incremental program changes. In: Jalote, P., Briand, L.C., van der Hoek, A. (eds.) 36th International Conference on Software Engineering, ICSE 2014, Hyderabad, India, 31 May –07 June 2014, pp. 288–298. ACM (2014). https://doi.org/10.1145/2568225.2568243
5. Bolduc, C.: Lessons learned: using a static analysis tool within a continuous integration system. In: 2016 IEEE International Symposium on Software Reliability Engineering Workshops (ISSREW), pp. 37–40. IEEE (2016)
6. Cousot, P., Cousot, R.: Static determination of dynamic properties of recursive programs. In: Neuhold, E. (ed.) Formal Descriptions of Programming Concepts, pp. 237–277. North-Holland Publishing Company, Amsterdam (1977)
7. Cousot, P., Cousot, R.: Comparing the Galois connection and widening/narrowing approaches to abstract interpretation. In: Bruynooghe, M., Wirsing, M. (eds.) PLILP 1992. LNCS, vol. 631, pp. 269–295. Springer, Heidelberg (1992). https://doi.org/10.1007/3-540-55844-6_142
8. Cousot, P., Cousot, R.: Modular static program analysis. In: Horspool, R.N. (ed.) CC 2002. LNCS, vol. 2304, pp. 159–179. Springer, Heidelberg (2002). https://doi.org/10.1007/3-540-45937-5_13
9. Emanuelsson, P., Nilsson, U.: A comparative study of industrial static analysis tools. Electron. Notes Theor. Comput. Sci. **217**, 5–21 (2008)
10. Frielinghaus, S.S., Seidl, H., Vogler, R.: Enforcing termination of interprocedural analysis. Formal Methods Syst. Design **53**(2), 313–338 (2018). https://doi.org/10.1007/s10703-017-0288-5
11. Garcia-Contreras, I., Morales, J.F., Hermenegildo, M.V.: Towards incremental and modular context-sensitive analysis. In: Technical Communications of the 34th International Conference on Logic Programming (ICLP 2018). OpenAccess Series in Informatics (OASIcs). Dagstuhl Press, July 2018. (Extended Abstract)
12. Hermenegildo, M.V., Puebla, G., Marriott, K., Stuckey, P.: Incremental analysis of constraint logic programs. ACM Trans. Program. Lang. Syst. **22**(2), 187–223 (2000)
13. Jones, N.D., Muchnick, S.S.: A flexible approach to interprocedural data flow analysis and programs with recursive data structures. In: DeMillo, R.A. (ed.) Conference Record of the Ninth Annual ACM Symposium on Principles of Programming Languages, Albuquerque, New Mexico, USA, January 1982, pp. 66–74. ACM Press (1982). https://doi.org/10.1145/582153.582161
14. Muthukumar, K., Hermenegildo, M.: Deriving a fixpoint computation algorithm for top-down abstract interpretation of logic programs. Technical report ACT-DC-153-90, Microelectronics and Computer Technology Corporation (MCC), Austin, TX, April 1990
15. Karbyshev, A.: Monadic parametricity of second-order functionals. Ph.D. thesis, Technical University Munich (2013). http://nbn-resolving.de/urn:nbn:de:bvb:91-diss-20130923-1144371-0-6
16. Le Charlier, B., Van Hentenryck, P.: A universal top-down fixpoint algorithm. Technical report CS-92-25. CS Department, Brown University (1992)
17. Liqat, U., et al.: Energy consumption analysis of programs based on XMOS ISA-level models. In: Gupta, G., Peña, R. (eds.) LOPSTR 2013. LNCS, vol. 8901, pp. 72–90. Springer, Cham (2014). https://doi.org/10.1007/978-3-319-14125-1_5
18. O'Hearn, P.W.: Continuous reasoning: scaling the impact of formal methods. In: Dawar, A., Grädel, E. (eds.) Proceedings of the 33rd Annual ACM/IEEE Symposium on Logic in Computer Science, LICS 2018, Oxford, UK, 09–12 July 2018, pp. 13–25. ACM (2018). https://doi.org/10.1145/3209108.3209109

19. Puebla, G., Hermenegildo, M.: Optimized algorithms for incremental analysis of logic programs. In: Cousot, R., Schmidt, D.A. (eds.) SAS 1996. LNCS, vol. 1145, pp. 270–284. Springer, Heidelberg (1996). https://doi.org/10.1007/3-540-61739-6_47
20. Ramalingam, G., Reps, T.W.: A categorized bibliography on incremental computation. In: Deusen, M.S.V., Lang, B. (eds.) Conference Record of the Twentieth Annual ACM SIGPLAN-SIGACT Symposium on Principles of Programming Languages, Charleston, South Carolina, USA, January 1993, pp. 502–510. ACM Press (1993). https://doi.org/10.1145/158511.158710
21. Ramalingam, G., Reps, T.W.: An incremental algorithm for a generalization of the shortest-path problem. J. Algorithms **21**(2), 267–305 (1996). https://doi.org/10.1006/jagm.1996.0046
22. Reps, T.W., Horwitz, S., Sagiv, S.: Precise interprocedural dataflow analysis via graph reachability. In: Cytron, R.K., Lee, P. (eds.) Conference Record of POPL 1995: 22nd ACM SIGPLAN-SIGACT Symposium on Principles of Programming Languages, San Francisco, California, USA, 23–25 January 1995, pp. 49–61. ACM Press (1995). https://doi.org/10.1145/199448.199462
23. Seidl, H., Vogler, R.: Three improvements to the top-down solver. In: Sabel, D., Thiemann, P. (eds.) Proceedings of the 20th International Symposium on Principles and Practice of Declarative Programming, PPDP 2018, Frankfurt am Main, Germany, 03–05 September 2018, pp. 21:1–21:14. ACM (2018). https://doi.org/10.1145/3236950.3236967
24. Sharir, M., Pnueli, A.: Two approaches to interprocedural data flow analysis. In: Muchnick, S., Jones, N. (eds.) Program Flow Analysis: Theory and Applications, pp. 189–233. Prentice-Hall, Englewood Cliffs (1981)
25. Vojdani, V., Apinis, K., Rõtov, V., Seidl, H., Vene, V., Vogler, R.: Static race detection for device drivers: the Goblint approach. In: Proceedings of the 31st IEEE/ACM International Conference on Automated Software Engineering, ASE 2016, pp. 391–402. ACM (2016). https://doi.org/10.1145/2970276.2970337

Correctly Slicing Extended Finite State Machines

Torben Amtoft[1(✉)], Kelly Androutsopoulos[2], and David Clark[3]

[1] Kansas State University, Manhattan, KS 66506, USA
tamtoft@ksu.edu
[2] Middlesex University, London NW4 4BT, UK
K.Androutsopoulos@mdx.ac.uk
[3] University College London, London WC1E 6BT, UK
david.clark@ucl.ac.uk

Abstract. We consider slicing extended finite state machines. Extended finite state machines (EFSMs) combine a finite state machine with a store and can model a range of computational phenomena, from high-level software to cyber-physical systems. EFSMs are essentially interactive, possibly non-terminating or with multiple exit states and may be nondeterministic, so standard techniques for slicing, developed for control flow graphs of programs with a functional semantics, are not immediately applicable.

This paper addresses the various aspects of correctness for slicing of EFSMs, and provides syntactic criteria that we prove are sufficient for our proposed notions of semantic correctness. The syntactic criteria are based on the "weak commitment" and "strong commitment" properties highlighted by Danicic et alia. We provide polynomial-time algorithms to compute the least sets satisfying each of these two properties. We have conducted experiments using widely-studied benchmark and industrial EFSMs that compare our slicing algorithms with those using existing definitions of control dependence. We found that our algorithms produce the smallest average slices sizes, 21% of the original EFSMs when "weak commitment" is sufficient and 58% when "strong commitment" is needed (to preserve termination properties).

Keywords: Extended finite state machines · Slicing

1 Introduction

Program slicing was introduced in 1979 by Weiser [33,34] to remove the parts of a program that are irrelevant in a given context. Given a slicing criterion <p, V>, where p is a program point and V is a subset of program variables, the slice consists of all the statements that may directly or indirectly affect the slicing criterion (determined by computing dependences on control flow graphs). Weiser's notion of slicing is executable (the slice is not just a closure of statements but can be compiled and run), static (all possible executions are considered) and

A. Di Pierro et al. (Eds.): Festschrift Hankin, LNCS 12065, pp. 149–197, 2020.
https://doi.org/10.1007/978-3-030-41103-9_6

backward (considers dependences backwards from the slicing criterion). Since then, there have been hundreds of papers on program slicing, and several surveys [28,31,35] that illustrate its wide application in many areas including compiler optimizations, debugging and reverse engineering.

Researchers have moved from considering only program slicing to model slicing [3,13,19,23], because a significant amount of software production is done with models, in particular specification and design. Models describe many types of information better than programs but can become large and unmanageable more quickly. Applications of slicing at the model level include model checking, comprehension, testing and reuse.

We consider slicing extended finite state machines. Extended finite state machines (EFSMs) and their variants (e.g. Harel's [15] and UML state machines) are now widely used to model dynamic behaviour of cyber physical systems [10], embedded systems [24] and product lines [11,23]. EFSMs combine a finite state machine with a store, i.e. a set of program variables that can be updated via commands attached to transitions between the atomic states. EFSMs can model a range of computational phenomena, from high-level software to cyber-physical systems. They are essentially interactive and may be nondeterministic and possibly without exit points (or having multiple such). For EFSMs it is more natural to let slices be sets of transitions rather than sets of nodes (atomic states) [20].

One might imagine, given the long history of both slicing and finite state machines, that the correct slicing of finite state machines is a closed question or, if not closed, a relatively uninteresting one. However, neither consideration is true when aiming to produce correct, executable slices, that is to say reduced "programs" for which any nodes not relevant to the restricted semantics have been removed and the "program" has been "rewired" to maintain executability (the slice can be compiled and run). What makes EFSMs particularly interesting from a theoretical perspective is that their interactive, event driven semantics offers a simplified laboratory in which to analyse the correctness of slicing in the presence of interaction with an environment.

Work on program slicing has largely focused on programs with a functional semantics, and notions of correctness assume interaction only at the initial input. But in the context of interaction with a dynamic environment, traditional slicing algorithms produce incorrect slices. To see this, consider the example program illustrated in Fig. 1. If the environment generates the input sequence 1, 2, 3, 4,..., the value of z in program P is 4 (as x = 1, y = 2, z = 3). If P is sliced with respect to <5, {z}>, line 2 will be removed as line 4 is not (data) dependent on it. For the same input sequence the value of z in the slice will be 3 (as x = 1, z = 2), which does not satisfy Weiser's correctness property.

In the context of model slicing, any work has to address interaction with environmental events. A critical line of research leading to this paper has been contributions on how to slice correctly when the input sequence is part of the implicit state and not described by any program variable.

Sivagurunathan et al. [29] address this issue by sketching a program transformation technique to be applied before the application of traditional slicing

Program P		Weiser's Static Slice P' of P w.r.t. (5,{z})	
1	read(x);	1	read(x);
2	read(y);	2	
3	read(z);	3	read(z);
4	z=x+z;	4	z=x+z;
5		5	

Fig. 1. A simple interactive program and its slice.

algorithms in order to make the implicit state explicit by modeling it with additional variables. The limitations in that work are that a suitably general set of transformation rules is not given and the programs to which it can be applied assume a single exit on which the definition of control dependence rests.

Ganapathy et al. [12] defined correctness for slicing reactive programs, where events (in particular generated events) are part of the slicing criterion. However, not all interactive systems generate events, and the drawback of their work is the large size of their slices: they include all states and transitions that reach the event of interest. Labbé et al. [21] presented polynomial algorithms for slicing Input/Output Symbolic Transition Systems (IOSTSs). However, they do not state correctness properties, and consider only a control dependence [26] that is sensitive to non-termination.

Korel et al. [20] considered slicing of EFSMs and proposed new correctness criteria, in particular they made a key observation: in order for the sliced EFSM to have behavior similar to the original EFSM, certain events must be allowed to be "idle" (the term "stuttering" is also used), that is, not trigger transitions. For example, in Fig. 1 we would have 3 kinds of events: `read-x`, `read-y`, and `read-z`; the sliced program would be idle on the event `read-y` and hence an input sequence `read-x(1), read-y(2), read-z(3)` would result in 1 and 3 being read, but 2 ignored. We build on the work of Korel et al. [20] and our approach can be viewed as a formalisation and validation of their correctness notion, but extended to handle even EFSMs with no (or with multiple) exit states.

Like EFSMs, modern programs may have no exit points (or multiple such) so as to program indefinitely non-terminating, reactive systems. This has required the development of new definitions of control dependence for programming languages and models [1,4,26,27]. Definitions of control dependence began to proliferate until Danicic and others extracted order out of chaos by focusing on the desirable *properties* of the resulting dependence relations rather than the control dependence definitions themselves [9]:

- *Weak Commitment Closure* (WCC) means that each node has *at most one* "next observable", where an observable is a node relevant to the slicing criterion.
- *Strong Commitment Closure* (SCC) in addition demands that from a node that has a "next observable", there is no way to infinitely avoid that observable.

EFSMs, being models rather than programs, also allow non-determinism. An early attempt to handle non-determinism by Hatcliff et alia considers a multi-threaded language but does not prove any correctness results [16].

Any correct algorithm for slicing, in formalisms based on finite state models, has to consider all of the above issues: interaction with the environment; non-determinism; and a suitable notion of control dependence. This paper is the first that successfully and generally deals with all these issues to develop algorithms that we can prove produce (the least) slices for EFSMs that satisfy certain correctness properties. In the process, we establish that the commitment closure approach to control dependence of Danicic et alia [9] is flexible enough to provide correctness for interactive programming languages with modern control structures and non-determinism.

The main contributions of this paper are:

1. A formalization (Sect. 3) of new definitions of correctness for slicing an EFSM (whose syntax and semantics are defined in Sect. 2) in the presence of interaction, non-determinism and non-termination. We require (Completeness) that the slice machine simulates the original machine, with some events now perhaps idle. In the other direction (Soundness), we cannot hope for a perfect simulation, but demand that for each "observable" step by the slice machine, either the original machine simulates it, perhaps inserting some events, or the original machine either *(i)* gets stuck, or *(ii)* loops (a possibility excluded by "strong soundness").
2. Syntactic criteria (Sect. 4) that allow us to prove (in Appendix A) the correctness properties from 1. It suffices to demand that the set of transitions in the slice satisfies two conditions: it must be closed under the well-known notion of data dependence, and it must have the abovementioned WCC property, first highlighted by Danicic et al. but in this work adapted (in a non-trivial way) to EFSMs; to get strong soundness, we need the SCC property (likewise adapted to EFSMs). Observe that, unlike the classical notions of control dependence [6,17,25], we can handle control flow graphs without an exit point (or with multiple such).
3. Two novel polynomial-time algorithms for computing least slices for EFSMs (Sect. 5); one does it wrt. WCC and another wrt. SCC. (We discovered them in 2013 as part of a very early version[1] of this work, whereas Danicic et al. presented in their groundbreaking work [9] only a very abstract algorithm—though in subsequent work they have presented algorithms that, while for different settings, are somewhat similar to our algorithms.)
4. A demonstration of the practical usefulness of the two algorithms over a suite of thirteen EFSMs, taken from textbook benchmark examples and real world production EFSM models, and experimental comparisons with existing control dependence definitions (Sects. 6 and 7).

[1] Published as Research Note RN/13/22 from University College London.

2 Extended Finite State Machines (EFSMs)

We shall now formally describe the kind of Extended Finite State Machines (EFSMs) we shall consider in this paper. An EFSM M is a finite state automaton where transitions have guards, may manipulate a common store, and may be triggered by external events. In Sect. 2.1 we shall formally define their syntax and in Sect. 2.3 their semantics, which is designed so as to also allow reasoning about a slice as defined in Sect. 2.2. In Sect. 3 we shall discuss what it means for a slice to be semantically correct, and propose suitable definitions of correctness.

2.1 Syntax of EFSMs

An EFSM is a tuple $(\hat{S}, \hat{T}, \hat{E}, \hat{V})$ where $\hat{S}$ is a finite set of states (one state may be designated as the initial), $\hat{T}$ is a finite set of transitions, $\hat{E}$ is a finite set of events (some of which may take a parameter), and $\hat{V}$ is a finite set of variables; variables and parameters are assumed to be eventually bound to scalar values such as integers or strings. We shall employ naming conventions and in particular use n or m to range over states, and use t or u to range over transitions.

A transition t (or u) $\in \hat{T}$ has a source state $\mathsf{S}(t) \in \hat{S}$, a target state $\mathsf{T}(t) \in \hat{S}$, and a label of the form $\tilde{e}[g]/a$, each component of which we shall now describe:

- $\tilde{e} = \mathsf{E}(t)$ is either ε (the transition is "spontaneous"), of the form e where e is an event in $\hat{E}$ that does not take a parameter, or of the form $e(v_b)$ where e is an event in $\hat{E}$ that expects a parameter which will be bound to v_b (not occurring in $\hat{V}$).
- $g = \mathsf{G}(t)$ is a guard (which must be true in order for t to be chosen), and a is an action which we may assume is either of the form $v := A$ with $v \in \hat{V}$ or `skip`; here g and A may refer to the variables in $\hat{V}$ and also to v_b if it exists.

We define $\mathsf{D}(t)$, the variables defined by t, as $\{v\}$ if a is $v := A$ and $\emptyset$ if a is `skip`; we define $\mathsf{A}(t)$, the (arithmetic) expression mentioned in t, as A if a is $v := A$ and 0 (arbitrarily) if a is `skip`. We define $\mathsf{U}(t)$, the variables used by t, as (with fv/bv denoting free/bound variables) $(fv(\mathsf{G}(t)) \cup fv(\mathsf{A}(t))) \setminus bv(\tilde{e})$ where $bv(e(v_b)) = \{v_b\}$ and $bv(e) = bv(\varepsilon) = \emptyset$. When depicting an EFSM, all parts of a label are optional: one may omit $\tilde{e}$ if it is ε, omit g if it is `true`, and omit a if it is `skip`. For examples of labels on transitions, see Fig. 3. A transition where all parts of its label can be omitted is called an ε-transition. Note that we do not allow for transitions that *produce* events.

Figure 2 depicts a non-trivial EFSM (with the initial state `S1`) that shall serve as our main running example. Since a main challenge of this paper is to provide a better understanding of control aspects, while employing a standard notion of data dependence (which is already well understood and rather non-controversial), we ignore the labels of the transitions.

We say that t is *self-looping* if $\mathsf{S}(t) = \mathsf{T}(t)$, and that u is a *successor* of t if $\mathsf{S}(u) = \mathsf{T}(t)$. In Fig. 2, `t11` is self-looping, and `t5` is a successor of `t1`.

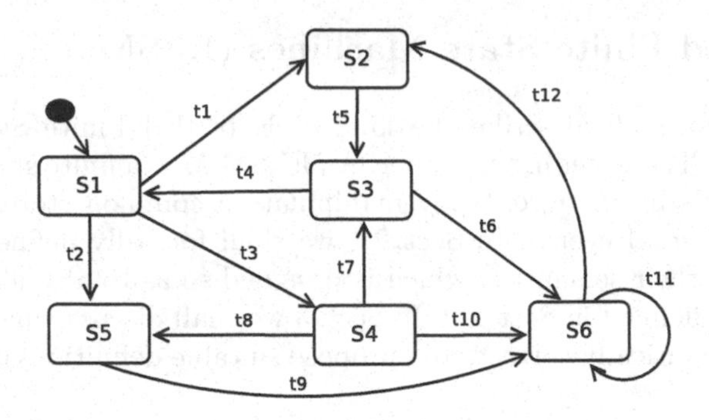

Fig. 2. Our running example EFSM.

We say that n is an *exit* state if no other node can be reached from n, that is, there is no transition t with $\mathsf{S}(t) = n$ and $\mathsf{T}(t) \neq n$. In Fig. 2, there is no exit state. An exit state n may, or may not, be the source and target of a self-looping transition.

We say that $[t_1..t_k]$ $(k \geq 0)$ is a path (of length k) if for all $j \in 1 \ldots k-1$, t_{j+1} is a successor of t_j. If $k \geq 1$, we say that $[t_1..t_k]$ is a path from $\mathsf{S}(t_1)$ to $\mathsf{T}(t_k)$ (if $k = 0$, then $[t_1..t_k]$ is a path from n to n for all n). We say that n occurs in $[t_1..t_k]$ if there exists $j \in 1 \ldots k$ such that $n = \mathsf{S}(t_j)$ or $n = \mathsf{T}(t_j)$. With L as set of transitions, we say that a path $[t_1..t_k]$ is outside L iff $t_j \notin L$ for all $j \in 1 \ldots k$. In Fig. 2, [`t5, t4, t3`] is a path from `S2` to `S4` where the states `S2`, `S3`, `S1` and `S4` occur but which is outside say {`t1,t6`}. We say that $[t_1..t_k..]$ is an infinite path from $\mathsf{S}(t_1)$ if t_{j+1} is a successor of t_j for all $j \geq 1$; we say that the path avoids n if $n \neq \mathsf{S}(t_j)$ for all $j \geq 1$. In Fig. 2, the path [`t4, t3, t7, t4, t3, t7, ...`] is an infinite path from `S3` that avoids all of `S2`, `S5`, and `S6`.

2.2 Slicing EFSMs

In general, a slice is a subpart of the original program. In our setting, this raises the question: should we consider a slice to be a subset of the *states* (together with the transitions between these states), or to be a subset of the *transitions* (together with the states involved in those transitions). We shall (as Korel et al. [20]) choose the latter option, as the former is a special case (where a slice contains a transition iff it contains its source and also its target), thus:

Definition 1 (Slice Set). *A slice set for an EFSM* $(\hat{S}, \hat{T}, \hat{E}, \hat{V})$ *is a subset of* $\hat{T}$.

We must demand that a slice set contains the given *slicing criterion*, that is, those transitions that are required to be preserved by the slice. But in order for also the context of such transitions to be "invariant" under the slice, and thus ensure certain kinds of semantic correctness (as defined in Sect. 3), we must in addition demand that the slice set includes the transitions that may "impact" the slicing criterion; this can be formalized as a demand that the slice set satisfies certain conditions (given in Sect. 4). We shall even aim for the slice set to be the least such set (in Sect. 5 we shall show how to achieve that).

Our definitions will often be implicitly parameterized with respect to a given EFSM, and with respect to a fixed slice set which is often called $\mathcal{L}$ while its members are called "observables". As will be formalized in Sect. 2.3, slicing amounts to keeping the transitions in $\mathcal{L}$ as they are, whereas transitions not in $\mathcal{L}$ are replaced by ε-transitions (all parts of the label are removed).

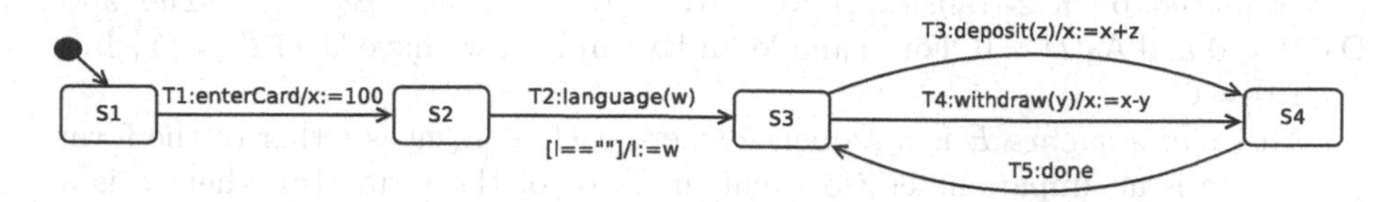

Fig. 3. A simplified EFSM for ATM operations.

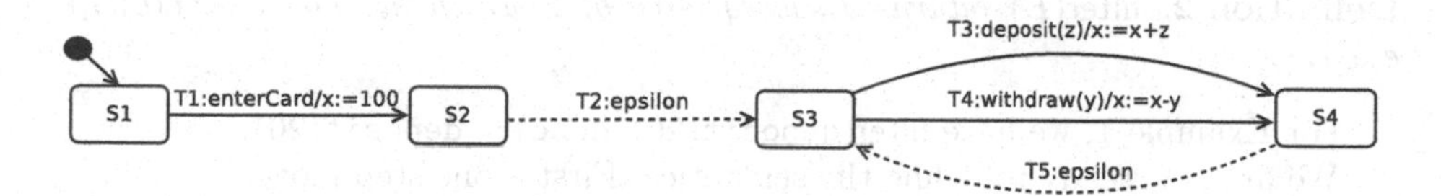

Fig. 4. The result of slicing the EFSM from Fig. 3 with respect to T1 and T3 and T4.

Example 1. *Consider the EFSM in Fig. 3 which models a simple ATM system, and assume that the slicing criterion consists of* T4. *Then the slice set $\mathcal{L}$ must also contain* T1 *and* T3 *(as these transitions provide definitions of the* x *which is used by* T4*). The result of slicing wrt. $\mathcal{L}$ is the EFSM depicted in Fig. 4.*

The presence of ε-transitions may enable the further optimization of slices [2]. For example, in Fig. 4, since S3 is reachable by an ε-transition from S2 and from S4, these three states may be merged. We shall not further address this optimization.

We use $\hat{E}_{\mathcal{L}}$ to denote the events relevant for the sliced EFSM:

$$\hat{E}_{\mathcal{L}} = \{e \in \hat{E} \mid \exists t \in \mathcal{L} : \mathsf{E}(t) = e \text{ or } \mathsf{E}(t) = e(v_b) \text{ for some } v_b\}.$$

For Example 1, we have $\hat{E}_{\mathcal{L}} = \{\texttt{enterCard}, \texttt{deposit}, \texttt{withdraw}\}$.

2.3 Semantics of EFSMs

We shall present a semantics that facilitates reasoning about slicing. Our development is relative to a given EFSM $(\hat{S}, \hat{T}, \hat{E}, \hat{V})$ and a fixed slice set $\mathcal{L}$. We shall use Example 1, where $\mathcal{L}$ contains T1 and T3 and T4, to illustrate the definitions.

A *configuration* C is a pair (n, s) where $n \in \hat{S}$ is a state and s is a store which maps variables $v \in \hat{V}$ to values. The domain of values is unspecified but we assume an expression language with $[\![A]\!]s$ denoting the value resulting from evaluating expression A in store s; similarly we assume a guard language with $[\![B]\!]s \in \{true, false\}$ denoting the value of the boolean expression B wrt. store s.

In our definitions, we shall use the subscript 1 to refer to the original EFSM, and subscript 2 to refer to the sliced EFSM. Thus $\mathsf{E}_1(t) = \mathsf{E}(t)$ and $\mathsf{G}_1(t) = \mathsf{G}(t)$, *etc.* If $t \in \mathcal{L}$ then $\mathsf{E}_2(t) = \mathsf{E}(t)$ and $\mathsf{G}_2(t) = \mathsf{G}(t)$, *etc.*, but otherwise (as then t is replaced by a ε-transition) we have $\mathsf{E}_2(t) = \varepsilon$ and $\mathsf{G}_2(t) = \texttt{true}$ and $\mathsf{D}_2(t) = \emptyset$ and $\mathsf{A}_2(t) = 0$. For example, in Example 1, we have $\mathsf{D}_1(\texttt{T2}) = \{\texttt{l}\}$ but $\mathsf{D}_2(\texttt{T2}) = \emptyset$.

An *event sequence* E is a sequence where each element is either of the form e where e is an unparameterized event in $\hat{E}$, or of the form $e(c)$ where e is a parameterized event in $\hat{E}$ and c is a value; the empty event sequence is denoted ε. To disregard events that are "irrelevant" for the sliced program, we use the (idempotent) function filter:

Definition 2. filter(E) *returns a subsequence of E which includes e or $e(c)$ iff $e \in \hat{E}_{\mathcal{L}}$.*

For Example 1, we have filter($\texttt{deposit}(20), \texttt{done}) = \texttt{deposit}(20)$.

We are now ready to define the semantics. First a one-step move:

Definition 3. *We write*

$$i \vdash t : (n, s) \xrightarrow{E} (n', s')$$

to denote that (n, s) in the i-semantics ($i = 1, 2$) through transition t moves to (n', s') while consuming the event sequence E (which will be either empty or a singleton). This happens when t is such that all the conditions listed below hold:

- $\mathsf{S}(t) = n$ *and* $\mathsf{T}(t) = n'$
- *either* $\mathsf{E}_i(t) = \varepsilon = E$ *or there exists* $e \in \hat{E}$ *such that either* $\mathsf{E}_i(t) = e = E$ *or (for some* c *and* v_b*)* $\mathsf{E}_i(t) = e(v_b)$ *and* $E = e(c)$
- $[\![g]\!]s = true$
- *if* $\mathsf{D}_i(t) = \emptyset$ *then* $s' = s$ *but if* $\mathsf{D}_i(t)$ *is a singleton* $\{v\}$ *then* $s' = s[v \mapsto [\![A]\!]s]$

where g and A are defined as follows: if $E = e(c)$ and $\mathsf{E}_i(t) = e(v_b)$ then $g = \mathsf{G}_i(t)[c/v_b]$ and $A = \mathsf{A}_i(t)[c/v_b]$ but otherwise $g = \mathsf{G}_i(t)$ and $A = \mathsf{A}_i(t)$.

In Example 1, when $E = \texttt{deposit}(20)$ we have (for $i = 1, 2$ and for all stores s):

$$i \vdash \texttt{T3} : (\texttt{S3}, s) \xrightarrow{E} (\texttt{S4}, s[\texttt{x} \mapsto s(\texttt{x}) + 20]).$$

A configuration is *stuck* (as is $(\texttt{S2}, s)$ in Example 1 if $s(\texttt{l})$ is non-empty) if no transitions apply:

Definition 4. *We say that (n, s) is i-stuck if for all t with $\mathsf{S}(t) = n$: $[\![\mathsf{G}_i(t)]\!]s = false$, or if $\mathsf{E}_i(t)$ is of the form $e(v_b)$: $[\![\mathsf{G}_i(t)[c/v_b]]\!]s = false$ for all c.*

As discussed in Sect. 1, we shall allow "idle" events: an EFSM will ignore an event for which there is no transition in the system, rather than block on it. This is formalized in the rule for multi-step moves:

Definition 5. *We write* $i \vdash \pi : C \xrightarrow{E} C'$ *iff with* $\pi = [t_1..t_{k-1}]$ $(k \geq 1)$ *there exists* $C_1, \ldots, C_k$ *with* $C = C_1$ *and* $C' = C_k$, *and* $E_1, \ldots, E_{k-1}$ *where* $E_1 \ldots E_{k-1}$ *is a subsequence of* E, *such that* $i \vdash t_j : C_j \xrightarrow{E_j} C_{j+1}$ *for all* $j \in 1 \ldots k-1$.

If the moves are all non-idle, i.e., $E = E_1 \ldots E_{k-1}$, *we write* $i \vdash_{ni} \pi : C \xrightarrow{E} C'$.

Example 2. *In Example 1, with* $E_1 = \texttt{deposit(30)}, \texttt{done}, \texttt{withdraw(10)}$ *and with*
$E_2 = \texttt{deposit(30)}, \texttt{withdraw(10)}$, *for all stores* s *we have*

$$1 \vdash_{ni} [\texttt{T3}, \texttt{T5}, \texttt{T4}] : (\texttt{S3}, s) \xrightarrow{E_1} (\texttt{S4}, s[\texttt{x} \mapsto s(\texttt{x}) + 20])$$

$$2 \vdash_{ni} [\texttt{T3}, \texttt{T5}, \texttt{T4}] : (\texttt{S3}, s) \xrightarrow{E_2} (\texttt{S4}, s[\texttt{x} \mapsto s(\texttt{x}) + 20])$$

but also, since E_2 *is a subsequence of* E_1:

$$2 \vdash [\texttt{T3}, \texttt{T5}, \texttt{T4}] : (\texttt{S3}, s) \xrightarrow{E_1} (\texttt{S4}, s[\texttt{x} \mapsto s(\texttt{x}) + 20]).$$

Since the path taken to get from one configuration to another is invisible to the environment, but the event sequence is visible, we often only depict the latter:

Definition 6. *We write* $i \vdash C \xrightarrow{E} C'$ *iff* $i \vdash \pi : C \xrightarrow{E} C'$ *for some path* π.
We write $i \vdash_{ni} C \xrightarrow{E} C'$ *iff* $i \vdash_{ni} \pi : C \xrightarrow{E} C'$ *for some path* π.

We are often interested in moves that are unobservable (wrt. a given $\mathcal{L}$):

Definition 7. *We write* $i \vdash C \xRightarrow{E} C'$ *iff* $i \vdash \pi : C \xrightarrow{E} C'$ *for some* π *outside* $\mathcal{L}$.
We write $i \vdash_{ni} C \xRightarrow{E} C'$ *iff* $i \vdash_{ni} \pi : C \xrightarrow{E} C'$ *for some* π *outside* $\mathcal{L}$.

Even if $1 \vdash_{ni} (n, s) \xRightarrow{E} (n', s')$ it may happen that $s' \neq s$ and $E \neq \varepsilon$. But we have

Fact 8. *If* $2 \vdash t : (n, s) \xrightarrow{E} (n', s')$ *with* $t \notin \mathcal{L}$ *then* $s' = s$ *and* $E = \varepsilon$.
If $2 \vdash (n, s) \xRightarrow{E} (n', s')$ *then* $s' = s$, *and if* $2 \vdash_{ni} (n, s) \xRightarrow{E} (n', s')$ *also* $E = \varepsilon$.

Of special interest is moves where all but the last step are unobservable:

Definition 9. *We write* $i \vdash^t_{ni} C \Rightarrow\!\!\xrightarrow{E} C'$ *if* $t \in \mathcal{L}$ *and there exists* C_0, *and* E_0, E' *with* $E = E_0 E'$, *such that* $i \vdash_{ni} C \xRightarrow{E_0} C_0$ *and* $i \vdash t : C_0 \xrightarrow{E'} C'$.

Example 3. *In Example 1 we have (for all s with* $s(\mathtt{l})$ *the empty string)*

$$1 \vdash^{\mathtt{T3}}_{ni} (\mathtt{S2}, s) \overset{E_1}{\Longrightarrow} (\mathtt{S4}, s[\mathtt{x} \mapsto s(\mathtt{x}) + 20, l \mapsto \mathtt{English}])$$

$$2 \vdash^{\mathtt{T3}}_{ni} (\mathtt{S2}, s) \overset{E_2}{\Longrightarrow} (\mathtt{S4}, s[\mathtt{x} \mapsto s(\mathtt{x}) + 20])$$

where $E_1 = \mathtt{language(English), deposit(20)}$ *and* $E_2 = \mathtt{deposit(20)}$.

It will often be the case that a slice set is *uniform* in that it contains either all the transitions that mention a given event, or none of them:

Definition 10 (Uniformity). *We say that a slice set* $\mathcal{L}$ *is* uniform *iff the following property holds for all* $e \in \hat{E}$*: if* $t_1, t_2 \in \hat{T}$ *are such that* $\mathsf{E}(t_1)$ *equals* e *or is of the form* $e(v_b)$*, and* $\mathsf{E}(t_2)$ *equals* e *or is of the form* $e(v_b)$*, then either* $t_1, t_2 \in \mathcal{L}$ *or* $t_1, t_2 \notin \mathcal{L}$*.*

We shall see in the next section that a uniform slice set allows for stronger correctness properties, since we have:

Fact 11. *Assume that* $\mathcal{L}$ *is uniform. If* $i \vdash_{ni} C \overset{E}{\Longrightarrow} C'$ *then* $\mathsf{filter}(E) = \varepsilon$*.*

3 Slicing an EFSM: What Does Correctness Mean?

In this section we shall develop definitions of what it means for slicing, with respect to a given slice set $\mathcal{L}$, to be semantically correct. We would like to capture two (yet rather informal) notions:

Completeness moves by the original EFSM can also be done by the sliced EFSM.
Soundness moves by the sliced EFSM can also be done by the original EFSM.

This suggests (assuming non-idle moves) that we aim for the following goals:

Completeness Attempt 1. *If* $1 \vdash_{ni} C \overset{E}{\longrightarrow} C'$ *then* $2 \vdash_{ni} C \overset{E}{\longrightarrow} C'$*.*

Soundness Attempt 1. *If* $2 \vdash_{ni} C \overset{E}{\longrightarrow} C'$ *then* $1 \vdash_{ni} C \overset{E}{\longrightarrow} C'$*.*

But these goals are too ambitious, for several reasons which will be detailed in the next subsections where we shall rewrite the goals so as to motivate the final versions, given in Sect. 3.3 (completeness) and Sect. 3.4 (soundness). We believe that the resulting goals capture what it should mean for slicing of EFSMs to be correct.

3.1 Event Sequences

Looking at Completeness Attempt 1 and Soundness Attempt 1, one may wonder that the *same* event sequence is used in the antecedent as in the consequent. And in fact, for soundness this demand is too strong, as we shall now show using Example 1. With $E_2 = \texttt{deposit(30)}, \texttt{withdraw(10)}$, we have (cf. Example 2)

$$2 \vdash_{\mathrm{ni}} [\texttt{T3}, \texttt{T5}, \texttt{T4}] : (\texttt{S3}, s) \xrightarrow{E_2} (\texttt{S4}, s[\texttt{x} \mapsto s(\texttt{x}) + 20]).$$

Yet $1 \vdash_{\mathrm{ni}} (\texttt{S3}, s) \xrightarrow{E_2} (\texttt{S4}, s')$ does *not* hold for any s'; in order for the original EFSM to execute E_2 we need to *pad* a $\texttt{done}$ event: with $E_1 = \texttt{deposit(30)}, \texttt{done}, \texttt{withdraw(10)}$, we have $\mathsf{filter}(E_1) = E_2$ and (cf. Example 2)

$$1 \vdash_{\mathrm{ni}} [\texttt{T3}, \texttt{T5}, \texttt{T4}] : (\texttt{S3}, s) \xrightarrow{E_1} (\texttt{S4}, s[\texttt{x} \mapsto s(\texttt{x}) + 20]).$$

This motivates the following modified goal:

Soundness Attempt 2. *If* $2 \vdash_{ni} C \xrightarrow{E_2} C'$ *then* E_2 *is a subsequence of an* E_1 *with* $1 \vdash_{ni} C \xrightarrow{E_1} C'$.
Moreover, if $\mathcal{L}$ *is uniform then* $\mathsf{filter}(E_1) = E_2$.

We shall now address completeness, and again look at Example 1 where with $E_1 = \texttt{deposit(30)}, \texttt{done}, \texttt{withdraw(10)}$ we have (cf. Example 2)

$$1 \vdash_{\mathrm{ni}} [\texttt{T3}, \texttt{T5}, \texttt{T4}] : (\texttt{S3}, s) \xrightarrow{E_1} (\texttt{S4}, s[\texttt{x} \mapsto s(\texttt{x}) + 20])$$

but we clearly do not have

$$2 \vdash_{\mathrm{ni}} [\texttt{T3}, \texttt{T5}, \texttt{T4}] : (\texttt{S3}, s) \xrightarrow{E_1} (\texttt{S4}, s[\texttt{x} \mapsto s(\texttt{x}) + 20])$$

since $\texttt{done} \notin \hat{E}_{\mathcal{L}}$. On the other hand, $\texttt{done}$ can be considered idle so (by Definition 5) we do have (cf. Example 2)

$$2 \vdash [\texttt{T3}, \texttt{T5}, \texttt{T4}] : (\texttt{S3}, s) \xrightarrow{E_1} (\texttt{S4}, s[\texttt{x} \mapsto s(\texttt{x}) + 20])$$

and (with $\mathsf{filter}(E_1) = \texttt{deposit(30)}, \texttt{withdraw(10)}$)

$$2 \vdash_{\mathrm{ni}} [\texttt{T3}, \texttt{T5}, \texttt{T4}] : (\texttt{S3}, s) \xrightarrow{\mathsf{filter}(E_1)} (\texttt{S4}, s[\texttt{x} \mapsto s(\texttt{x}) + 20]).$$

This motivates the following modified goal:

Completeness Attempt 2. *If* $1 \vdash_{ni} C \xrightarrow{E} C'$ *then* $2 \vdash C \xrightarrow{E} C'$.
Moreover, if $\mathcal{L}$ *is uniform then* $2 \vdash_{ni} C \xrightarrow{\mathsf{filter}(E)} C'$.

We shall now explain why we need to demand uniformity in the last parts of Completeness Attempt 2 and Soundness Attempt 2. For that purpose, assume for a moment that Example 1 has been redesigned so as to replace the event $\texttt{done}$

by the event $\texttt{deposit(y)}$ where $\texttt{y}$ is ignored; thus the slice set $\mathcal{L} = \{\texttt{T1}, \texttt{T3}, \texttt{T4}\}$ is not uniform since $\texttt{deposit}$ occurs both in $\texttt{T3}$ and $\texttt{T5}$.

First observe that with $E = \texttt{deposit}(10), \texttt{deposit}(20)$ we have

$$1 \vdash_{\mathrm{ni}} [\texttt{T5}, \texttt{T3}] : (\texttt{S4}, s) \xrightarrow{E} (\texttt{S4}, s[\texttt{x} \mapsto s(\texttt{x}) + 20])$$

and certainly we also have

$$2 \vdash [\texttt{T5}, \texttt{T3}] : (\texttt{S4}, s) \xrightarrow{E} (\texttt{S4}, s[\texttt{x} \mapsto s(\texttt{x}) + 20])$$

(but we also have say $2 \vdash_{\mathrm{ni}} [\texttt{T5}, \texttt{T3}, \texttt{T5}, \texttt{T3}] : (\texttt{S4}, s) \xrightarrow{E} (\texttt{S4}, s[\texttt{x} \mapsto s(\texttt{x}) + 30])$).

Without the requirement of $\mathcal{L}$ being uniform, Completeness Attempt 2 would require (as $\mathsf{filter}(E) = E$) that

$$2 \vdash_{\mathrm{ni}} (\texttt{S4}, s) \xrightarrow{E} (\texttt{S4}, s[\texttt{x} \mapsto s(\texttt{x}) + 20])$$

but this is clearly *not* possible as a non-idle execution of the sliced program on E will add 30 to $\texttt{x}$.

Next observe that we have

$$2 \vdash_{\mathrm{ni}} [\texttt{T5}] : (\texttt{S4}, s) \xrightarrow{\varepsilon} (\texttt{S3}, s)$$

and certainly we also have

$$1 \vdash_{\mathrm{ni}} [\texttt{T5}] : (\texttt{S4}, s) \xrightarrow{\texttt{deposit}(20)} (\texttt{S3}, s).$$

Without the requirement of $\mathcal{L}$ being uniform, Soundness Attempt 2 would require $\mathsf{filter}(\texttt{deposit}(20)) = \varepsilon$ which is clearly *not* the case.

3.2 Relevant Variables

Our tentative correctness demands are still too restrictive, as they require the sliced EFSM to produce *exactly* the same store as the original. But since slicing removes transitions that are not "relevant", the sliced EFSM is likely (as in Example 3) to disagree with the original EFSM on variables that are not *relevant*:

Definition 12. *For a transition t or a state n, we define its relevant variables as follows:*

1. *We say that v is relevant for t wrt. $\mathcal{L}$, written $v \in Rv_{\mathcal{L}}(t)$, iff there exists $t' \in \mathcal{L}$ such that $v \in \mathsf{U}(t')$, and there exists a path $[t_1..t_k]$ with $k \geq 1$ and $t = t_1$ and $t' = t_k$ such that for all $j \in 1 \ldots k-1$, $v \notin \mathsf{D}(t_j)$.*
2. *We say that v is relevant for a state n wrt. $\mathcal{L}$, written $v \in Rv_{\mathcal{L}}(n)$, iff there exists t with $\mathsf{S}(t) = n$ such that $v \in Rv_{\mathcal{L}}(t)$.*

We write $Rv(t)$ and $Rv(n)$), rather than $Rv_{\mathcal{L}}(t)$ and $Rv_{\mathcal{L}}(n)$, when $\mathcal{L}$ is clear from context.

In Example 1 (with $\mathcal{L} = \{\texttt{T1, T3, T4}\}$), `x` is relevant for `T3` and `T4` since `x` is used there, and also for `T2` and `T5`, but `x` is not relevant for `T1` as `x` is defined there but not used. Thus `x` is relevant for all states except `S1`.

Fact 13. *If $t \in \mathcal{L}$ then $\mathsf{U}(t) \subseteq Rv_{\mathcal{L}}(t)$.*

We have motivated the following attempt to phrase completeness:

Completeness Attempt 3. *If $1 \vdash_{ni} C \xrightarrow{E} C_1'$ then $2 \vdash C \xrightarrow{E} C_2'$ for some C_2' with $C_1' \; Q \; C_2'$.*

Moreover, if $\mathcal{L}$ is uniform then $2 \vdash_{ni} C \overset{\mathsf{filter}(E)}{\longrightarrow} C_2'$.

Here we use Q to relate configurations that are in the same state and whose stores agree on the relevant variables of that state:

Definition 14. $(n_1, s_1) \; Q \; (n_2, s_2)$ *iff* $n_1 = n_2$ *and* $s_1(v) = s_2(v)$ *for all* $v \in Rv(n_1)$.

In order to make that result compose over a sequence of moves, we need to relax the assumptions and allow also the input stores to disagree on irrelevant variables:

Completeness Attempt 4. *If $1 \vdash_{ni} C_1 \xrightarrow{E} C_1'$ and $C_1 \; Q \; C_2$ then $2 \vdash C_2 \xrightarrow{E} C_2'$ for some C_2' with $C_1' \; Q \; C_2'$.*

Moreover, if $\mathcal{L}$ is uniform then $2 \vdash_{ni} C_2 \overset{\mathsf{filter}(E)}{\longrightarrow} C_2'$.

Similarly, we can take another stab at soundness:

Soundness Attempt 3. *If $2 \vdash_{ni} C_2 \xrightarrow{E_2} C_2'$ and $C_1 \; Q \; C_2$ then E_2 is a subsequence of some E_1 such that $1 \vdash_{ni} C_1 \xrightarrow{E_1} C_1'$ for some C_1' with $C_1' \; Q \; C_2'$.*

Moreover, if $\mathcal{L}$ is uniform then $\mathsf{filter}(E_1) = E_2$.

3.3 Completeness

We are now ready to give our final version of completeness, where we can actually assume that the sliced EFSM goes through the same path of transitions as the original:

Desideratum 1 (Completeness). *If $1 \vdash_{ni} \pi : \; C_1 \xrightarrow{E} C_1'$ and $C_1 \; Q \; C_2$ then $2 \vdash \pi : \; C_2 \xrightarrow{E} C_2'$ for some C_2' with $C_1' \; Q \; C_2'$.*

Moreover, if $\mathcal{L}$ is uniform then $2 \vdash_{ni} \pi : \; C_2 \overset{\mathsf{filter}(E)}{\longrightarrow} C_2'$.

3.4 Soundness

Before giving our final version of soundness, we must deal with one crucial issue: if the sliced EFSM does an ε-transition, with $2 \vdash t_0 : (n, s) \xrightarrow{\varepsilon} (n_1, s)$ where $t_0 \notin \mathcal{L}$, there may not exist (n_2, s_2) (and E) such that $1 \vdash t_0 : (n, s) \xrightarrow{E} (n_2, s_2)$, since $[\![\mathsf{G}(t_0)]\!]s$ may be false.

In that case, the original EFSM may still eventually "catch up" with the sliced EFSM, by doing some unobservable loop that eventually results in a store s' such that $[\![\mathsf{G}(t_0)]\!]s'$ is true. Restricting our attention (unlike previous attempts) to moves with *exactly one* observable step, this suggests:

Soundness Attempt 4. *If* $2 \vdash^{t}_{ni} C_2 \overset{E_2}{\Rightarrow\!\!\rightarrow} C_2'$ *and* $C_1\ Q\ C_2$ *then* $1 \vdash^{t}_{ni} C_1 \overset{E_1 E_2}{\Rightarrow\!\!\rightarrow} C_1'$ *for some* C_1', E_1 *with* $C_1'\ Q\ C_2'$*. Moreover, if* $\mathcal{L}$ *is uniform then* $\mathsf{filter}(E_1) = \varepsilon$.

But it may also happen that the original EFSM gets *stuck* or *loops* from (n, s), as we shall now illustrate by Example 1 where for all s and z we have

$$2 \vdash^{\mathtt{T3}}_{\mathrm{ni}} (\mathtt{S2}, s) \overset{\mathtt{deposit(z)}}{\Rightarrow\!\!\rightarrow} (\mathtt{S4}, s') \text{ with } s' = s[\mathtt{x} \mapsto \mathtt{x} + z].$$

If $s(\mathtt{l})$ is the empty string then for all w we do indeed have for some s' with $s'(\mathtt{x}) = s(\mathtt{x}) + z$:

$$1 \vdash^{\mathtt{T3}}_{\mathrm{ni}} (\mathtt{S2}, s) \overset{\mathtt{language(w),deposit(z)}}{\Rightarrow\!\!\rightarrow} (\mathtt{S4}, s').$$

But if $s(\mathtt{l}) \neq$ "" then the original EFSM gets stuck at (S2, s), and by extension from (S1, s), before reaching even the source of T3.

Definition 15. *[Stuck from] We say that the original EFSM gets stuck from* (n, s)*, avoiding the slice set* $\mathcal{L}$*, iff there exists* (n', s') *which is 1-stuck (cf. Definition 4) such that for some* π *and* E *we have* $1 \vdash \pi : (n, s) \xrightarrow{E} (n', s')$ *where neither* n'*, nor the source of a transition in* π*, is also the source of a transition in* $\mathcal{L}$ *(thus* π *is outside* $\mathcal{L}$*).*

Now imagine that at S2 we add a self-looping transition with guard $\mathtt{l}$ = "English", and look at what happens to a configuration $C_2 = (\mathtt{S2}, s)$ with $s(\mathtt{l})$ = "English". If the action of the added transition is $\mathtt{l}$:= "", then the original EFSM will indeed eventually catch up on the sliced EFSM on C_2, but if the action of the added transition is skip, then the original EFSM will do an infinite loop and never reach even the source of T3—in which case we say that the original EFSM loops from C_2, avoiding {T3}, as formalized by:

Definition 16. *[Loop from] We say that the original EFSM loops from* (n, s)*, avoiding the slice set* $\mathcal{L}$*, if for all* $j \geq 1$ *there exists* t_j *whose source is not the source of a transition in* $\mathcal{L}$ *such that for all* $k \geq 1$*, for some* E_k, C_k*:* $1 \vdash [t_1..t_k] : C \xrightarrow{E_k} C_k$.

In Fig. 3, the original EFSM loops from (S3,s) avoiding T1 and T2, but not avoiding T3 and not avoiding T5.

We have motivated a final version of soundness, called *weak soundness*:

Desideratum 2 (Weak Soundness). *If* $2 \vdash_{ni}^{t} C_2 \overset{E_2}{\Rightarrow\rightarrow} C_2'$ *and* $C_1 \; Q \; C_2$ *then either*

1. $1 \vdash_{ni}^{t} C_1 \overset{E_1 E_2}{\Rightarrow\rightarrow} C_1'$ *for some* C_1', E_1 *with* $C_1' \; Q \; C_2'$ *(and if* $\mathcal{L}$ *is uniform then* $\mathsf{filter}(E_1) = \varepsilon$*), or*
2. *the original EFSM gets stuck from* C_1 *avoiding* $\mathcal{L}$*, or*
3. *the original EFSM loops from* C_1 *avoiding* $\mathcal{L}$*.*

While this requirement allows for several kinds of behavior (hence the term "weak"), keep in mind what it rules out:

- that all moves from C_1 will eventually result in an observable step, but
- each such move has a transition other than t as its first observable step.

In some situations, it may not be desirable to slice away loops, for example if we want to ensure that certain temporal properties are preserved. If this is the case, we shall go for *strong* soundness:

Desideratum 3 (Strong Soundness). *If* $2 \vdash_{ni}^{t} C_2 \overset{E_2}{\Rightarrow\rightarrow} C_2'$ *and* $C_1 \; Q \; C_2$ *then either*

1. $1 \vdash_{ni}^{t} C_1 \overset{E_1 E_2}{\Rightarrow\rightarrow} C_1'$ *for some* C_1', E_1 *with* $C_1' \; Q \; C_2'$ *(and if* $\mathcal{L}$ *is uniform then* $\mathsf{filter}(E_1) = \varepsilon$*), or*
2. *the original EFSM gets stuck from* C_1 *avoiding* $\mathcal{L}$*.*

It may seem a natural final step to propose a "very strong soundness" requirement that would not allow the original program to get stuck. To ensure that, it appears we would need the slice set to contain all transitions with guards that are not always true. While this may be of interest in some applications, it may be considered orthogonal to our development which shows how Weak/Strong Soundness follows from general principles (first discovered by [9]).

Even though the requirements for Weak Soundness and Strong Soundness mention only moves with exactly one observable steps, they can in the natural way be extended to requirements about sequences of such moves (as the postcondition coincides with the precondition with both using Q).

4 Slicing an EFSM: How to Get Correctness?

We shall develop conditions on the slice set $\mathcal{L}$ that ensure completeness and/or strong/weak soundness. In Sect. 4.1 we introduce the standard notion of data dependence which suffices for completeness. For soundness, in addition to data dependence we need two crucial conditions: in Sect. 4.3 we introduce the condition "weak commitment closed" ($\mathcal{WCC}$) which suffices for weak soundness, and in Sect. 4.4 we introduce the condition "strong commitment closed" ($\mathcal{SCC}$) which ensures strong soundness; both are expressed using the notion of "next observable" introduced in Sect. 4.2.

4.1 Data Dependence

Our definition is standard except that it relates transitions rather than states:

Definition 17 (Data Dependence). *We say that t' is data dependent on t, written $t \rightarrow_{\mathrm{dd}} t'$, iff there exists a variable $v \in \mathsf{D}(t) \cap \mathsf{U}(t')$ and a path $[t_1..t_k]$ ($k \geq 0$) from $\mathsf{T}(t)$ to $\mathsf{S}(t')$ such that for all $j \in 1 \ldots k$, $v \notin \mathsf{D}(t_j)$.*

For example, in Fig. 3, `T4` is data dependent on `T1` and `T3`, and also on itself which is uninteresting for the purpose of computing a slice set that is closed under $\rightarrow_{\mathrm{dd}}$:

Definition 18. *Say $\mathcal{L}$ is closed under $\rightarrow_{\mathrm{dd}}$ iff $t \in \mathcal{L}$ whenever $t \rightarrow_{\mathrm{dd}} t'$ and $t' \in \mathcal{L}$.*

Being closed under data dependence is sufficient for completeness (cf. Desideratum 1):

Theorem 1 (Completeness). *Assume that $\mathcal{L}$ is closed under $\rightarrow_{\mathrm{dd}}$. If*

- $1 \vdash_{ni} \pi : C_1 \xrightarrow{E} C_1'$ *and*
- $C_1 \; Q \; C_2$

then there exists C_2' such that

- $2 \vdash \pi : C_2 \xrightarrow{E} C_2'$ *and*
- $C_1' \; Q \; C_2'$.

Moreover, if $\mathcal{L}$ is uniform then $2 \vdash_{ni} \pi : C_2 \xrightarrow{\mathsf{filter}(E)} C_2'$.

The proof is in Appendix A.

4.2 Next Observable

Following recent trends in the theoretical foundation of slicing [1,9] we shall employ the concept of "next observable" which allows us to express the classical notion of "control dependence" in a way that doesn't require the existence of a unique exit state. In our setting, where a slice set consists of transitions rather than states, we need to phrase "next observable" in a somewhat different way:

Definition 19. *For a slice set $\mathcal{L}$, for each state n we define $obs_{\mathcal{L}}(n)$ (written $obs(n)$ when $\mathcal{L}$ is given by the context) as the set of states n' such that*

- *there exists $t \in \mathcal{L}$ with $\mathsf{S}(t) = n'$, and*
- *there exists a path outside $\mathcal{L}$ from n to n'.*

For example, consider the EFSM in Fig. 2, and assume that $\mathcal{L} = \{\mathtt{t5}, \mathtt{t6}\}$. Then for all n we have $obs_{\mathcal{L}}(n) \subseteq \{\mathsf{S}(\mathtt{t5}), \mathsf{S}(\mathtt{t6})\} = \{\mathtt{S2}, \mathtt{S3}\}$. In particular, $obs_{\mathcal{L}}(\mathtt{S1}) = \{\mathtt{S2}, \mathtt{S3}\}$ whereas $obs_{\mathcal{L}}(\mathtt{S2}) = \{\mathtt{S2}\}$ (since there is no path outside $\mathcal{L}$ from `S2` to `S3`) but $obs_{\mathcal{L}}(\mathtt{S3}) = \{\mathtt{S2}, \mathtt{S3}\}$ (since [`t4`, `t1`] is a path outside $\mathcal{L}$ from `S3` to `S2`).

4.3 Weak Commitment Closure

In order to obtain (weak) soundness, we in general cannot allow a set $obs_{\mathcal{L}}(n)$ to contain *two* (or more) states. To see this, consider Fig. 2 with $\mathcal{L} = \{\mathtt{t5}, \mathtt{t6}\}$ so that $obs_{\mathcal{L}}(\mathtt{S1}) = \{\mathtt{S2}, \mathtt{S3}\}$. Then the sliced EFSM may move to S3 (through ε-transitions) and perform the observable transition t6, while such a move may be impossible for the original EFSM (for example if G(t3) is false) which could instead move to S2 (if G(t1) is true) and perform t5. This motivates the notion of "weak commitment closed" ($\mathcal{WCC}$):

Definition 20 (WCC). *We say that $\mathcal{L}$ satisfies $\mathcal{WCC}$ iff for each state n, $obs_{\mathcal{L}}(n)$ is either empty or a singleton.*

We see that in Fig. 2, the set $\{\mathtt{t5}, \mathtt{t6}\}$ does *not* satisfy $\mathcal{WCC}$.

Satisfying $\mathcal{WCC}$ (and being closed under data dependence) is sufficient for weak soundness (cf. Desideratum 2):

Theorem 2 (Weak Soundness). *Assume that $\mathcal{L}$ is closed under $\rightarrow_{\mathrm{dd}}$ and satisfies $\mathcal{WCC}$.*

If $2 \vdash^t_{ni} C_2 \overset{E_2}{\Longrightarrow} C_2'$ and $C_1\ Q\ C_2$ then there are 3 possibilities:

1. *$1 \vdash^t_{ni} C_1 \overset{E_1 E_2}{\Longrightarrow} C_1'$ for some C_1', E_1 with $C_1'\ Q\ C_2'$ (and if $\mathcal{L}$ is uniform then $\mathsf{filter}(E_1) = \varepsilon$), or*
2. *the original EFSM gets stuck from C_1 avoiding $\mathcal{L}$ (cf. Definition 15), or*
3. *the original EFSM loops from C_1 avoiding $\mathcal{L}$ (cf. Definition 16).*

The proof is in Appendix A.

4.4 Strong Commitment Closure

In order to obtain strong soundness, we cannot allow a state n to be part of an infinite path that avoids $obs_{\mathcal{L}}(n)$. To see this, consider Fig. 2 but this time with $\mathcal{L} = \{\mathtt{t9}\}$. This trivially satisfies $\mathcal{WCC}$ as $obs_{\mathcal{L}}(n)$ will always be either $\emptyset$ or $\{\mathtt{S5}\}$, in particular $obs_{\mathcal{L}}(\mathtt{S1}) = \{\mathtt{S5}\}$ but there is an infinite path from S1 that avoids S5. Thus the sliced EFSM may move to S5 and perform the observable transition t9, while such a move may be impossible for the original EFSM (if say G(t2) and G(t3) are both false) which could instead (for certain values of the store) cycle infinitely between S1, S2, and S3. This would violate strong soundness, and motivates the notion of "strong commitment closed" ($\mathcal{SCC}$):

Definition 21 (SCC). *We say that $\mathcal{L}$ satisfies $\mathcal{SCC}$ iff for each state n, either*

- *$obs_{\mathcal{L}}(n)$ is empty, or*
- *$obs_{\mathcal{L}}(n)$ is a singleton n', and there is no infinite path from n that avoids n'.*

Clearly a slice set that satisfies $\mathcal{SCC}$ will also satisfy $\mathcal{WCC}$.

Satisfying $\mathcal{SCC}$ (and being closed under data dependence) is sufficient for strong soundness (cf. Desideratum 3):

Theorem 3 (Strong Soundness). *Assume $\mathcal{L}$ is closed under $\rightarrow_{\mathrm{dd}}$ and satisfies $\mathcal{SCC}$.*

If $2 \vdash^t_{ni} C_2 \stackrel{E_2}{\Longrightarrow} C'_2$ and $C_1 \; Q \; C_2$ then there are 2 possibilities:

1. $1 \vdash^t_{ni} C_1 \stackrel{E_1 E_2}{\Longrightarrow} C'_1$ for some C'_1, E_1 with $C'_1 \; Q \; C'_2$ (and if $\mathcal{L}$ is uniform then $\mathsf{filter}(E_1) = \varepsilon$*), or*
2. the original EFSM gets stuck from C_1 avoiding $\mathcal{L}$.

The proof is in Appendix A.

5 Computing Least Slices

For a given EFSM, there may be many slice sets that satisfy $\mathcal{WCC}$ and are closed under data dependence.

Example 4. *Consider the EFSM given in Fig. 5. If the slicing criterion is transition t8, and t8 is data dependent on t3 and t6 (but no other data dependences exist), then a superset of {t3, t6, t8} is closed under data dependence, and will satisfy $\mathcal{WCC}$ iff it contains t2 and t5. (That is, any of t1, t4 or t7 may or may not be there, so there are 8 possible supersets.)*

However, among all the sets that satisfy $\mathcal{WCC}$ and are closed under data dependence, there will always be a *least* one (that is, one which is a subset of all other such sets). This follows since in Sect. 5.1 we shall present an algorithm that we can prove always returns that least set. In the above example, this will be $\{t2, t3, t5, t6, t8\}$ which is constructed as follows: we first close $\{t8\}$ under data dependence which yields $\{t3, t6, t8\}$ (in bold in Fig. 5), so that the observable states (the sources of these transitions) are $\{S3, S5, S7\}$ (filled in Figure 5); we then do a backwards search from these states and see that $S2$ has *two* next observables ($S3$ and $S5$) so we need to add $t2$ and $t5$, after which no more transitions need to be added. Note that even though also $S1$ has two next observables, it is important that $S2$ is considered first, as otherwise $t1$ is needlessly added.

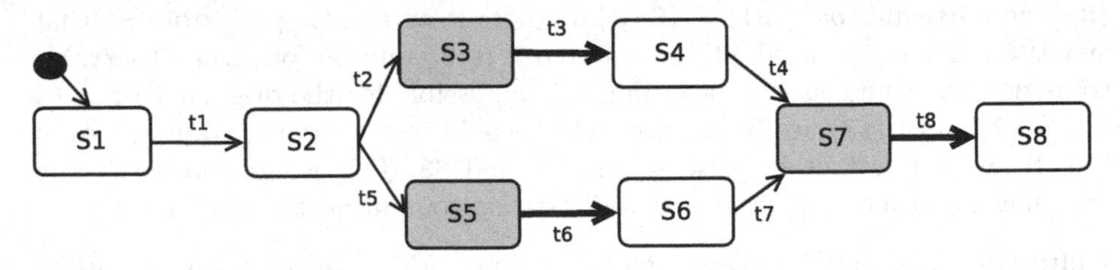

Fig. 5. An EFSM where the current slice set $\{t3, t6, t8\}$ (in bold) does not satisfy $\mathcal{WCC}$, as the corresponding set of observable states $\{S3, S5, S7\}$ (filled) allows some states ($S1$ and $S2$) to have *two* next observables.

Similarly, for a given EFSM there may be many slice sets that satisfy $\mathcal{SCC}$ and are closed under data dependence.

Example 5. *Consider the EFSM given in Fig. 6. If the slicing criterion is transition* t4 *(and there are no data dependences), then the following supersets satisfy SCC:* $\{t3, t4\}$, $\{t1, t3, t4\}$, $\{t2, t3, t4\}$, *and* $\{t1, t2, t3, t4\}$.

In Sect. 5.2, we shall present an algorithm that always returns the *least* such set. In the above example, this will be $\{t3, t4\}$ which is constructed as follows: with the initial observable state being $S3$ (filled in Fig. 6), we again do a backwards search from $S3$ and see that $S2$ can avoid $S3$ and hence we need to add $t3$, after which no more transitions need to be added. Note that even though also $S1$ can avoid $S3$, it is important that $S2$ is considered first, as otherwise $t1$ is needlessly added.

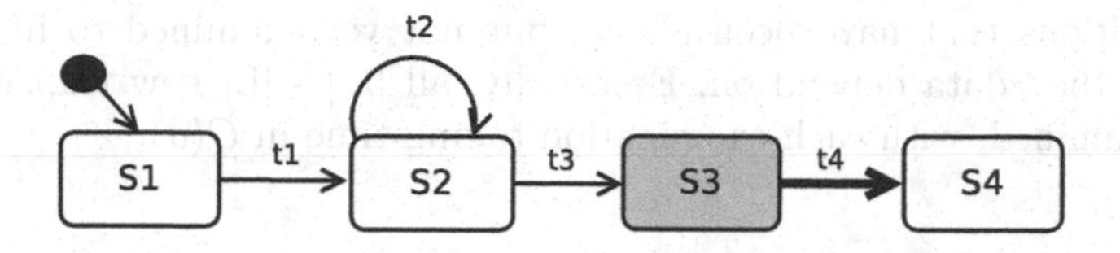

Fig. 6. An EFSM where the current slice set $\{t4\}$ (in bold) does not satisfy *SCC*, as the corresponding set of observable states $\{S3\}$ (filled) allows some states ($S1$ and $S2$) to *avoid* that observable.

When analyzing the complexity of our algorithms, we assume that sets of states and sets of transitions are represented as bitmaps, allowing one element to be added in constant time. To prepare for our algorithms, we now address how to ensure data dependence; we shall assume a pre-computed table `DD` such that `DD`(t, u) is true iff $t \rightarrow_{\mathrm{dd}} u$ holds.

Lemma 22. *The table* `DD` *can be computed in time* $O(a^2)$ *where a is the number of transitions in the EFSM.*

To do so, for each transition t where $\mathsf{D}(t)$ is non-empty and thus a singleton v, we first mark all nodes except $\mathsf{T}(t)$ as unvisited and then do a depth-first search from $\mathsf{T}(t)$ to find how far this definition "propagates"; the procedure `Reaches` finds the transitions u that use v and which can be reached through transitions that do not redefine v.

We can use the table `DD` to add transitions while preserving the property of being closed under data dependence:

Lemma 23. *Given a table* `DD` *such that* `DD`(t, u) *holds iff* $t \rightarrow_{\mathrm{dd}} u$*, we can write a function* `DDclose` *that given a slice set* L *closed under data dependence, and another slice set* L_2 *with* $L \cap L_2 = \emptyset$*, returns the least slice set* L' *that contains* $L \cup L_2$ *and is closed under data dependence. The running time of that function is in* $O(a \cdot (|L'| - |L|))$ *where a is the number of transitions in the EFSM.*

Procedure. Reaches(n, v)

```
  ; /* Finds the transitions that use v and are reachable from n through
  an acyclic path that does not redefine v. */
1 foreach u with S(u) = n do
2     if v ∈ U(u) then
3         output u
4     if v ∉ D(u) and T(u) has not yet been visited then
5         mark T(u) as visited
6         Reaches(T(u), v)
```

The function `DDclose` is defined below. It works by maintaining a queue Q of transitions that have been added, but not yet examined to include any transitions they data depend on. Eventually, all |L'| - |L| new transitions will have be examined, with each examination taking time in $O(a)$.

Procedure. DDclose(L, $L2$)

```
1 Q := L2
2 L := L ∪ L2
3 while Q ≠ ∅ do
4     remove an element, t2, from Q
5     foreach t1 with (t1, t2) ∈ DD do
6         if t1 ∉ L then
7             add t1 to Q and to L
8 return L
```

5.1 Computing Least $\mathcal{WCC}$-Satisfying Slices

We present an algorithm (Algorithm 1) that for a given slice set `L` returns the least superset of `L` that satisfies $\mathcal{WCC}$ and is closed under data dependence. The algorithm works by adding transitions to `L` until no new transitions need to be added. In each iteration, the algorithm computes in `B` the states that are sources of transitions in `L`, and does from `B` a backwards breadth-first search through transitions not in `L`, with `V` being the states that have been visited so far. For each $n \in$ `V`, the array entry $X[n]$ is defined as the state in `B` that can be reached (through transitions not in `L`) from n; that state will be a next observable of n. The current frontier of the search is called `C`, the exploration of which builds up `Cnew`, the next frontier. If there is a transition $t \notin$ `L` to a node $m \in C$, from a node n that already belongs to `V` but with $X[n] \neq X[m]$, we detect that n has *two* next observables and we infer that t has to be included in the slice set (as motivated in Example 4 and justified by the correctness result).

Algorithm 1. Computes the least $\mathcal{WCC}$-closed slice.

Input: An EFSM M; a set $\mathcal{L}$ of transitions in M
Output: the least $\rightarrow_{dd}$-closed superset of $\mathcal{L}$ that satisfies $\mathcal{WCC}$

```
1  L := DDclose(∅, 𝓛)
2  repeat
          /* L is closed under →dd, and is a subset of any →dd-closed
       superset of 𝓛 that satisfies 𝓦𝓒𝓒 */
3      Lnew := ∅
4      B := {n | ∃t ∈ L : n = S(t)}
5      foreach n ∈ B do
6          X[n] := n
7      V, C := B
8      while C ≠ ∅ and Lnew = ∅ do
9          Cnew := ∅
10         foreach m ∈ C do
11             foreach transition t ∉ L with T(t) = m do
12                 n := S(t)
13                 if n ∈ V then
14                     if X[n] ≠ X[m] then
15                         Lnew := Lnew ∪ {t}
16                 else
17                     V := V ∪ {n}
18                     Cnew := Cnew ∪ {n}
19                     X[n] := X[m]
20         C := Cnew
21     L := DDclose(L,Lnew)
22 until Lnew = ∅
```

Example 6. *Let us apply Algorithm 1 to the EFSM in Fig. 2, to find the least set that satisfies $\mathcal{WCC}$ and contains t6 and t9.*

In the first iteration, $B = \{S3, S5\}$. Since S3 can be reached from S4 by t7, and S5 can be reached from S4 by t8, there is a conflict at S4 so we add t7 (or t8) to L.

In subsequent iterations, with $B = \{S3, S4, S5\}$, we add t8 (due to the conflict at S4), as well as t2 and t3 (due to a conflict at S1).

We now have $B = \{S1, S3, S4, S5\}$, and may add t4 due to a conflict at S3. The next iteration adds t1 since from S1 one can reach S3 through the path $[t1, t5]$. The following iteration adds t10 since from S4 one can reach S3 through the path $[t10, t12, t5]$.

We are left with $\mathcal{L} = \{t1, t2, t3, t4, t6, t7, t8, t9, t10\}$, illustrated in Fig. 7 at which point no new transitions can be added, since $\mathsf{S}(t5)$ does not have other observables than S3. And $\mathcal{L}$ does indeed satisfy $\mathcal{WCC}$: for all states n we have $obs_{\mathcal{L}}(n) = \{n\}$, except $obs_{\mathcal{L}}(S2) = obs_{\mathcal{L}}(S6) = \{S3\}$.

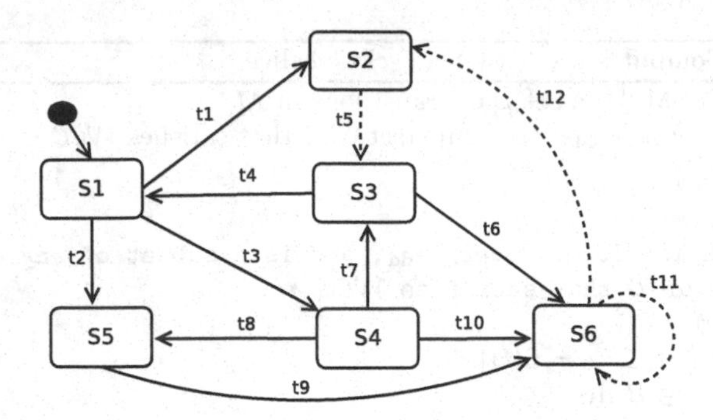

Fig. 7. Applying Algorithm 1 for computing the least WCC-closed set containing $t6$ and $t9$ to our running example (Fig. 2). Transitions in $\mathcal{L}$ are shown using solid lines, and those not with dotted lines.

Theorem 4 (WCC Algorithm Complexity). *Algorithm 1 (including the construction of the table* `DD`*) can be implemented to run in time $O(a^2)$ where a is the number of transitions.*

Proof. By Lemma 22, the table `DD` can be computed in time $O(a^2)$. By Lemma 23, the total time of the calls to `DDclose` is in $O(a^2)$.

The outer loop eventually terminates; it continues only when `Lnew` $\neq \emptyset$ and as `Lnew` is disjoint from `L` by construction, the assignment `L := DDclose(L, Lnew)` will strictly increase the size of `L` but this cannot go on forever as the number of transitions is finite. In particular, we see that the outer loop iterates $O(a)$ times.

It is thus sufficient to show that each iteration of the outer loop, except for the call to `DDclose`, runs in time $O(a)$.

Towards this goal, first observe that the `while` loop will always terminate, as eventually no more states can be added to `V` in which case `C` will be empty and the loop thus exit (if it hasn't done already). In particular, each node is processed at most once by the outer **foreach** loop, and each transition is processed at most once by the inner **foreach** loop. As the body of the inner **foreach** loop executes in constant time, this yields the claim.

In Appendix B, we prove:

Theorem 5 (WCC Algorithm Correctness). *Assuming that the table* `DD` *correctly computes data dependence, Algorithm 1 returns in* `L` *the least superset of $\mathcal{L}$ that is closed under data dependence and satisfies $\mathcal{WCC}$.*

5.2 Computing Least $\mathcal{SCC}$-Closed Slices

We present Algorithm 2 that extends Algorithm 1 to compute the least superset of $\mathcal{L}$ that satisfies $\mathcal{SCC}$ (and is closed under data dependence): as we explore each state n from which a state m' in `B` is reachable, we check if there from n

is an infinite path that avoids m'; if that is the case, we can add to L the first transition in the path from n to m'.

Algorithm 2 is identical to Algorithm 1 except that two lines (13 and 14) have been added. The algorithm employs a pre-computed table `LoopAvoids` such that `LoopAvoids(`n, m`)` is true iff there is an infinite path from n that avoids m.

Algorithm 2. Computes the least $\mathcal{SCC}$-closed slice

Input: An EFSM M; a set $\mathcal{L}$ of transitions in M
Output: the least $\rightarrow_{dd}$-closed superset of $\mathcal{L}$ that satisfies $\mathcal{SCC}$

```
1  L := DDclose(∅, 𝓛)
2  repeat     /* L is closed under →dd, and a subset of any →dd-closed
   superset of 𝓛 that satisfies SCC */
3      Lnew := ∅
4      B := {n | ∃t ∈ L : n = S(t)}
5      foreach n ∈ B do
6          X[n] := n
7      V, C := B
8      while C ≠ ∅ and Lnew = ∅ do
9          Cnew := ∅
10         foreach m ∈ C do
11             foreach transition t ∉ L with T(t) = m do
12                 n := S(t)
13                 if LoopAvoids(n, X[m]) then        /* See Algorithm 3 */
14                     Lnew := Lnew ∪ {t}
15                 if n ∈ V then
16                     if X[n] ≠ X[m] then
17                         Lnew := Lnew ∪ {t}
18                 else
19                     V := V ∪ {n}
20                     Cnew := Cnew ∪ {n}
21                     X[n] := X[m]
22         C := Cnew
23     L := DDclose(L,Lnew)
24 until Lnew = ∅
```

The table `LoopAvoids` can be constructed by Algorithm 3. For each m, the transitions involving m are ignored, and a depth-first search is made; then `LoopAvoids(`n, m`)` is set to true iff a back edge is reachable from n.

Fact 24. *For all n, m,* `LoopAvoids(`n, m`)` *is true iff there is an infinite path from n that avoids m.*

Algorithm 3. Constructing the table `LoopAvoids`, with procedure DFS.

```
1  foreach state m do
2      foreach state n do
3          LoopAvoids(n, m) := false
4      T_m := the transitions that do not involve m
5      foreach state n except m do
6          color[n] := white
7      while (exists n: color[n] = white) do
8          let n be a state with color[n] = white
9          call DFS(n);

10 Procedure DFS(n)
11     color[n] := gray
12     foreach t ∈ T_m with S(t) = n do
13         n' := T(t)
14         if color[n'] = gray then
15             LoopAvoids(n, m) := true
16         else if color[n'] = black then
17             if LoopAvoids(n', m) then
18                 LoopAvoids(n, m) := true
19         else if color[n'] = white then
20             DFS(n')
21             if LoopAvoids(n', m) then
22                 LoopAvoids(n, m) := true
23     color[n] := black
```

Example 7. *Let us apply Algorithm 2 to the EFSM in Fig. 2, to find the least set that satisfies SCC and contains t9.*

In the first iteration, $B = \{S5\}$. Since t2 has source $S1$ and target $S5$, and there is an infinite path from $S1$ that avoids $S5$, we add t2. Similarly, since t8 has source $S4$ and target $S5$, and there is an infinite path from $S4$ that avoids $S5$, we add t8.

Now, $B = \{S1, S4, S5\}$. Since $S4$ can be reached from $S1$ by t3, we add t3. Since t4 has target $S1$ and there is an infinite path from $S3$ that avoids $S1$, we add t4.

Now, $B = \{S1, S3, S4, S5\}$. Since $S3$ can be reached from $S4$ by t7, we add t7. Since $S3$ can be reached from $S1$ by the path $[t1, t5]$, we add t1. Since $S3$ can be reached from $S6$ by the path $[t12, t5]$, and there is an infinite path from $S6$ (via t11) that avoids $S3$, we add t12.

Now, $B = \{S1, S3, S4, S5, S6\}$. Since $S6$ can be reached from $S3$ by t6, and from $S4$ by t10, we add t6 and t10.

We are left with $\mathcal{L} = \{t1, t2, t3, t4, t6, t7, t8, t9, t10, t12\}$ at which point no new transitions can be added. And $\mathcal{L}$ does indeed satisfy SCC: $obs_{\mathcal{L}}(S1) = \{S1\}$; $obs_{\mathcal{L}}(S2) = \{S3\}$; $obs_{\mathcal{L}}(S3) = \{S3\}$; $obs_{\mathcal{L}}(S4) = \{S4\}$; $obs_{\mathcal{L}}(S5) = \{S5\}$; $obs_{\mathcal{L}}(S6) = \{S6\}$ but no infinite path from $S2$ avoids $S3$ (Fig. 8).

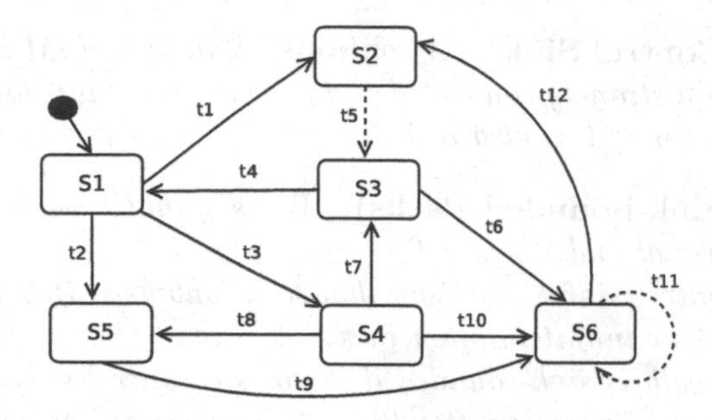

Fig. 8. Applying Algorithm 2 to our running example (Fig. 2) for computing the least set that satisfies SCC that contains $t9$. Transitions in $\mathcal{L}$ are shown using solid lines, and those not in $\mathcal{L}$ with dotted lines.

Theorem 6 (SCC Algorithm Complexity). *Algorithm 2, including the construction of the tables* DD *and* LoopAvoids, *can be implemented to run in time $O(a^2)$ where a is the number of transitions.*

Proof. To construct the table LoopAvoids, Algorithm 3 will for each m use time in $O(a)$, and hence (since $|\hat{S}| \in O(a)$) have a total running time in $O(a^2)$. Next Algorithm 2 can use that table, and will also run in time $O(a^2)$ (by an analysis almost identical to the one given in the proof of Theorem 4 for Algorithm 1).

In Appendix B, we prove:

Theorem 7 (SCC Algorithm Correctness). *Assuming that the table* DD *correctly computes data dependence, Algorithm 2 (using Algorithm 3) returns in* L *the least superset of $\mathcal{L}$ that is closed under data dependence and satisfies $\mathcal{SCC}$.*

6 Comparing Control Dependence Definitions for EFSMs

There have been several previous attempts to define suitable notions of control dependence for EFSMs. These approaches, none of which have allowed a proof that the resulting slices are in some sense "correct", will in this section be described, and illustrated on various examples while comparing to $\mathcal{WCC}/\mathcal{SCC}$-based slicing.

6.1 Preliminary Definitions

Definition 25 (Post-Dominance [20]). *Given a machine with exactly one exit state (reachable from all other states), and two states Y and Z, we say that Z post-dominates Y iff Z is in every path from Y to the exit state.*

Definition 26 (Maximal Path). *A path in an EFSM is maximal iff it terminates in an exit state, or is infinite.*

Definition 27 (Control Sink). *A control sink in an EFSM is a set of transitions $\mathcal{K}$ that forms a strongly connected component such that, for each transition t in $\mathcal{K}$, each successor of t is also in $\mathcal{K}$.*

Definition 28 (Sink-bounded Paths). *We say that a path π joins a control sink $\mathcal{K}$ if π is maximal and $\mathcal{K} \cap \pi \neq \emptyset$.*

If π joins a control sink $\mathcal{K}$, we say that π is unbiased if π is finite or if all transitions in $\mathcal{K}$ occur infinitely often in π.

We say that a path is sink-bounded if it joins a control sink and is unbiased.

With some abuse of notation (for historical reasons), we shall say that π is an unfair sink-bounded path iff π joins a control sink (even if π is unbiased).

6.2 Definitions for EFSMs with Exactly One Exit State

Korel et al. [20] present a definition of control dependence for EFSMs that

- is insensitive to non-termination
- is phrased in terms of post-dominance, similar to what is standard for static backward slicing of programs [25,33]
- is thus not applicable if there are multiple exit states, or none (in which case the machine is likely designed to be potentially non-terminating).

Definition 29 (Insensitive Control Dependence (ICD) [20]). *Transition T_k is control dependent on transition T_i if:*

1. *source(T_k) post-dominates target(T_i), and*
2. *source(T_k) does not post-dominate source(T_i).*

6.3 Definitions for EFSMs with Arbitrary Number of Exit States

Ranganath et al. [26] were the first to propose notions of control dependence, one that is sensitive to non-termination and one which is not, that apply to programs with multiple or none exit points. Androutsopoulos et al. [4] adapted these definitions for EFSMs, yielding what is known as Non-Termination Sensitive Control Dependence (NTSCD) and Non-Termination Insensitive Control Dependence (NTICD), and also introduced Unfair Non-Termination Insensitive Control Dependence (UNTICD) that overcame a limitation of NTICD.

To express those definition, we first present a general definition, parametrized on the set *PATH* of paths we consider:

Definition 30 (Control Dependence (CD)). *A transition T_j is control dependent on a transition T_i, written $T_i \xrightarrow{CD} T_j$, iff:*

1. *for all paths $\pi \in PATH$ from target(T_i), also source(T_j) belongs to π;*
2. *there exists a path $\pi \in PATH$ from source(T_i)) such that source(T_j) does not belong to π.*

We may then instantiate as follows: with PATH the set of

- maximal paths, we get *NTSCD*;
- sink-bounded paths, we get *NTICD*;
- unfair sink-bounded paths, we get *UNTICD*.

NTICD cannot compute control dependences within control sinks. UNTICD overcomes this limitation by considering also paths that are not unbiased, allowing for extra control dependences within control sinks to be introduced; these dependences, however, are the same as for NTSCD.

Control dependence definitions for EFSMs are classified, in Table 1, according to whether they are "weak", that is not sensitive to non-termination (thus aiming only at something like "weak soundness"), or "strong", that is sensitive to non-termination (thus aiming at something like "strong soundness").

Table 1. Control dependence definitions for EFSMs

Weak control dependence	ICD [20], NTICD [4], UNTICD [4], WCC
Strong control dependence	NTSCD [4], SCC

6.4 Examples of Slicing Wrt Various Definitions

Let us first consider the EFSM in Fig. 9 where there are no data dependencies. Here S5 is the unique exit state. Apart from `start`, all other states form a strongly connected group, and each have a transition (invoked in case of "error") to S5; moreover, S2, S3 and S4 have self-transitions.

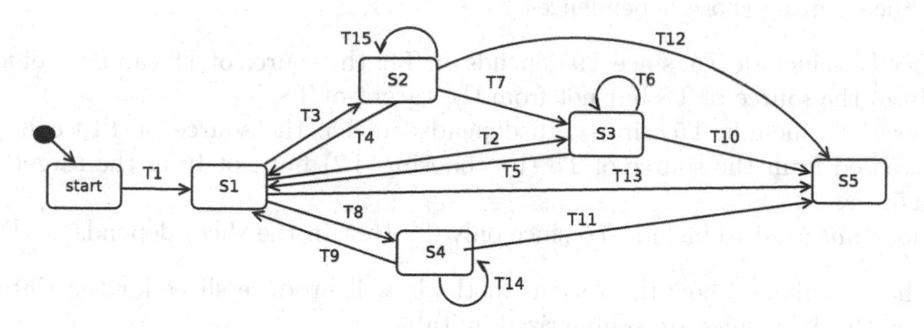

Fig. 9. An EFSM where NTSCD and SCC produce slices of the same size.

Consider the result of slicing the EFSM in Fig. 9 wrt. criterion T2. WCC will add no other transitions (as only S1 can be a next observable), whereas SCC will add T4, T5, T9 (since from S2, S3, S4 there are infinite paths that avoid S1) and then (as now all of S1, S2, S3, S4 may be next observables) also T3, T7, T8. SCC thus yields the slice {T2, T3, T4, T5, T7, T8, T9}. Also NTSCD, as well as NTICD and UNTICD (note that a path is sink-bounded iff it ends in S5), will produce that slice, since T2 depends on T4 and on T5 and on T9, and T4/T5/T9 depends on T3/T7/T8.

We conclude that for this example, WCC outperforms all other "weak" definitions.

Let us next consider the EFSM depicted in Fig. 10 which illustrates, among other things, why it may happen that NTSCD produces a smaller slice than SCC but one which fails to satisfy our correctness property.

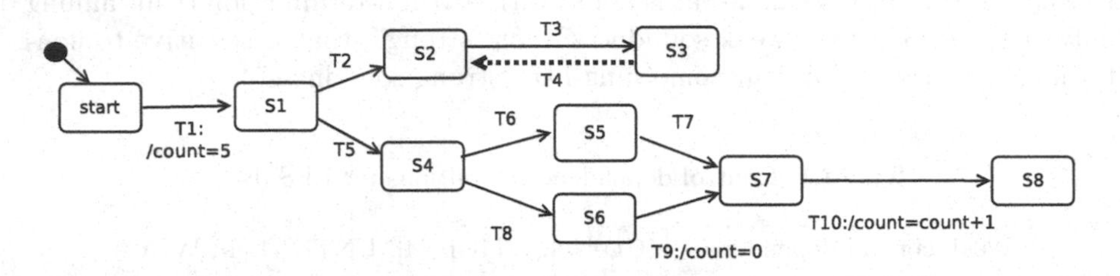

Fig. 10. An EFSM used to illustrate the difference between slices produced using SCC and NTSCD.

With T10 our slicing criterion, data dependence forces T9 and T1 to be included in the slice. For WCC and SCC, in order for S4 to have a unique next observable, we then need to include T6 and T8. For SCC, since there is an infinite path from S1 that avoids S4, we need to include T5.

Concerning NTSCD, NTICD and UNTICD, observe that in this case those dependences equal each other, since the maximal paths equal the ("fair" or not) sink-bounded paths: both are those that end in S8, or alternate between S2 and S3. Slicing using those dependences

- needs to include T8, since T9 depends on T8: the source of T9 can be avoided from the source of T8 but not from the target of T8;
- needs to include T5, since T10 depends on T5: the source of T10 can be avoided from the source of T5 (by choosing T2) but not from the target of T5;
- does *not* need to include T6 since only T7 (not in the slice) depends on T6.

We have explained how the various methods will produce slices for the EFSM in Fig. 10; the results are summarized in Table 2.

Thus the slice produced by SCC includes both T6 and T8, rather than just T8; intuitively, this is because they both determine (or control) whether the execution of T10 will be preceded by execution of T9. If we slice away T6, that is replace it by an ε-transition, the sliced EFSM may bypass T9 and then the value of `count` at S8 will be 6. But if T6 has a strong guard, this behavior can*not* be replicated by the original EFSM (which may instead choose T8 and then the value of `count` at S8 will be 1). To slice away T6, as done by NTSCD, will thus violate Strong Soundness, as presented in this paper.

Table 2. Slices produced, by various methods, for the EFSM in Fig. 10 with respect to T10.

Control dependence	Slice wrt. T10
WCC	$\{T1, T6, T8, T9, T10\}$
SCC	$\{T1, T5, T6, T8, T9, T10\}$
NTSCD/NTICD/UNTICD	$\{T1, T5, T8, T9, T10\}$

Observe that if in Fig. 10 we remove (the dotted) transition T4 then the slices produced by the various methods remain the same, except that SCC no longer will include T5 since now there is no infinite path from S1 (that avoids S4).

6.5 Simple Quantitative Comparison of Various Approaches

Let us consider an EFSM for modeling door control (a subcomponent of an elevator control system [30]) which is shown in Fig. 11. The door component controls the elevator door, i.e. it opens the door, waits for the passengers to enter or leave the elevator (by using a timer) and finally shuts the door. The data dependenceis, computed using Definition 17, for the door controller are given in Table 3. We adopt a short-hand notation when describing dependences, e.g., T1 $\rightarrow_{dd}$ T2, T3 denotes T1 $\rightarrow_{dd}$ T2 and T1 $\rightarrow_{dd}$ T3.

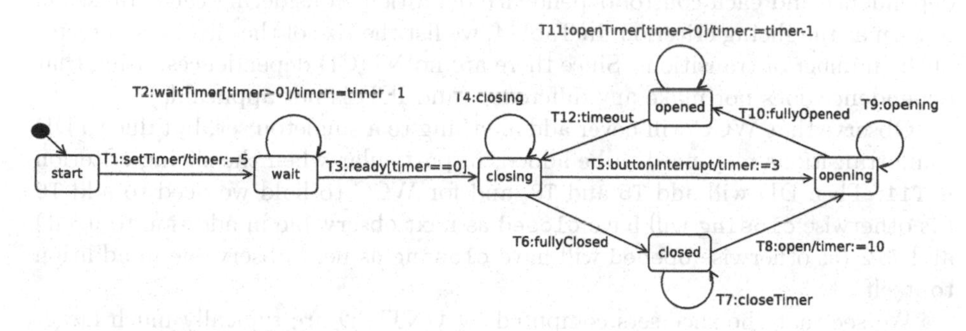

Fig. 11. An EFSM specification for the door controller of the elevator system.

Table 3 also lists the various control dependences computed for the door controller. ICD cannot be applied as there is no (unique) exit state. On the other hand, there is a "control sink" containing all transitions except T1, T2 and T3.

NTICD has no dependences because there are no non-termination insensitive control dependences outside of the control sink and it cannot compute any control dependences within control sinks. UNTICD has dependences within the control sink, due to the self-looping transitions T4, T7, T9, T11. NTSCD has the same dependences within the control sink as UNTICD, as is always the case (as shown in [4]), but because of the self-looping transition T2 in addition some outside: T3 $\rightarrow_{\text{NTSCD}}$ T4, T5, T6.

Table 3. Data dependence (DD) and control dependence for the door controller (Fig. 11).

DD	T1 $\rightarrow_{\text{dd}}$ T2, T3	T2 $\rightarrow_{\text{dd}}$ T2, T3	T11 $\rightarrow_{\text{dd}}$ T11
	T5 $\rightarrow_{\text{dd}}$ T11	T8 $\rightarrow_{\text{dd}}$ T11	
ICD	Cannot be applied as no unique exit state		
NTICD	No dependences		
UNTICD	T5 → T9, T10	T6 → T7, T8	T8 → T9, T10
	T10 → T11, T12	T12 → T4, T5, T6	
NTSCD	T3 → T4, T5, T6	T5 → T9, T10	T6 → T7, T8
	T8 → T9, T10	T10 → T11, T12	T12 → T4, T5, T6

We cannot describe WCC and SCC as in Table 3 because WCC and SCC do not relate individual transitions, but describe properties that a slice set must satisfy. For ICD, NTICD, UNTICD and NTSCD, the slice set is the transitive closure, with respect to a slicing criterion, of that control dependence (with data dependence added).

We have computed the slice sets for the door controller EFSM, using data dependence and each control dependence definition, considering each transition in turn as the slicing criterion. In Table 4, we list the size of the slice sets in terms of the number of transitions. Since there are no NTICD dependences, using that dependence does not make any difference (and ICD is not applicable).

Observe that WCC will never add anything to a singleton set, but due to DD some transitions may need to be added, in particular when the slicing criterion is T11. Then DD will add T5 and T8, and for WCC to hold we need to add T6 (as otherwise `closing` will have `closed` as next observable in addition to itself) and T12 (as otherwise `opened` will have `closing` as next observable in addition to itself).

We see that the slice sets computed by UNTICD are typically much larger than those computed by WCC, with the slice sets computed by NTSCD even larger, and (in this case) always equal to those computed by SCC. We shall briefly explain each of these computations, again using T11 as the slicing criterion. In all cases, DD will cause T5 and T8 to be added to the slice.

For UNTICD, we see from Table 3 that we need to add T6, T10, and T12. The presence of T10 is the difference between UNTICD and WCC, and arises because UNTICD is sensitive to non-termination inside the control sink; from the source of T10 it is possible, due to the self-looping transition T9, to avoid the source of T12.

For NTSCD, we see from Table 3 that (due to the presence of T5) we need to add also T3, and by data dependence even T1 and T2. (The only transitions not included in the slice are thus the self-looping T4, T7, and T9.)

For the least set satisfying SCC, observe that it must satisfy WCC and hence contain {T5, T6, T8, T11, T12}. With that slice set, the next observable of opening is opened, but (due to T9) there is an infinite path from opening that avoids opened, so T10 must be added. Also, the next observable of wait is closing, but (due to T2) there is an infinite path from wait that avoids closing, so T3 must be added, and by DD also T1 and T2. The resulting slice thus equals the slice for NTSCD.

Table 4. For each kind of control dependence, and each transition t in the EFSM of the door controller (Fig. 11), we use that control dependence to slice with respect to a slicing criterion that consists of only t, and list the size (*i.e.*, the number of transitions) of the resulting slice.

Slicing criterion	T1	T2	T3	T4	T5	T6	T7	T8	T9	T10	T11	T12
DD, or NTICD + DD	1	2	3	1	1	1	1	1	1	1	3	1
WCC (+ DD)	1	2	3	1	1	1	1	1	1	1	5	1
UNTICD (+ DD)	1	2	3	6	5	5	6	5	6	5	6	5
SCC or NTSCD (+ DD)	1	2	3	9	8	8	9	8	9	8	9	8

7 Empirical Evaluation

We shall now explore the practical usefulness of using WCC and SCC to slice EFSMs. Our measure will be the size of the slices produced, since if slices are small compared to the original models then they provide advantages, such as aiding users in understanding the models. We thus pose the following research questions:

RQ1. What is the typical size of a backward slice produced using WCC or SCC?
RQ2. How do the sizes of backward slices produced by WCC and SCC compare to those produced using existing control dependence definitions for EFSMs?

We use thirteen EFSMs, described in Table 5, as subjects for our experiments. The EFSMs are drawn from a variety of sources. The first six, which we collectively label as group A, were used in the research of model-based slicing with traditional dependence analysis [20]; each has a unique exit state (reachable from all other states). The last seven EFSMs have no exit states; we have divided them into two groups, labeled B and C:

- group B consists of four machines, where TCP [36] is extracted from a SDL specification, and so is TCSbin which is a proprietary real-world model from Motorola; when extracting into EFSMs, we have done some simplifications, including the omission of history states. DoorController has already been described (Sect. 6); INRES is an example of an EFSM specified protocol as described in [8].

- group C consists of Alarm, Tamagotchi and WaterPump which are flattened state machines from the study of clustering finite state machines [14]; the transitions in these machines contain only events (that is, not guards nor actions).

A visual inspection of the EFSMs in group A reveals that their structure varies even though they all have a unique exit state; of the EFSMs in group A, only FuelPump has a CFG-like structure (*i.e.*, almost all transitions belong to an cycle-free path from the initial state to the exit state) while the rest of the EFSMs in group A contain a few strongly connected components.

All of the EFSMs from group B and C, apart from DoorController (as in Fig. 11), consist of a large control sink (of which all transitions are part).

Table 5. EFSMs used in our experiments.

EFSMs	$\|\hat{S}\|$	$\|\hat{T}\|$	$\|\hat{V}\|$	EXIT	Description
ATM	9	23	8	Yes	Automated Teller Machine [20]
Cashier	12	21	10	Yes	Cashier Machine [20]
CruiseControl	5	17	18	Yes	Cruise Control System [20]
FuelPump	13	25	12	Yes	Fuel Pump System [20]
PrinTok	11	89	5	Yes	Print Token [20]
VendingMachine	7	28	7	Yes	Vending Machine system [20]
INRES	8	18	6	No	INRES protocol [8]
DoorController	6	12	1	No	Door Controller of the Lift System [30]
TCP	12	57	31	No	TCP Standard(RFC793) [36]
TCSbin	24	65	61	No	Telephony Protocol (Motorola)
Alarm	10	25	0	No	Alarm Clock [14]
WaterPump	11	14	0	No	Water Pump Controller [14]
Tamagotchi	15	38	0	No	Virtual Pet [14]
Total	143	432	162		

7.1 Implementation and Tool

In [5], a tool was developed in Python to automatically compute slices (both forward and backward) for EFSMs by taking closures of data dependence, and any of the NTSCD, NTICD and UNTICD dependences. In this paper, to observe the effect of different control dependences we shall consider only backward slices.

For an EFSM M the tool will first compute, among the transitions in M, the data dependences, using Definition 17, and then the control dependences, using (based on the user's choice) one of the three control dependence definitions: NTSCD, NTICD, UNTICD. These dependences will be represented as a dependence graph, which is a directed graph whose nodes represent transitions in M

and whose edges represent either data or control dependences between transitions. The slicing algorithm marks all backwardly reachable transitions from the slicing criterion c, *i.e.*, the transitive closure of c wrt. control and data dependence All marked transitions will be in the slice set $\mathcal{L}$; all unmarked transitions become ε-transitions.

We have extended the tool described in [5] by adding Python implementations of the WCC and SCC algorithms. For an EFSM M the tool will first compute, using Definition 17, the data dependences among the transitions in M and produce the corresponding dependence graph. Then the slice set $\mathcal{L}$ will initially consist of the slicing criterion c and any transitions that are data dependent on c, *i.e.*, that are backwardly reachable from c in the dependence graph. Then according to the WCC or SCC algorithm as defined in Algorithms 1 and 2, transitions will be added to the slice. For any new transitions that are added to the slice, the dependence graph will be checked and any backwardly reachable transitions from the new transitions will be added. This process will continue until no new transitions are added in $\mathcal{L}$; all transitions not in $\mathcal{L}$ become ε-transitions.

An algorithm for removing ε-transitions is described in [2]; it is an adaptation of Ilie and Yu's NFA minimisation algorithm [18]. This algorithm has been used with either NTSCD, NTICD, UNTICD to produce slices that are amorphous, in that they are not sub-graphs of the original. In fact, in some cases the number of transitions can be larger than the original, though empirical results in [2] show that in practice this did not occur. This algorithm could be applied to the slices produced with Algorithm 1 (for WCC) or Algorithm 2 (for SCC), to remove ε-transitions and merge equivalent states.

For the empirical evaluation, we consider the size of the slice sets for WCC and SCC, respectively, as well as for the slice sets for the three control dependences: NTSCD, NTICD, UNTICD. We do not consider the slice sizes of ICD, since ICD is applicable only for EFSMs with a unique exit state (group A), and for such EFSMs it is shown [4] that ICD is a special case of UNTICD and of NTICD (*i.e.*, ICD, UNTICD and NTICD produce the same slices for EFSMs that have a unique exit state).

7.2 Metrics

We shall now formalise the metrics used for measuring slice sizes for a given EFSM M with transitions $\hat{T}$. With m a slicing method, and c a slicing criterion, we shall define $\mathcal{L}(M, m, c)$ as the slice set (computed as sketched in Sect. 7.1) given by:

- if m is WCC or SCC, $\mathcal{L}(M, m, c)$ is the least set that satisfies that property, is closed under data dependence, and contains c;
- if m is NTSCD or NTICD or UNTICD, $\mathcal{L}(M, m, c)$ is the least set that is closed under that dependence and under data dependence, and contains c.

We shall define $\mathsf{ratio}(M, m, c)$, the *slicing ratio* of method m (on M) wrt. c, as the percentage of transitions that get included in the slice:

$$\mathsf{ratio}(M, m, c) = \frac{|\mathcal{L}(M, m, c)|}{|\hat{T}|}.$$

The number of possible slicing criteria is exponential in $|\hat{T}|$, but we shall expect slicing criteria to be "small", and restrict our attention to slicing criteria of a certain size; we define $\mathsf{avg\text{-}ratio}(M, m, sc)$ as the average (*i.e.*, mean) slicing ratio, taken over all slicing criteria of size sc. That is:

$$\mathsf{avg\text{-}ratio}(M, m, sc) = \frac{\sum_{c \in \mathcal{P}(\hat{T}) \ | \ |c| = sc} \mathsf{ratio}(M, m, c)}{\frac{|\hat{T}|!}{sc! \ (|\hat{T}| - sc)!}}.$$

In particular, we have:

$$\mathsf{avg\text{-}ratio}(M, m, 1) = \frac{\sum_{t \in \hat{T}} \mathsf{ratio}(M, m, \{t\})}{|\hat{T}|}.$$

With X the set of all EFSM models considered in this section (thus $|X| = 13$), we finally define $\mathsf{total\text{-}avg\text{-}ratio}(m, sc)$, the *average singleton slicing ratio* of method m for slicing criteria of size sc, as

$$\mathsf{total\text{-}avg\text{-}ratio}(m, sc) = \frac{\sum_{M \in X} \mathsf{avg\text{-}ratio}(M, m, sc)}{|X|}.$$

7.3 Results Concerning RQ1: Typical Slices Using WCC or SCC

To give a tentative answer to the first research question, we compute for each EFSM M (of the thirteen given): $\mathsf{avg\text{-}ratio}(M, \text{SCC}, 1)$, $\mathsf{avg\text{-}ratio}(M, \text{WCC}, 1)$, and $\mathsf{avg\text{-}ratio}(M, \text{WCC}, 2)$. (We could also have computed $\mathsf{avg\text{-}ratio}(M, \text{SCC}, 2)$ but even for a singleton slicing criteria, SCC produces slices that are not much smaller than the original EFSM.) The results, to be discussed in the following paragraphs, are given in Table 6, whose last line lists $\mathsf{total\text{-}avg\text{-}ratio}(\text{SCC}, 1)$, $\mathsf{total\text{-}avg\text{-}ratio}(\text{WCC}, 1)$, and $\mathsf{total\text{-}avg\text{-}ratio}(\text{WCC}, 2)$.

Discussing the Results for WCC. The results for WCC are depicted in Table 6. We first look at the case (second rightmost column) where the slicing criterion is a single transition, and observe that then the average slicing ratio for many EFSMs is very low (for FuelPump, PrinTok, Alarm, WaterPump and Tamagotchi, even less than 10%). On closer inspection, this is not surprising, since if the transition that forms the slicing criterion is not data dependent on anything then WCC will cause no other transitions to be added and the resulting slice will have size 1; indeed, 62.26% of all slices computed are of size one. (In particular, if the EFSM uses *no* variables, as is the case for the three EFSMs in group C (Alarm, Tamagotchi and WaterPump), then *all* slices will have size 1.)

Table 6. Average slicing ratios computed using WCC or SCC and data dependence for the thirteen EFSMs. The variable *sc* refers to the number of transitions used as a slicing criterion.

Group	EFSMs	SCC (sc = 1)	WCC (sc = 1)	WCC (sc = 2)
A	ATM	33.46%	16.45%	29.11%
	Cashier	41.95%	14.74%	28.71%
	CruiseControl	46.37%	41.86%	58.43%
	FuelPump	28%	8.8%	19.11%
	PrinTok	60.45%	6.02%	43.02%
	VendingMachine	79.97%	32.40%	71.68%
B	INRES	60.19%	25.31%	40.96%
	DoorController	56.94%	13.19%	23.36%
	TCP	54.05%	41.36%	61.02%
	TCSbin	83.6%	42.84%	70.94%
C	Alarm	92.32%	4%	47.25%
	WaterPump	40.31%	7.14%	25.67%
	Tamagotchi	82.48%	2.63%	34.78%
Total average		58.46%	21.11%	42.62%

We therefore perform a second experiment (rightmost column in Table 6) where we consider slicing criteria that consist of *two* transitions. As expected, this results in substantially larger slices, but typically still containing less than half of the original transitions.

Discussing the Results for SCC (with Singleton Slicing Criteria). The results for SCC are also depicted in Table 6. Since a slice set that satisfies SCC also satisfies WCC, no SCC slice can be smaller than the corresponding WCC slice, and in fact, the typically SCC slice is significantly larger. In particular, only 3.01% of all SCC slices are of size 1 (recall that for WCC the number is 62.26%).

For CruiseControl, however, the average SCC slice is only slightly larger than the average WCC slice. In fact, a closer inspection reveals that the slices are identical, except when the slicing criterion consists of a certain transition (T17); in that situation, transitions need to be added because of an infinite path that avoids the next observable.

Considering Median Slicing Ratio (versus Average). One could argue that when measuring slicing ratios, the *median* is more interesting than the average (mean), as the latter may be severely impacted by a few extreme cases. We shall therefore also compute the median of the slicing ratios, but (in this paper) only when the slicing criterion is a singleton; we implement that by first sorting the $|\hat{T}|$ slicing ratios and then *(i)* picking the $(|\hat{T}|+1)/2$ largest when $|\hat{T}|$ is odd, and *(ii)* picking the mean of the $|\hat{T}|/2$ largest and the $|\hat{T}|/2+1$ largest when $|\hat{T}|$ is even.

The results are depicted in Table 7. Comparing with Table 6, we see that

- for SCC, the median slicing ratio is rather close to the average slicing ratio;
- for WCC, the median slicing ratio is often much smaller than the average slicing ratio.

These results are rather unsurprising, given that we saw (Sect. 7.3) that most WCC slices are very small; hence the median is also very small whereas the few large slices boost the average.

Table 7. The **median** of slicing ratios computed using WCC or SCC and data dependence is given for each of the thirteen EFSMs (with the slicing criterion always a singleton).

Group	EFSMs	SCC (sc = 1)	WCC (wc = 1)
A	ATM	34.78%	13.04%
	Cashier	42.86%	9.52%
	CruiseControl	29.41%	29.41%
	FuelPump	24%	4%
	PrinTok	60.67%	1.12%
	VendingMachine	82.14%	3.57%
B	INRES	61.11%	5.56%
	DoorController	66.67%	8.33%
	TCP	52.63%	52.63%
	TCSbin	84.21%	81.54%
C	Alarm	96%	4%
	WaterPump	42.86%	7.14%
	Tamagotchi	84.21%	2.63%

A third way of summarizing the data, not considered in this work, would be to compute the *geometric* mean, rather than the arithmetic mean.

7.4 Results Concerning RQ2(a): Comparing WCC to Previous Control Dependences

To give a tentative answer to the first part of the second research question, comparing WCC to previous definitions of non-termination *in*sensitive control dependence, we compute for each EFSM M (of the thirteen given): avg-ratio(M, NTICD, 1) and avg-ratio(M, UNTICD, 1); we even compute avg-ratio(M, DD, 1) where DD generates slices using only data dependence. The results, together with the corresponding results for WCC (taken from Table 6), are given in Table 8, whose last line lists total-avg-ratio(WCC, 1), total-avg-ratio(NTICD, 1), total-avg-ratio(UNTICD, 1), and total-avg-ratio(DD, 1).

Table 8. Average slicing ratios, for singleton slicing criteria, for methods insensitive to non-termination.

Group	EFSMs	WCC	NTICD	UNTICD	DD
A	ATM	16.45%	23.25%	23.25%	14.18%
	Cashier	14.74%	67.57%	67.57%	11.34%
	CruiseControl	41.86%	70.24%	70.24%	31.49%
	FuelPump	8.8%	9.28%	9.28%	7.52%
	PrinTok	6.02%	60.45%	60.45%	1.88%
	VendingMachine	32.40%	34.69%	34.69%	18.88%
B	INRES	25.31%	15.74%	60.19%	15.74%
	DoorController	13.19%	11.81%	38.19%	11.81%
	TCP	41.36%	6.77%	42.97%	6.77%
	TCSbin	42.84%	5.61%	82.06%	5.61%
C	Alarm	4%	4%	92.32%	4%
	WaterPump	7.14%	7.14%	40.31%	7.14%
	Tamagotchi	2.63%	2.63%	82.48%	2.63%
Total Average		21.11%	24.55%	54.15%	10.69%

For the EFSMs in group A, each with a unique exit state, UNTICD coincides with NTICD (as the final transition that leads to the exit state forms a trivial control sink where "fairness" holds vacuously). On the other hand, WCC appears to produce slices that are somewhat smaller, at least on average. On closer inspection, we found that only in the VendingMachine are the sizes of some of the slices produced using NTICD smaller than those produced using WCC. This is because WCC includes "order" dependences, which NTICD/UNTICD don't (see Fig. 10 and Table 2 for an example).

For the EFSMs in group B or C, the control sinks are typically large, and as UNTICD coincides with (the non-termination sensitive) NTSCD within a control sink, UNTICD produces large slices. On the other hand, NTICD (that considers only "fair" paths within control sinks) will add no transitions to the slices within control sinks (apart from what has already been added by DD). For the EFSMs in group B, it is thus no wonder that NTICD produces smaller slices than WCC does (e.g. see Table 4 for the size of the slices produced for the DoorController with respect to transition T11), but one should keep in mind that no correctness property has ever been proven, or even stated, for NTICD.

For the EFSMs in group C, having no variables and hence no data dependences, WCC as well as NTICD gives slices of size 1 when given a slicing criterion of size 1.

We have carried out further investigations where we focus on non-trivial slices, and therefore remove from consideration all slicing criteria that result in a slice of size one (in particular, only EFSMs from groups A and B were taken into account). Doing so, we found that the total average slicing ratio for WCC is 51.04% (still, 20.24% of all WCC slices are equal to those computed using only DD). We can compare this total to the results from [5] (which also does not consider slices of size 1, in particular not the EFSMs from group C, but does consider all the EFSMs from groups A and B); there it is found that NTICD produces an average slicing ratio of 49.48% (but remember that for group B this ratio does not mean much) while UNTICD produces an average slicing ratio of 66.83%.

7.5 Results Concerning RQ2(b): Comparing SCC to Previous Control Dependences

To give a tentative answer to the second part of the second research question, comparing SCC to previous definitions of non-termination *sensitive* control dependence, we compute avg-ratio(M, NTSCD, 1) for each EFSM M (of the thirteen given). The results, together with the corresponding results for SCC (taken from Table 6) and UNTICD (taken from Table 8), are given in Table 9, whose last line lists total-avg-ratio(SCC, 1), total-avg-ratio(NTSCD, 1), and total-avg-ratio(UNTICD, 1).

Table 9. Average slicing ratios, for singleton slicing criteria, for methods sensitive to non-termination.

Group	EFSMs	SCC	NTSCD	UNTICD
A	ATM	33.46%	32.70%	23.25%
	Cashier	41.95%	71.66%	67.57%
	CruiseControl	46.37%	70.24%	70.24%
	FuelPump	28%	26.72%	9.28%
	PrinTok	60.45%	60.45%	60.45%
	VendingMachine	79.97%	79.97%	34.69%
B	INRES	60.19%	60.19%	60.19%
	DoorController	56.94%	56.94%	38.19%
	TCP	54.05%	42.97%	42.97%
	TCSbin	83.6%	82.06%	82.06%
C	Alarm	92.32%	92.32%	92.32%
	WaterPump	40.31%	40.31%	40.31%
	Tamagotchi	82.48%	82.48%	82.48%
Total average		58.46%	61.46%	54.15%

First observe that since UNTICD coincides with NTSCD within a control sink, it is not surprising that UNTICD often equals NTSCD (but otherwise produces smaller slices). Next we compare SCC to NTSCD and see that on average, SCC produces much smaller slices for two EFSMs (Cashier and Cruise-Control), but for four EFSMs produces slices that are slightly (except for TCP) larger; for the remaining EFSMs, the (average) slicing ratios are identical. Moreover, SCC has a slightly smaller total average slicing ratio than NTSCD.

It may seem discomforting that SCC sometimes produces slices that are larger than those produced by NTSCD. But keep in mind that no correctness property has been proven, or even stated, for NTSCD in the context of EFSMs. In the context of programs expressed as CFGs, the control dependence with that name [26] has been proven correct, but for irreducible CFGs only when applied together with a kind of "order dependence" as described in [27]. The properties WCC and SCC have been designed to include the notion of order dependence.

8 Conclusion

We have proposed algorithms, running in low polynomial time, for slicing EFSMs. The algorithms are based on notions (a "weak" and a "strong") of *commitment closure* adapted from the work by Danicic et al. [9] who originally phrased them for programs represented as control-flow graphs. We have proposed new semantic definitions of "correct slices" that integrate interaction, non-determinism and non-termination, and proved that these criteria are satisfied by slices produced by our algorithms.

We have conducted experiments using both benchmark and real world production EFSMs to measure the practical usefulness of our slicing algorithms and to compare them with slices computed using existing definitions of control dependence. Slicing wrt. "weak" commitment closure (WCC) will often significantly reduce the size of the EFSMs (the average relative slice size is 21% if the slicing criterion is a singleton), while slicing wrt. "strong" commitment closure will often give a modest size reduction (the average relative slice size is 58%). In both cases, this is typically smaller than what one gets by using algorithms based on previous definitions with similar aims, but not always—this reflects that we found that previous algorithms do not always satisfy the correctness properties presented in this paper.

To establish a firm foundation for our approach, we would like to eventually use a proof assistant to formally verify our algorithms, as was recently done by Léchenet et al. [22] for the computation of weak commitment closure (but not yet handling data dependence) in the original work by Danicic et.al. [9]. Also Wasserrab [32] and Blazy et al. [7] have formally verified algorithms for slicing.

To model bidirectional interaction with the environment, we would like to consider EFSMs that not just consume events but also *generate* them, and investigate which new notions are needed in order to accommodate the slicing of such EFSMs.

In order to make slicing even more useful, it is desirable to generate smaller slices than we currently do. To do so, without compromising correctness, recall

(Sect. 4.3) that WCC was motivated as a sufficient condition for weak soundness, *no matter the values of the guards* (similarly for SCC wrt. strong soundness). While simple to implement, this is very conservative. We would like to investigate a more liberal version of WCC that while still sufficient for weak soundness may allow a node to have more than one observable, as long as certain guards have certain values. Equipped with a suitable data flow analysis, we believe this will result in significantly fewer transitions being "sucked" into a slice.

A Proofs of Completeness and (Weak/Strong) Soundness

We shall now prove the correctness theorems from Sect. 4, but shall first establish some generally applicable results.

A.1 General Results

Lemma 31. *Assume that $\mathcal{L}$ is closed under $\rightarrow_{\text{dd}}$, and that $t \notin \mathcal{L}$. With $n = \mathsf{S}(t)$ and $n' = \mathsf{T}(t)$ we then have $Rv_{\mathcal{L}}(n') \subseteq Rv_{\mathcal{L}}(n)$, and if $i \vdash t : (n, s) \xrightarrow{E} (n', s')$ for some $i \in \{1, 2\}$ then $s(v) = s'(v)$ for all $v \in Rv(n')$.*

Proof. Consider $v \in Rv(n')$. There exists a path $[t_1..t_k]$ with $k \geq 1$ from n' such that $t_k \in \mathcal{L}$ with $v \in \mathsf{U}(t_k)$ and such that $v \notin \mathsf{D}(t_j)$ for all $j \in 1 \ldots k-1$.

Assume, to get a contradiction, that $v \in \mathsf{D}(t)$. Then $t \rightarrow_{\text{dd}} t_k$, which together with $t_k \in \mathcal{L}$ and $t \notin \mathcal{L}$ contradicts $\mathcal{L}$ being closed under $\rightarrow_{\text{dd}}$.

Hence $v \notin \mathsf{D}(t)$. Thus the path $[t, t_1..t_k]$ will establish $v \in Rv(n)$, and if $i \vdash t : (n, s) \xrightarrow{E} (n', s')$ then $s'(v) = s(v)$ (no matter whether $i = 1$ or $i = 2$).

Lemma 32. *Assume that for $t \in \mathcal{L}$ we have*

$1 \vdash t : (n, s_1) \xrightarrow{E} (n', s'_1)$ *and*
$2 \vdash t : (n, s_2) \xrightarrow{E} (n', s'_2)$

where $s_1(v) = s_2(v)$ for all $v \in Rv(n)$. Then $s'_1(v) = s'_2(v)$ for all $v \in Rv(n')$.

Proof. Let $v \in Rv(n')$ be given, so as to show $s'_1(v) = s'_2(v)$. We split into two cases.

The first case is when $v \notin \mathsf{D}(t)$. Then $v \in Rv(n)$, implying the desired equality $s'_1(v) = s_1(v) = s_2(v) = s'_2(v)$.

The second case is when $v \in \mathsf{D}(t)$, where we must prove that $[\![A]\!]s_1 = [\![A]\!]s_2$ where $A = \mathsf{A}(t)[c/v_b]$ if E is of the form $e(c)$, and $A = \mathsf{A}(t)$ otherwise. But this follows from the fact that for all $w \in fv(A) = fv(\mathsf{A}(t)) \setminus \{v_b\}$ we have $s_1(w) = s_2(w)$, since $w \in \mathsf{U}(t)$ and thus (as $t \in \mathcal{L}$) also $w \in Rv(t) \subseteq Rv(n)$.

A.2 Proving Completeness

We shall first establish some intermediate results:

Lemma 33. *Assume that $\mathcal{L}$ is closed under $\rightarrow_{\mathsf{dd}}$. If with $t \notin \mathcal{L}$ we have*

- $1 \vdash t : (n, s_1) \xrightarrow{E_1} (n', s_1')$ *and*
- $s_1(v) = s_2(v)$ *for all* $v \in Rv(n)$

then

- $2 \vdash t : (n, s_2) \xrightarrow{\varepsilon} (n', s_2)$ *and*
- $s_1'(v) = s_2(v)$ *for all* $v \in Rv(n')$.

where $\mathsf{filter}(E_1) = \varepsilon$ *if* $\mathcal{L}$ *is uniform.*

Proof. Since $t \notin \mathcal{L}$, we see that $\mathsf{G}_2(t) = \mathit{true}$ and $\mathsf{E}_2(t) = \varepsilon$ and $\mathsf{D}_2(t) = \emptyset$ which yields the first claim. For the second claim, consider $v \in Rv(n')$: since $t \notin \mathcal{L}$, we infer by Lemma 31 that $s_1(v) = s_1'(v)$, and that $v \in Rv(n)$ which by assumption implies $s_1(v) = s_2(v)$; hence we can conclude that $s_1'(v) = s_2(v)$. The last claim follows from Fact 11.

Lemma 34. *If with* $t \in \mathcal{L}$ *we have*

- $1 \vdash t : (n, s_1) \xrightarrow{E} (n', s_1')$ *and*
- $s_1(v) = s_2(v)$ *for all* $v \in Rv(n)$

then there exists s_2' *such that*

- $2 \vdash t : (n, s_2) \xrightarrow{E} (n', s_2')$ *and*
- $s_1'(v) = s_2'(v)$ *for all* $v \in Rv(n')$.

Proof. Our assumptions entail $n = \mathsf{S}(t)$, $n' = \mathsf{T}(t)$, $\mathsf{G}_1(t) = \mathsf{G}_2(t) = \mathsf{G}(t)$, and $[\![g]\!]s_1 = \mathit{true}$ where $g = \mathsf{G}(t)[c/v_b]$ if E is of the form $e(c)$, and $g = \mathsf{G}(t)$ otherwise. For an arbitrary $w \in \mathit{fv}(\mathsf{G}(t)) \setminus \{v_b\}$ we infer by Fact 13 that $w \in Rv(t) \subseteq Rv(n)$, implying $s_1(w) = s_2(w)$. Hence also $[\![g]\!]s_2 = \mathit{true}$, implying that there exists s_2' such that $2 \vdash t : (n, s_2) \xrightarrow{E} (n', s_2')$. That $s_1'(v) = s_2'(v)$ for all $v \in Rv(n')$ now follows from Lemma 32.

Theorem 1 (Completeness): Assume that $\mathcal{L}$ is closed under $\rightarrow_{\mathsf{dd}}$. If

- $1 \vdash_{\mathrm{ni}} \pi : C_1 \xrightarrow{E} C_1'$ and
- $C_1 \; Q \; C_2$

then there exists C_2' such that

- $2 \vdash \pi : C_2 \xrightarrow{E} C_2'$ and
- $C_1' \; Q \; C_2'$.

Moreover, if $\mathcal{L}$ is uniform then $2 \vdash_{\mathrm{ni}} \pi : C_2 \xrightarrow{\mathsf{filter}(E)} C_2'$.

Proof. We do induction in the length of π. If π is empty, then $C_1' = C_1$ and $E_1 = \varepsilon$, and the claim is trivial with $C_2' = C_2$.

If π is a singleton t, we split into two cases: if $t \in \mathcal{L}$ then the claim follows from Lemma 34; if $t \notin \mathcal{L}$ then the claim follows from Lemma 33.

Now assume that π can be written $\pi = \pi''\pi'$ with π' and π'' not empty. It is easy to see that there exists E'' and E' with $E = E''E'$ such that $1 \vdash_{\mathrm{ni}} \pi'' : C_1 \xrightarrow{E''} C_1''$ and $1 \vdash_{\mathrm{ni}} \pi' : C_1'' \xrightarrow{E'} C_1'$. Inductively on π'', there exists C_2'' with $C_1''\ Q\ C_2''$ such that $2 \vdash \pi'' : C_2 \xrightarrow{E''} C_2''$; inductively on π', there then exists C_2' with $C_1'\ Q\ C_2'$ such that $2 \vdash \pi' : C_2'' \xrightarrow{E'} C_2'$. But then it is easy to see that we have the desired $2 \vdash \pi : C_2 \xrightarrow{E} C_2'$. Finally, if $\mathcal{L}$ is uniform, we inductively get $2 \vdash_{\mathrm{ni}} \pi'' : C_2 \xrightarrow{\mathsf{filter}(E'')} C_2''$ and $2 \vdash_{\mathrm{ni}} \pi' : C_2'' \xrightarrow{\mathsf{filter}(E')} C_2'$ and thus clearly also $2 \vdash_{\mathrm{ni}} \pi : C_2 \xrightarrow{\mathsf{filter}(E'')\mathsf{filter}(E')} C_2'$ which yields the claim since $\mathsf{filter}(E) = \mathsf{filter}(E'')\mathsf{filter}(E')$.

A.3 Proving Soundness

As discussed in Sect. 3.4, we cannot quite hope for the converse of Theorem 1 in that the original EFSM may get stuck, or loop, rather than reach the next observable state. This is formalized by the following result:

Lemma 35. *Let $obs(n) = \{m\}$. Given s, one of the 3 cases below applies:*

1. *there exists E and s' such that $1 \vdash_{ni} (n, s) \overset{E}{\Longrightarrow} (m, s')$*
2. *the original EFSM gets stuck from (n, s) avoiding $\mathcal{L}$*
3. *the original EFMS loops from (n, s) avoiding $\mathcal{L}$.*

Proof. Consider the following iterative algorithm, incrementally constructing n_j, s_j, E_j for $j \geq 0$. With $n_0 = n$, $s_0 = s$ and $E_0 = \varepsilon$, the invariant is that

$$1 \vdash_{\mathrm{ni}} (n, s) \overset{E_j}{\Longrightarrow} (n_j, s_j) \text{ with } n_0..n_{j-1} \text{ not the source of a transition in } \mathcal{L}.$$

This trivially holds for $j = 0$. For each iteration (with $j \geq 0$), there are 3 possible actions: if $n_j = m$ we exit the loop, and we have established case 1 with $E = E_j$ and $s' = s_j$.

Otherwise, if $n_j \neq m$, we can infer that n_j is not the source of a transition in $\mathcal{L}$, since if $n_j = \mathsf{S}(t)$ with $t \in \mathcal{L}$ then $n_j \in obs(n) = \{m\}$. There are now two possibilities:

- if there exists t with $\mathsf{S}(t) = n_j$ such that $[\![\mathsf{G}(t)]\!]s_j = true$, or such that $[\![\mathsf{G}_i(t)[c/v_b]]\!]s = true$ for some c if $\mathsf{E}_i(t)$ is of the form $e(v_b)$, we define t_{j+1} as one such t (say the "smallest"). Then there exists n_{j+1} and s_{j+1} and E_j' such that

$$1 \vdash t_j : (n_j, s_j) \overset{E_j'}{\to} (n_{j+1}, s_{j+1})$$

  Let $E_{j+1} = E_j E_j'$. We then increment j by one and repeat the loop; the invariant will be maintained since $t \notin \mathcal{L}$ and n_j is not the source of a transition in $\mathcal{L}$.

– otherwise, we exit the loop, concluding that (n_j, s_j) is 1-stuck which establishes case 2.

If we never exit the loop, this will establish case 3.

Theorem 2 (Weak Soundness): Assume that $\mathcal{L}$ is closed under $\rightarrow_{dd}$ and satisfies $\mathcal{WCC}$. If $2 \vdash^t_{ni} C_2 \stackrel{E_2}{\Longrightarrow} C'_2$ and $C_1 \; Q \; C_2$ then there are 3 possibilities:

1. $1 \vdash^t_{ni} C_1 \stackrel{E_1 E_2}{\Rrightarrow} C'_1$ for some C'_1, E_1 with $C'_1 \; Q \; C'_2$ (and if $\mathcal{L}$ is uniform then $\mathsf{filter}(E_1) = \varepsilon$), or
2. the original EFSM gets stuck from C_1 avoiding $\mathcal{L}$ (cf. Definition 15), or
3. the original EFSM loops from C_1 avoiding $\mathcal{L}$ (cf. Definition 16).

Proof. Let $C_2 = (n, s_2)$ and $C'_2 = (n', s'_2)$ and $C_1 = (n, s_1)$. From Definition 9 and Fact 8 we see that $t \in \mathcal{L}$ and that for some m: $2 \vdash_{ni} (n, s_2) \stackrel{\varepsilon}{\Longrightarrow} (m, s_2)$ and $2 \vdash t : (m, s_2) \stackrel{E_2}{\rightarrow} (n', s'_2)$. Thus $m \in obs(n)$, and from the $\mathcal{WCC}$ property we infer $obs(n) = \{m\}$. Lemma 35 now tells us that either

1. there exists E_1 and s''_1 such that $1 \vdash_{ni} (n, s_1) \stackrel{E_1}{\Longrightarrow} (m, s''_1)$, or
2. the original EFSM gets stuck from (n, s_1) avoiding $\mathcal{L}$, or
3. the original EFSM loops from (n, s_1) avoiding $\mathcal{L}$.

If case 2 or case 3 holds, we are done. We thus assume that case 1 holds (and from Fact 11 we see that if $\mathcal{L}$ is uniform then $\mathsf{filter}(E_1) = \varepsilon$).

By assumption, we have $s_1(v) = s_2(v)$ for all $v \in Rv(n)$. By repeated application of Lemma 31, we infer that $Rv(m) \subseteq Rv(n)$ and that $s''_1(v) = s_1(v)$ for all $v \in Rv(m)$. Thus $s''_1(v) = s_2(v)$ for all $v \in Rv(m)$, which establishes

$$(m, s''_1) \; Q \; (m, s_2).$$

From $t \in \mathcal{L}$ we have $\mathsf{G}_1(t) = \mathsf{G}_2(t) = \mathsf{G}(t)$, and from $2 \vdash t : (m, s_2) \stackrel{E_2}{\rightarrow} (n', s'_2)$ we know that $[\![g]\!]s_2 = true$ where $g = \mathsf{G}(t)[c/v_b]$ if E_2 is of the form $e(c)$, and $g = \mathsf{G}(t)$ otherwise. For an arbitrary $w \in fv(\mathsf{G}_1(t)) \setminus \{v_b\}$ we infer by Fact 13 that $w \in Rv(t) \subseteq Rv(m)$, implying $s''_1(w) = s_2(w)$. Hence also $[\![g]\!]s''_1 = true$, implying that there exists s'_1 such that $1 \vdash t : (m, s''_1) \stackrel{E_2}{\rightarrow} (n', s'_1)$, and thus with $C'_1 = (n', s'_1)$ we have $1 \vdash^t_{ni} C_1 \stackrel{E_1 E_2}{\Rrightarrow} C'_1$. That $C'_1 \; Q \; C'_2$, that is $s'_1(v) = s'_2(v)$ for all $v \in Rv(n')$, now follows from Lemma 32.

If $\mathcal{L}$ satisfies not just $\mathcal{WCC}$ but also $\mathcal{SCC}$, we can rule out case 3, which establishes

Theorem 3 (Strong Soundness): Assume $\mathcal{L}$ is closed under $\rightarrow_{dd}$ and satisfies $\mathcal{SCC}$.

If $2 \vdash^t_{ni} C_2 \stackrel{E_2}{\Longrightarrow} C'_2$ and $C_1 \; Q \; C_2$ then there are 2 possibilities:

1. $1 \vdash^t_{ni} C_1 \stackrel{E_1 E_2}{\Rrightarrow} C'_1$ for some C'_1, E_1 with $C'_1 \; Q \; C'_2$ (and if $\mathcal{L}$ is uniform then $\mathsf{filter}(E_1) = \varepsilon$), or
2. the original EFSM gets stuck from C_1 avoiding $\mathcal{L}$.

B Proofs of Correctness for Algorithms Computing Least Slices

Theorem 5 (WCC Algorithm Correctness): Assuming that the table DD correctly computes data dependence, Algorithm 1 returns in L the least superset of $\mathcal{L}$ that is closed under data dependence and satisfies $\mathcal{WCC}$.

Proof. We shall first state a number of loop invariants for the inner loop; they are expressed in terms of k, the number of iterations so far. We shall use C^k to denote the value of C after k iterations; similarly we shall write V^k and X^k (it is convenient to let $V^{-1} = \emptyset$). For $k \geq 0$ and each state n we define $obs_L^k(n)$ as the set of states n' such that there exists t in L with $\mathsf{S}(t) = n'$, and there exists a cycle-free path of length $\leq k$ outside L from n to n'. We observe that $n' \in obs_L^0(n)$ iff $n' = n$ and n in B, that $obs_L^k(n) \subseteq obs_L^{k'}(n)$ if $k \leq k'$, and that $n' \in obs_L(n)$ iff there exists $k \geq 0$ such that $n' \in obs_L^k(n)$.

The **while** loop invariants are, with $k \geq 0$:

1. $V^k = \{n \mid X^k[n] \text{ is defined }\}$
2. V^k is the disjoint union of V^{k-1} and C^k
3. for all $n \in V^k$, $X^k[n] \in obs_L^k(n)$
4. for all n, if $obs_L^k(n) \neq \emptyset$ then $n \in V^k$
5. if $\mathtt{Lnew}^k = \emptyset$ then $obs_L^k(n) = \{X^k[n]\}$ for all $n \in V^k$
6. if L' satisfies $\mathcal{WCC}$ with $\mathtt{L} \subseteq L'$ then $\mathtt{Lnew} \subseteq L'$

It is easy to see that all invariants hold at loop entry, with $k = 0$. We shall now argue that the invariants are maintained by the body of the while loop; in particular, that they hold for $k+1$ where we can assume that they hold for k.

1. This follows easily from inspecting the code.
2. This follows easily from inspecting the code.
3. The claim is obvious if $X^{k+1}[n] = X^k[n]$ as then $X^k[n] \in obs_L^k(n)$ (as invariant 3 holds before the iteration) and thus $X^{k+1}[n] \in obs_L^{k+1}(n)$.
   Next consider the situation where $X^{k+1}[n]$ assumes the value of $X^k[m]$ because there exists $t \notin \mathtt{L}$ with $\mathsf{T}(t) = m$ and $\mathsf{S}(t) = n$ where $m \in C^k$. Since invariant 3 holds before the iteration, we know that $X^k[m] \in obs_L^k(m)$. As it is easy to see (as $t \notin \mathtt{L}$) that $obs_L^k(m) \subseteq obs_L^{k+1}(n)$, this yields the desired $X^{k+1}[n] \in obs_L^{k+1}(n)$.
4. We assume that $obs_L^{k+1}(n) \neq \emptyset$. If also $obs_L^k(n) \neq \emptyset$, we know (since invariant 4 holds) that $n \in V^k$ and thus $n \in V^{k+1}$.
   We can thus assume that $obs_L^k(n) = \emptyset$, and infer that there exists $t \notin \mathtt{L}$ with $\mathsf{S}(t) = n$ such that with $m = \mathsf{T}(t)$ it holds that $obs_L^k(m)$ is non-empty, but $obs_L^{k-1}(m) = \emptyset$. From invariants 4, 1, 3 and 2 we infer $m \in V^k$ and even $m \in C^k$. But then the iteration will add n to V so that $n \in V^{k+1}$.
5. We assume that Lnew remains empty; since we have already established invariant 3, our task is to prove that if $n' \in obs_L^{k+1}(n)$ then $n' = X^{k+1}[n]$. If $n' \in obs_L^k(n)$, we know (from invariants 4 and 5) that $n' = X^k[n]$ and thus also $n' = X^{k+1}(n)$.

So assume that $n' \notin obs_L^k(n)$. The situation is that there exists $t \notin \mathtt{L}$ with $n = \mathsf{S}(t)$ such that with $m = \mathsf{T}(t)$ we have $n' \in obs_L^k(m)$ but $n' \notin obs_L^{k-1}(m)$. From $n' \in obs_L^k(m)$, and `Lnew` being empty before the iteration (as otherwise the loop would exit), we see (as invariants 4 and 5 hold) that $m \in V^k$ and $X^k[m] = n'$ and $obs_L^k(m) = \{n'\}$. From $n' \notin obs_L^{k-1}(m)$ we infer that $obs_L^{k-1}(m) = \emptyset$, and thus (from invariants 3 and 2) we have $m \in C^k$. Thus t is considered during the iteration, and since `Lnew` remains empty, we infer that $X^{k+1}[n] = n'$ (as $X[n]$ will be assigned either due to t, or some other transition being considered before t).

6. Let t be a member of $\mathtt{Lnew}^{k+1}$. With $n = \mathsf{S}(t)$ and $m = \mathsf{T}(t)$, we have $n \in V^{k+1}$ and $m \in C^k$, and there exists n' and m' with $n' \neq m'$ such that $n' = X^{k+1}[n]$ and $m' = X^k[m]$. From invariant 3 we see that $n' \in obs_L^{k+1}(n)$ and that $m' \in obs_L^k(m)$ which implies $m' \in obs_L^{k+1}(n)$. There thus exists a cycle-free path π_1 from n to n', and a cycle-free path π_2 from n to m'.
These paths can have nothing in common. For assume, to get a contradiction, that some node n'' occurs on both paths. Then there exists j with $j \leq k$ such that $obs_L^j(n'')$ is not a singleton. From invariant 5 we infer that $\mathtt{Lnew}^j$ is non-empty, but then the loop would exit after j iterations, which is a contradiction.
Now assume that L' contains `L` and satisfies $\mathcal{WCC}$. In particular, $obs_{L'}(n)$ has to be at most a singleton. But since $obs_{L'}(n)$ has to contain a node from π_1, and also a node from π_2, this is only possible if $obs_{L'}(n) = \{n\}$. But this can only happen if $t \in L'$.

After having stated and proved the invariants for the **while** loop, let us now address the invariant for the **repeat** loop. It obviously holds initially. We shall now argue it is maintained by an iteration, which is obvious for the part about being closed under data dependence (due to the call of `DDclose` at the end of the body).

Now let L' be a superset of $\mathcal{L}$ that is closed under data dependence and satisfies $\mathcal{WCC}$. Our goal is to show that after an iteration, $\mathtt{L} \subseteq L'$. From the invariant, we know that $\mathtt{L} \subseteq L'$ holds before the iteration. From invariant 6 for the **while** loop, we see that $\mathtt{Lnew} \subseteq L'$. By the specification of `DDclose`, we infer that $\mathtt{DDclose(L,Lnew)} \subseteq L'$. But this is what we aimed to prove.

We are left with proving that `L` does indeed satisfy $\mathcal{WCC}$ when the **repeat** loop exits, with $\mathtt{Lnew} = \emptyset$. Let the last iteration have k iterations of the **while** loop; thus $C^k = \emptyset$ but $C^j \neq \emptyset$ for $j < k$.

Now assume, to get a contradiction, that `L` does *not* satisfy $\mathcal{WCC}$.

There thus exists i and m such that $obs_L^i(m)$ has at least two elements. We can assume that i is the least such number, that is: for $j < i$ and all n, $obs_L^j(n)$ is at most a singleton. We infer that there exists n' such that $n' \in obs_L^i(m)$, and that there exists m' with $m' \neq n'$ such that $m' \in obs_L^i(m)$ but $m' \notin obs_L^{i-1}(m)$. There thus exists $m_0..m_i$ with $m_i = m$ and $m_0 = m'$, and a cycle-free path through $m_i...m_0$ from m to m', such that for all $j \in 0 \ldots i$ it holds that $m' \in obs_L^j(m_j)$ but $m' \notin obs_L^{j-1}(m_j)$. For $j < i$ we know, as $obs_L^j(m_j)$ is at most a singleton, that $obs_L^j(m_j) = \{m'\}$, and thus $obs_L^{j-1}(m_j) = \emptyset$ from which we infer that $m_j \in V^j \setminus V^{j-1} = C^j$. Since $C^j \neq \emptyset$ for $j < i$, and $C^k = \emptyset$, we infer that $i \leq k$. Thus $obs_L^k(m)$ has at least two elements. But this conflicts with invariant 5 for the **while** loop, giving the desired contradiction.

Theorem 7 (SCC Algorithm Correctness): Assuming that the table DD correctly computes data dependence, Algorithm 2 (using Algorithm 3) returns in L the least superset of $\mathcal{L}$ that is closed under data dependence and satisfies $\mathcal{SCC}$.

Proof. The proof is quite similar to the proof of Theorem 5; we shall only list the features to be added.

First, we need to add the following invariant for the **while** loop:

$$\text{If } \texttt{Lnew}^k = \emptyset \text{ and } n' \in X^k[n] \text{ then no infinite path from } n \text{ avoids } n' \quad (1)$$

To prove this, first observe that if $n = n'$, and hence when $k = 0$, the claim is trivial. We may thus assume that $k > 0$ and $n' \neq n$. By invariant 3 we have $n' \in obs_L^k(n)$. Thus there exists $t \notin \texttt{L}$ with $n = \mathsf{S}(t)$ such that with $m = \mathsf{T}(t)$ we have $n' \in obs_L^{k-1}(m)$ and by invariant 4 thus $m \in V^{k-1}$. We infer that $m \in C^j$ for some $j \leq k - 1$ and hence the $(j + 1)$th iteration will examine t. Since $\texttt{Lnew}^{j+1} = \emptyset$, we infer that LoopAvoids(n, n') does not hold, and hence (by Fact 24) there is no infinite path from n that avoids n'.

Also, we need to modify Invariant 6 for the **while** loop into:

$$\text{if } L' \text{ satisfies } \mathcal{SCC} \text{ with } \texttt{L} \subseteq L' \text{ then } \texttt{Lnew} \subseteq L' \quad (2)$$

To see that this invariant is maintained, assume that L' contains L and satisfies $\mathcal{SCC}$ (and thus also $\mathcal{WCC}$), and that t is a member of $\texttt{Lnew}^{k+1}$; we must show that $t \in L'$.

If t was added by line 17 of Algorithm 2, we can reason as in the proof of Theorem 7 to show that $t \in L'$.

We can therefore assume that t was added by Line 14. With $n = \mathsf{S}(t)$ and $m = \mathsf{T}(t)$ and $m' = X^k[m]$, so that there is a cycle-free path π outside L from n to m' that starts with t, there thus is a infinite path from n that avoids m'.

Assume, to get a contradiction, that $t \notin L'$. Then $obs_{L'}(n)$ (which obviously cannot be empty) would consist of a node in π different from n, say n'. There exists $i \leq k$ such that $m' = X^i[n']$, and as $\texttt{Lnew}^i = \emptyset$ we see by (1) that no infinite path from n' avoids m'. But then there is an infinite path from n that avoids n', as otherwise no infinite path from n would avoid m'. Since $obs_{L'}(n) = \{n'\}$, this contradicts L' satisfying $\mathcal{SCC}$. We conclude that $t \in L'$, as desired.

Having proved that (2) and (1) are indeed extra invariants for the **while** loop, we can now proceed as in the proof of Theorem 5 to show the correctness of the invariant listed in Algorithm 2 for the **repeat** loop.

We are left with proving that L does indeed satisfy $\mathcal{SCC}$ when the **repeat** loop exits, with $\texttt{Lnew} = \emptyset$. Let the last iteration have k iterations of the **while** loop; thus $C^k = \emptyset$ but $C^j \neq \emptyset$ for $j < k$.

Now assume, to get a contradiction, that L does *not* satisfy $\mathcal{SCC}$. There can be two reasons for this: if L does not satisfy $\mathcal{WCC}$, we can show as in the proof of Theorem 5 that this conflicts with invariant 5 for the **while** loop.

We can thus assume that L satisfies $\mathcal{WCC}$ but *not* $\mathcal{SCC}$. Then there exists n and n' such that $n' \in obs_L(n)$, and yet there is an infinite path from n that avoids n'. Let i be the smallest number such that $n' \in obs_L^i(n)$. There thus exists $n_0..n_i$ with $n_i = n$ and $n_0 = n'$, and a cycle-free path through $n_i...n_0$ from n to n', such that for all $j \in 0 \ldots i$ it holds that $n' \in obs_L^j(n_j)$ but $n' \notin obs_L^{j-1}(n_j)$. For $j \leq i$ we know, as $obs_L^j(n_j)$ is at most a singleton, that $obs_L^j(n_j) = \{n'\}$, and thus $obs_L^{j-1}(n_j) = \emptyset$ from which we infer that $n_j \in V^j \setminus V^{j-1} = C^j$. Since $C^j \neq \emptyset$ for $j \leq i$, and $C^k = \emptyset$, we infer that $i < k$. Thus $n' \in obs_L^k(n)$, and hence (by invariant 5 for the **while** loop) $n' = X^k[n]$. But then an infinite path from n that avoids n' conflicts with (1), giving the desired contradiction.

References

1. Amtoft, T.: Slicing for modern program structures: a theory for eliminating irrelevant loops. Inf. Process. Lett. **106**(2), 45–51 (2008). https://doi.org/10.1016/j.ipl.2007.10.002
2. Androutsopoulos, K., Clark, D., Harman, M., Hierons, R.M., Li, Z., Tratt, L.: Amorphous slicing of extended finite state machines. IEEE Trans. Softw. Eng. **39**(7), 892–909 (2013)
3. Androutsopoulos, K., Clark, D., Harman, M., Krinke, J., Tratt, L.: State-based model slicing: a survey. ACM Comput. Surv. **45**(4), 53:1–53:36 (2013)
4. Androutsopoulos, K., Clark, D., Harman, M., Li, Z., Tratt, L.: Control dependence for extended finite state machines. In: Chechik, M., Wirsing, M. (eds.) FASE 2009. LNCS, vol. 5503, pp. 216–230. Springer, Heidelberg (2009). https://doi.org/10.1007/978-3-642-00593-0_15
5. Androutsopoulos, K., Gold, N., Harman, M., Li, Z., Tratt, L.: A theoretical and empirical study of EFSM dependence. In: 25th IEEE International Conference on Software Maintenance (ICSM 2009), Edmonton, Alberta, Canada, 20–26 September 2009, pp. 287–296. IEEE Computer Society (2009)
6. Ball, T., Horwitz, S.: Slicing programs with arbitrary control-flow. In: Fritzson, P.A. (ed.) AADEBUG 1993. LNCS, vol. 749, pp. 206–222. Springer, Heidelberg (1993). https://doi.org/10.1007/BFb0019410
7. Blazy, S., Maroneze, A., Pichardie, D.: Verified validation of program slicing. In: Proceedings of the 2015 Conference on Certified Programs and Proofs, CPP 2015, pp. 109–117. ACM, New York (2015). https://doi.org/10.1145/2676724.2693169. http://doi.acm.org/10.1145/2676724.2693169
8. Bourhfir, C., Dssouli, R., Aboulhamid, E., Rico, N.: Automatic executable test case generation for extended finite state machine protocols. In: Kim, M., Kang, S., Hong, K. (eds.) Testing of Communicating Systems. ITIFIP, pp. 75–90. Springer, Boston, MA (1997). https://doi.org/10.1007/978-0-387-35198-8_6
9. Danicic, S., Barraclough, R.W., Harman, M., Howroyd, J.D., Kiss, A., Laurence, M.R.: A unifying theory of control dependence and its application to arbitrary program structures. Theoret. Comput. Sci. **412**(49), 6809–6842 (2011)
10. Derler, P., Lee, E.A., Sangiovanni-Vincentelli, A.L.: Modeling cyber-physical systems. Proc. IEEE **100**(1), 13–28 (2012). https://doi.org/10.1109/JPROC.2011.2160929
11. Hafemann Fragal, V., Simao, A., Mousavi, M.R.: Validated test models for software product lines: featured finite state machines. In: Kouchnarenko, O., Khosravi, R.

(eds.) FACS 2016. LNCS, vol. 10231, pp. 210–227. Springer, Cham (2017). https://doi.org/10.1007/978-3-319-57666-4_13
12. Ganapathy, V., Ramesh, S.: Slicing synchronous reactive programs. Electron. Notes Theoret. Comput. Sci. **65**(5), 50–64 (2002). SLAP 2002, Synchronous Languages, Applications, and Programming (Satellite Event of ETAPS 2002)
13. Gold, N.E., Binkley, D., Harman, M., Islam, S., Krinke, J., Yoo, S.: Generalized observational slicing for tree-represented modelling languages. In: Proceedings of the 2017 11th Joint Meeting on Foundations of Software Engineering, ESEC/FSE 2017, pp. 547–558. ACM, New York (2017)
14. Hall, M., McMinn, P., Walkinshaw, N.: Superstate identification for state machines using search-based clustering. In: Pelikan, M., Branke, J. (eds.) Proceedings of the 12th Annual Conference on Genetic and Evolutionary Computation (GECCO 2010), pp. 1381–1388. ACM (2010)
15. Harel, D., Politi, M.: Modeling Reactive Systems with Statecharts: The Statemate Approach, 1st edn. McGraw-Hill Inc., New York (1998)
16. Hatcliff, J., Corbett, J., Dwyer, M., Sokolowski, S., Zheng, H.: A formal study of slicing for multi-threaded programs with JVM concurrency primitives. In: Cortesi, A., Filé, G. (eds.) SAS 1999. LNCS, vol. 1694, pp. 1–18. Springer, Heidelberg (1999). https://doi.org/10.1007/3-540-48294-6_1
17. Hatcliff, J., Dwyer, M.B., Zheng, H.: Slicing software for model construction. High.-Order Symb. Comput. **13**(4), 315–353 (2000)
18. Ilie, L., Yu, S.: Reducing NFAs by invariant equivalences. Theoret. Comput. Sci. **306**(1–3), 373–390 (2003)
19. Kamischke, J., Lochau, M., Baller, H.: Conditioned model slicing of feature-annotated state machines. In: Proceedings of the 4th International Workshop on Feature-Oriented Software Development, FOSD 2012, pp. 9–16. ACM, New York (2012)
20. Korel, B., Singh, I., Tahat, L., Vaysburg, B.: Slicing of state-based models. In: IEEE International Conference on Software Maintenance (ICSM 2003), pp. 34–43. IEEE Computer Society Press, Los Alamitos, California, USA, September 2003 (2003)
21. Labbé, S., Gallois, J.: Slicing communicating automata specifications: polynomial algorithms for model reduction. Form. Asp. Comput. **20**(6), 563–595 (2008). https://doi.org/10.1007/s00165-008-0086-3
22. Léchenet, J.-C., Kosmatov, N., Le Gall, P.: Fast computation of arbitrary control dependencies. In: Russo, A., Schürr, A. (eds.) FASE 2018. LNCS, vol. 10802, pp. 207–224. Springer, Cham (2018). https://doi.org/10.1007/978-3-319-89363-1_12
23. Lity, S., Morbach, T., Thüm, T., Schaefer, I.: Applying incremental model slicing to product-line regression testing. In: Kapitsaki, G.M., Santana de Almeida, E. (eds.) ICSR 2016. LNCS, vol. 9679, pp. 3–19. Springer, Cham (2016). https://doi.org/10.1007/978-3-319-35122-3_1
24. Marwedel, P.: Embedded and cyber-physical systems in a nutshell. DAC.COM Knowl. Cent. Article no. 20 (2010)
25. Podgurski, A., Clarke, L.A.: A formal model of program dependences and its implications for software testing, debugging, and maintenance. IEEE Trans. Softw. Eng. **16**(9), 965–979 (1990)
26. Ranganath, V.P., Amtoft, T., Banerjee, A., Dwyer, M.B., Hatcliff, J.: A new foundation for control-dependence and slicing for modern program structures. In: Sagiv, M. (ed.) ESOP 2005. LNCS, vol. 3444, pp. 77–93. Springer, Heidelberg (2005). https://doi.org/10.1007/978-3-540-31987-0_7

27. Ranganath, V.P., Amtoft, T., Banerjee, A., Hatcliff, J., Dwyer, M.B.: A new foundation for control dependence and slicing for modern program structures. ACM Trans. Program. Lang. Syst. **29**(5) (2007). http://doi.acm.org/10.1145/1275497.1275502
28. Silva, J.: A vocabulary of program slicing-based techniques. ACM Comput. Surv. **44**(3), 12:1–12:41 (2012)
29. Sivagurunathan, Y., Harman, M., Danicic, S.: Slicing, I/O and the implicit state. In: Proceedings of the Third International Workshop on Automatic Debugging: Linköping, Sweden, AADEBUG 1997, pp. 59–67. Linköping Electronic Articles in Computer and Information Science, May 1997
30. Strobl, F., Wisspeintner, A.: Specification of an elevator control system - an autofocus case study. Technical report. TUM-I9906, Technische Universität München (1999)
31. Tip, F.: A survey of program slicing techniques. J. Program. Lang. **3**(3), 121–189 (1995)
32. Wasserrab, D.: From formal semantics to verified slicing. Ph.D. thesis, Karlsruher Institut für Technologie (2010)
33. Weiser, M.: Program slicing. IEEE Trans. Softw. Eng. **10**(4), 352–357 (1984)
34. Weiser, M.D.: Program slices: formal, psychological, and practical investigations of an automatic program abstraction method. Ph.D. thesis, University of Michigan, Ann Arbor, MI (1979)
35. Xu, B., Qian, J., Zhang, X., Wu, Z., Chen, L.: A brief survey of program slicing. ACM SIGSOFT Softw. Eng. Notes **30**(2), 1–36 (2005)
36. Zaghal, R.Y., Khan, J.I.: EFSM/SDL modeling of the original TCP standard (RFC793) and the congestion control mechanism of TCP Reno. Technical report. TR2005-07-22, Networking and Media Communications Research Laboratories, Department of Computer Science, Kent State University, July 2005 (2005)

7. [illegible] A new [illegible] (2007). https://doi.org/10.1007/[illegible]
8. [illegible] (2001)
9. [illegible] and the implicit state [illegible] In: [illegible] International [illegible] (2007) [illegible]
10. [illegible] systems [illegible] (2019)
11. [illegible]
12. [illegible] (2010)
13. [illegible] 10 [illegible]
14. [illegible] (2019)
15. [illegible] survey [illegible] (2016)
16. [illegible] the original TCP [illegible] Technical Report [illegible] Networking and [illegible] Communications Research [illegible] University [illegible] 2005 [illegible]

Security

Secure Guarded Commands

Flemming Nielson(✉) and Hanne Riis Nielson

Department of Mathematics and Computer Science,
Technical University of Denmark, 2800 Kgs. Lyngby, Denmark
{fnie,hrni}@dtu.dk

Abstract. We develop a lightweight approach to information flow control that interacts with the use of cryptographic schemes. The language is a version of Dijkstra's Guarded Commands language extended with parallelism, communication and symmetric cryptography. Information flow is modelled using security labels that are sets of hashed symmetric keys expressing the capabilities needed for access to data. In essence, encryption is used to encapsulate the protection offered by the information flow policy. We develop a type system aimed at tracking explicit, implicit, bypassing and correlation flows arising due to the parallel processes and the internal non-determinism inherent in Guarded Commands. The development is facilitated by the parallel processes having disjoint memories and is illustrated on a multiplexer scenario previously addressed using content-dependent information flow policies.

1 Introduction

Motivation. It is widely reported that The President of the European Commission, Jean-Claude Juncker, in his State of the Union address of September 2017 stated that *Cybersecurity is the second emergency in Europe, after climate change and before immigration.* The work of Chris Hankin has developed from the study of functional languages and λ-calculi [9], over the study of concurrent languages of the KLAIM variety [2,3,20] to have a clear focus on how to achieve cyber security [7].

The development of a strong European Cyber Security perimeter needs to take a *proactive* approach and to progress in at least two directions. One direction is the development of strong and powerful tools rooted in sound theoretical approaches based on type systems, program analysis, model checking, and theorem proving [10,17]. The other direction is the development of *teachable* approaches to secure software development so as to ensure that they can be applied by the majority of graduates with degrees in computer science or computer engineering.

Our development of the on-line tools at FormalMethods.dk focuses on the latter, supported by a text book on Formal Methods [13] and accompanying slides and videos, with a view to make information flow accessible to a wider audience [14].

A. Di Pierro et al. (Eds.): Festschrift Hankin, LNCS 12065, pp. 201–215, 2020.
https://doi.org/10.1007/978-3-030-41103-9_7

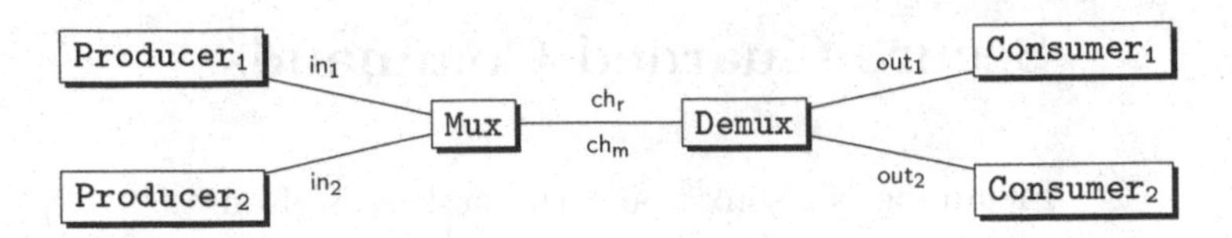

Fig. 1. Multiplexer scenario.

Contribution. The present paper extends our treatment of information flow in the presence of non-determinism [14] with parallelism, communication and symmetric cryptography.

In our view, the presence of non-determinism and parallelism is essential for modelling realistic examples where perhaps a number of deterministic programs jointly interact or interact with a database. Within a Trusted Computing Base, as might be a database system or a multi-tasking operating system, it is standard to enforce security using information flow security labels (including those of the Decentralised Label Model [11]), as they are adequate for controlling information flow and leakage within the Trusted Computing Base. Once data needs to be exchanged between different Trusted Computing Base it becomes essential to use cryptography (symmetric and asymmetric) to protect data that needs to flow along potentially insecure channels. By incorporating all of these constructs into the same programming language we are able to model the concurrent interaction between a number of Trusted Computing Bases.

Our technical development includes formulating a type system incorporating information flow security labels and symmetric cryptography and showing how the use of encryption allows to bypass the policies of information flow when communicating between Trusted Computing Bases. Also within a given Trusted Computing Base it is often necessary to bypass the security policy for properly controlled data, and the terms sanitisation [8], declassification [11], or endorsement [11] are used for this. In our development we explore the idea of using symmetric keys for controlling this as well. The type system builds on traditional approaches [18,19] but is able to deal with non-determinism, parallelism, communication, symmetric cryptography and sanitisation.

Throughout the paper we consider a simple example of two producers wanting to communicate with two consumers. Communication takes place over potentially insecure channels and involves a multiplexer and a demultiplexer. This is illustrated in Fig. 1. The challenge is to ensure that messages from the producer $\texttt{Producer}_i$ ends up at the consumer $\texttt{Consumer}_i$ (rather than $\texttt{Consumer}_{3-i}$) for both $i = 1$ and $i = 2$. We manage to do so in a simpler way than our previous treatments based on content-dependent information flow policies [12,16].

2 Syntax

In this section we extend Dijkstra's Guarded Commands language [6] with parallelism, communication and symmetric cryptography. We present the syntax

programs:	$P ::= E\ \mathtt{par}\ D_1\, C_1\ \Box\ \cdots\ \Box\ D_n\, C_n\ \mathtt{rap}$	$(n > 0)$
environment:	$E ::= \mathtt{chan}[\ c_1 : t_1; \cdots ; c_n : t_n\ ]$	$(n \geq 0)$
declarations:	$D ::= \mathtt{var}[\ x_1 : t_1\ell_1; \cdots ; x_n : t_n\ell_n\ ]$	$(n \geq 0)$
commands:	$C ::= x := e \mid c!e \mid c?x \mid C_1 ; C_2$	
	$\mid\ \mathtt{if}\ e_1 \rightarrow C_1\ \Box \ldots \Box\ e_n \rightarrow C_n\ \mathtt{fi}$	$(n > 0)$
	$\mid\ \mathtt{do}\ e_1 \rightarrow C_1\ \Box \ldots \Box\ e_n \rightarrow C_n\ \mathtt{od}$	$(n > 0)$
	$\mid\ \mathtt{sum}\ C_1\ \Box \ldots \Box\ C_n\ \mathtt{mus}$	$(n > 0)$
	$\mid\ \mathtt{decrypt}\ e_1\ \mathtt{as}\ x{:}t\,\ell\ \mathtt{using}\ e_2\ \mathtt{in}\ C$	
expressions:	$e ::= n \mid s \mid k \mid x \mid e_1 + e_2 \mid \cdots \mid$	
	$\mathtt{true} \mid e_1 = e_2 \mid \cdots \mid e_1 \wedge e_2 \mid \cdots$	
	$\mathtt{san}(e_1, e_2) \mid \mathtt{enc}(e_1, t, \ell, e_2)$	
types:	$t ::= \mathtt{int} \mid \mathtt{bool} \mid \mathtt{string} \mid \mathtt{crypt} \mid \mathtt{key} \mid \cdots$	
security labels:	$\ell ::= \{\langle k_1 \rangle, \cdots, \langle k_n \rangle\}$	$(n \geq 0)$

Fig. 2. Syntax of Secure Guarded Commands.

and introduce a variant of a multiplexer scenario previously addressed using content-dependent information flow policies [12,16].

The main syntactic category of *Secure Guarded Commands* is that of programs (denoted P). A program

$$E\ \mathtt{par}\ D_1\, C_1\ \Box\ \cdots\ \Box\ D_n\, C_n\ \mathtt{rap}$$

consists of an environment (E) defining the communication channels used in the program and a number of parallel processes, each having a local declaration (D_i) of variables and a command (C_i). The syntax is summarised in Fig. 2 and explained below.

The commands include those of Dijkstra's Guarded Commands so we have the basic command of assignment ($x := e$) in addition to sequencing ($C_1 ; C_2$) and constructs for conditionals ($\mathtt{if}\ e_1 \rightarrow C_1\ \Box \ldots \Box\ e_n \rightarrow C_n\ \mathtt{fi}$) and iteration ($\mathtt{do}\ e_1 \rightarrow C_1\ \Box \ldots \Box\ e_n \rightarrow C_n\ \mathtt{od}$). On top of this we introduce basic commands for output ($c!e$) and input ($c?x$) over a channel (c) and a command performing an 'external' non-deterministic choice among commands ($\mathtt{sum}\ C_1\ \Box \ldots \Box\ C_n\ \mathtt{mus}$); it will typically be the case that each C_i in $\mathtt{sum}\ C_1\ \Box \ldots \Box\ C_n\ \mathtt{mus}$ takes the form $c!e;C$ or $c?x;C$ but we do not need to impose this. Finally, we have a decryption command ($\mathtt{decrypt}\ e_1\ \mathtt{as}\ x{:}t\,\ell\ \mathtt{using}\ e_2\ \mathtt{in}\ C$) to be explained further below.

An expression may be a number (n), a string (s), a symmetric key (k), a variable (x), an arithmetic operation (e.g. $e_1 + e_2$), a truth value (e.g. $\mathtt{true}$), a relational operation (e.g. $e_1 = e_2$), or a logical operation (e.g. $e_1 \wedge e_2$). An expression may also be a sanitisation construct ($\mathtt{san}(e_1, e_2)$) for bypassing the security policy, and here e_1 is the data of interest, and e_2 is the symmetric key used as a 'witness' of the power to bypass the security policy. Finally, an

```
Mux :

var[x : crypt{ }]
do true →
     sum in₁ ? x ; chr ! 1 ; chm ! x
      □  in₂ ? x ; chr ! 2 ; chm ! x
     mus
od
```

```
Demux :

var[yr : int{ }; ym : crypt{ }]
do true →
     chr ? yr ; chm ? ym ;
     if yr = 1 → out₁ ! ym
      □ yr = 2 → out₂ ! ym
     fi
od
```

```
Producer_i :

var[m_i : string{⟨K0⟩, ⟨K_i⟩}]
do true →
     m_i := ··· ;
     in_i ! enc(m_i, string, {⟨K0⟩, ⟨K_i⟩}, K_i)
od
```

```
Consumer_i :

var[z_i : crypt{ };
    u_i : string{⟨K0⟩, ⟨K_i⟩};
    v_i : string{⟨K0⟩}]
do true →
     out_i ? z_i ;
     decrypt z_i as u_i : string{⟨K0⟩, ⟨K_i⟩}
     using K_i in v_i := san(u_i, K_i) ; ···
od
```

Fig. 3. Secure Guarded Commands for the multiplexer scenario ($i = 1, 2$).

expression may be an encryption ($\texttt{enc}(e_1, t, \ell, e_2)$), where once again e_1 is the data of interest, and e_2 is the symmetric key used to protect the data.

Types may be base types like integers (`int`), booleans (`bool`), strings (`string`), encryptions (`crypt`) or keys (`key`). Security labels will be *sets* of hashed symmetric keys (written $\langle k \rangle$ for a key k). The intention is that the identity of a parallel process is exhibited by the set of symmetric keys that it possesses, where each symmetric key is seen as defining a group of all those processes able to use it, and we use hashes of symmetric keys to protect the symmetric keys themselves.

We leave the syntax of channels, variables, numbers, strings, and symmetric keys unspecified.

Example 1. The system of Fig. 1 intends to let the two processes $\texttt{Producer}_1$ and $\texttt{Consumer}_1$ communicate securely over a potentially insecure channel between a multiplexer `Mux` and a demultiplexer `Demux`, and similarly to let the two processes $\texttt{Producer}_2$ and $\texttt{Consumer}_2$ communicate securely.

To achieve this it is assumed that symmetric keys $\texttt{K}_i$ (for $i = 1, 2$) are already established between the processes $\texttt{Producer}_i$ and $\texttt{Consumer}_i$; they will be used for encrypting the messages to be sent over the potentially insecure channel. Furthermore it is assumed that there is a symmetric key $\texttt{K}_0$ shared between all producers and consumers. The system may be written as:

```
chan[in₁ : crypt; in₂ : crypt; chr : int; chm : crypt; out₁ : crypt; out₂ : crypt]
par Producer₁ □ Producer₂ □ Mux □ Demux □ Consumer₁ □ Consumer₂
rap
```

The declarations and commands for the processes are given in Fig. 3. Each of the processes makes use of a few local variables and otherwise consists of an infinite loop specified using the do-construct together with the guard true.

The multiplexer uses the sum-construct to perform a non-deterministic choice between which producer to serve. This is an 'external' non-deterministic choice as it is made dependent on which of the two constituent commands is able to communicate. The local variables have empty security labels because the channels are considered potentially insecure. Based on the channel over which the message is received the multiplexer sends a destination over the channel $\mathtt{ch_r}$ and the payload over the channel $\mathtt{ch_m}$.

The demultiplexer also uses local variables with empty security labels. Based on the destination and payload received it will make use of the if-construct to decide which consumer a message is intended for and forward the payload along the appropriate channel.

Each producer $\mathtt{Producer}_i$ encrypts its message using the enc-construct and the symmetric key $\mathtt{K}_i$ already established with the consumer $\mathtt{Consumer}_i$. The message is given the security label $\{\langle\mathtt{K}_0\rangle, \langle\mathtt{K}_i\rangle\}$, rather than just $\{\langle\mathtt{K}_i\rangle\}$, so as to indicate that all the producers and consumers maintain some level of joint trust as expressed by their joint knowledge of $\mathtt{K}_0$.

Each consumer $\mathtt{Consumer}_i$ uses the decrypt-construct to extract the message using the key shared with the producer $\mathtt{Producer}_i$. In order to allow for a more widespread use of the received message it will sanitise (ideally only part of) the message by removing the hashed key $\langle\mathtt{K}_i\rangle$ so as to leave sanitised information in the variable $\mathtt{v}_i$. This variable is still linked to the hashed key $\langle\mathtt{K}_0\rangle$ so that it has not become entirely public. □

3 Semantics

We define the semantics of Secure Guarded Commands using operational semantics and will have a number of semantic judgements to this effect.

The processes of a program have disjoint memories so they can only exchange values by communicating over the channels. More precisely this means that for each process we will have a local memory assigning values to the variables of interest and we shall be based on synchronous communication.

Expressions are evaluated with respect to a memory σ that assigns values to all variables of interest and the semantic judgement takes the form

$$\sigma \vdash e \rhd v$$

The details are provided by the axiom schemes and rules of Fig. 4. Evaluation is undefined if the expression accesses a variable for which the memory does not assign a value. Note that the value of $\mathtt{san}(e_1, e_2)$ is the same as the value of e_1. For encryption we include a fresh nonce so as to ensure that $\mathtt{enc}(e_1, t, \ell, e_2) = \mathtt{enc}(e_1, t, \ell, e_2)$ evaluates to false as this is often required for a cryptographic scheme to be trustworthy.

$$\frac{}{\sigma \vdash n \rhd n} \quad \frac{}{\sigma \vdash s \rhd s} \quad \frac{}{\sigma \vdash k \rhd k} \quad \frac{}{\sigma \vdash \mathbf{true} \rhd \mathbf{tt}}$$

$$\frac{\sigma \vdash e_1 \rhd v_1 \quad \sigma \vdash e_2 \rhd v_2}{\sigma \vdash e_1 + e_2 \rhd v_1 + v_2} \quad \frac{\sigma \vdash e_1 \rhd v_1 \quad \sigma \vdash e_2 \rhd v_2}{\sigma \vdash e_1 = e_2 \rhd v_1 = v_2} \quad \frac{\sigma \vdash e_1 \rhd v_1 \quad \sigma \vdash e_2 \rhd v_2}{\sigma \vdash e_1 \wedge e_2 \rhd v_1 \wedge v_2}$$

$$\frac{}{\sigma \vdash x \rhd \sigma(x)} \text{ if } \sigma(x) \text{ defined} \qquad \frac{\sigma \vdash e_1 \rhd v_1}{\sigma \vdash \mathbf{san}(e_1, e_2) \rhd v_1}$$

$$\frac{\sigma \vdash e_1 \rhd v_1 \quad \sigma \vdash e_2 \rhd v_2}{\sigma \vdash \mathbf{enc}(e_1, t, \ell, e_2) \rhd \langle v_1, t, \ell, v_2, n \rangle} \text{ if } v_2 \text{ is a key and } n \text{ is fresh}$$

Fig. 4. Semantics of expressions.

A command will be interpreted relative to a memory σ that assigns values to all variables in the command of interest. The judgement has the form

$$(C, \sigma) \rightarrow^{\phi} (C', \sigma')$$

where the superscript ϕ indicates whether the action is silent (τ), an input ($c\,?\,v$) or an output ($c\,!\,v$). Here we allow C and C' to range both over commands and the special symbol $\surd$ indicating a terminated configuration. The details are provided by the axiom schemes and rules of Fig. 5.

Note that this semantics is purely non-deterministic and does not make use of a scheduler. If needed, we would model a scheduler by explicitly modifying the guards in guarded commands. Doing so would influence the results of the information flow type system to be developed in Sect. 5. This would be *our* way of modelling an attacker that might collude with a scheduler.

Furthermore note that the decryption construct acts as a skip construct if the message cannot be decrypted as required; it would be possible to incorporate an error command to be executed instead.

For processes the judgement has the form

$$(E \;\mathtt{par}\; D_1\, C_1 \cdots D_n\, C_n \;\mathtt{rap}, \sigma_1 \cdots \sigma_n) \rightarrow (E \;\mathtt{par}\; D_1\, C_1' \cdots D_n\, C_n' \;\mathtt{rap}, \sigma_1' \cdots \sigma_n')$$

where once more we allow C and C' to range both over commands and the special symbol $\surd$ indicating a terminated configuration.

We can either let one of the constituent processes perform a silent step or let two constituent processes produce matching input and output actions. Omitting the details needed to deal with output and input processes occurring in a different order, the details are provided by the axiom schemes and rules of Fig. 6. The use of symmetric communication, as opposed to asymmetric communication, is entirely a matter of choice. Note that if one of the processes terminates then the corresponding component in the configuration will contain $\surd$ and it will not be able to evolve further.

The side conditions of the rules insist that the domain of local memories (written $\mathsf{dom}(\sigma_i)$) includes the local variables declared in the program (written

$$\frac{\sigma \vdash e \rhd v}{(x := e, \sigma) \to^{\tau} (\surd, \sigma[x \mapsto v])} \text{ if } \sigma(x) \text{ is defined}$$

$$\frac{(C_1, \sigma) \to^{\phi} (C_1', \sigma')}{(C_1 \,; C_2, \sigma) \to^{\phi} (C_1' \,; C_2, \sigma')} \text{ if } C_1' \neq \surd \qquad \frac{(C_1, \sigma) \to^{\phi} (\surd, \sigma')}{(C_1 \,; C_2, \sigma) \to^{\phi} (C_2, \sigma')}$$

$$\frac{\sigma \vdash e_i \rhd \mathsf{tt}}{(\mathbf{if}\ e_1 \to C_1 \,\square \cdots \square\, e_n \to C_n\ \mathbf{fi}, \sigma) \to^{\tau} (C_i, \sigma)}$$

$$\frac{\sigma \vdash e_i \rhd \mathsf{tt}}{(\mathbf{do} \cdots \square\, e_i \to C_i \,\square \cdots \mathbf{od}, \sigma) \to^{\tau} (C_i \,; \mathbf{do} \cdots \square\, e_i \to C_i \,\square \cdots \mathbf{od}, \sigma)}$$

$$\frac{\sigma \vdash e_1 \rhd \mathsf{ff} \quad \cdots \quad \sigma \vdash e_n \rhd \mathsf{ff}}{(\mathbf{do}\ e_1 \to C_1 \,\square \cdots \square\, e_n \to C_n\ \mathbf{od}, \sigma) \to^{\tau} (\surd, \sigma)}$$

$$\frac{(C_i, \sigma) \to^{\phi} (C_i', \sigma')}{(\mathbf{sum}\ C_1 \,\square \cdots \square\, C_n\ \mathbf{mus}, \sigma) \to^{\phi} (C_i', \sigma')}$$

$$\frac{\sigma \vdash e \rhd v}{(c!e, \sigma) \to^{c!v} (\surd, \sigma)} \qquad \frac{}{(c?x, \sigma) \to^{c?v} (\surd, \sigma[x \mapsto v])} \text{ if } \sigma(x) \text{ is defined}$$

$$\frac{\sigma \vdash e_1 \rhd v_1 \quad \sigma \vdash e_2 \rhd v_2}{(\mathbf{decrypt}\ e_1\ \mathbf{as}\ x{:}t\,\ell\ \mathbf{using}\ e_2\ \mathbf{in}\ C, \sigma) \to^{\tau} (C, \sigma[x \mapsto v_0])} \text{ if } \begin{cases} v_1 = \langle v_0, t, \ell, v_2, n \rangle \\ \sigma(x) \text{ defined} \end{cases}$$

$$\frac{\sigma \vdash e_1 \rhd v_1 \quad \sigma \vdash e_2 \rhd v_2}{(\mathbf{decrypt}\ e_1\ \mathbf{as}\ x{:}t\,\ell\ \mathbf{using}\ e_2\ \mathbf{in}\ C, \sigma) \to^{\tau} (\surd, \sigma)} \text{ unless } \begin{cases} v_1 = \langle v_0, t, \ell, v_2, n \rangle \\ \quad \text{for some } v_0, n \\ \sigma(x) \text{ defined} \end{cases}$$

Fig. 5. Semantics of commands.

$$\frac{(C_i, \sigma_i) \to^{\tau} (C_i', \sigma_i')}{\begin{array}{l}(E\ \mathbf{par} \cdots D_i\, C_i \cdots \mathbf{rap}, \cdots \sigma_i \cdots) \\ \quad \to (E\ \mathbf{par} \cdots D_i\, C_i' \cdots \mathbf{rap}, \cdots \sigma_i' \cdots)\end{array}} \text{ if } \mathsf{dom}(\sigma_i) \supseteq \mathsf{dom}(\mathsf{env}(D_i))$$

$$\frac{(C_i, \sigma_i) \to^{c!v} (C_i', \sigma_i') \qquad (C_j, \sigma_j) \to^{c?v} (C_j', \sigma_j')}{\begin{array}{l}(E\ \mathbf{par} \cdots D_i\, C_i \,\square\, D_j\, C_j \cdots \mathbf{rap}, \cdots \sigma_i \sigma_j \cdots) \\ \quad \to (E\ \mathbf{par} \cdots D_i\, C_i' \,\square\, D_j\, C_j' \cdots \mathbf{rap}, \cdots \sigma_i' \sigma_j' \cdots)\end{array}} \text{ if } \begin{cases} \mathsf{dom}(\sigma_i) \supseteq \mathsf{dom}(\mathsf{env}(D_i)) \\ \mathsf{dom}(\sigma_j) \supseteq \mathsf{dom}(\mathsf{env}(D_j)) \\ c \in \mathsf{dom}(\mathsf{env}(E)) \end{cases}$$

Fig. 6. Semantics of systems.

$\mathsf{dom}(\mathsf{env}(D_i)))$ and that the channel used for communication is indeed declared in the program (written $\mathsf{dom}(\mathsf{env}(E)))$. Here the environments and declarations give rise to the following mappings

$$\mathsf{env}(\mathbf{chan}[\ c_1 : t_1; \cdots ; c_n : t_n\ ]) = [c_1 \mapsto (t_1, \{\ \}) \cdots c_n \mapsto (t_n, \{\ \})]$$
$$\mathsf{env}(\mathbf{var}[\ x_1 : t_1\ell_1; \cdots ; x_n : t_n\ell_n\ ]) = [x_1 \mapsto (t_1, \ell_1) \cdots x_n \mapsto (t_n, \ell_n)]$$

Note that for channels the security label is $\{\}$ reflecting that the channels are potentially insecure; it would be possible also to have channels considered to be more secure and provide a label distinct from $\{\}$ for those channels.

Furthermore note that each parallel process has its own local memory so that there is no possibility of shared variables.

4 The Information Flow Landscape

As a preparation for developing the information flow type system in Sect. 5 we shall present a brief overview of the kinds of flows that we are concerned about. This overview borrows from [14] although the method for dealing with them is different. In this section we shall assume that there is a security environment ϱ giving the security label of each variable mentioned.

Explicit Flows. In the command $\mathtt{y} := \mathtt{x}$ there is a direct and *explicit* flow from $\mathtt{x}$ to $\mathtt{y}$. For this flow to be admissible we shall assume that the security label $\varrho(\mathtt{x})$ of $\mathtt{x}$, and the security label $\varrho(\mathtt{y})$ of $\mathtt{y}$, satisfy that $\varrho(\mathtt{x}) \subseteq \varrho(\mathtt{y})$.

Implicit Flows. In the guarded command $\mathtt{x} = 0 \to \mathtt{y} := 0$, as might appear in $\mathtt{if}\ \mathtt{x} = 0 \to \mathtt{y} := 0\ \mathtt{fi}$ or $\mathtt{do}\ \mathtt{x} = 0 \to \mathtt{y} := 0\ \mathtt{od}$, there is a direct and *implicit* flow from $\mathtt{x}$ to $\mathtt{y}$. For this flow to be admissible we shall assume that $\varrho(\mathtt{x}) \subseteq \varrho(\mathtt{y})$.

Bypassing Flows. In $\mathtt{y} := 0;\ \mathtt{if}\ \mathtt{x} = 0 \to \mathtt{skip}\ [\!]\ \mathtt{true} \to \mathtt{y} := 1\ \mathtt{fi}$ there are no explicit flows from $\mathtt{x}$ to $\mathtt{y}$ and also no implicit flows. However, it is still the case that the final value of $\mathtt{y}$ might reveal something about $\mathtt{x}$ if one is able to run the program many times and observe the different non-deterministic outcomes. So we say that there is a *bypassing* flow from $\mathtt{x}$ to $\mathtt{y}$. For this flow to be admissible we shall assume that $\varrho(\mathtt{x}) \subseteq \varrho(\mathtt{y})$.

Correlation Flows. Bypassing flows capture some of the power of non-determinism but not all of it. In $\mathtt{if}\ \mathtt{true} \to \mathtt{y} := 0; \mathtt{x} := 0\ [\!]\ \mathtt{true} \to \mathtt{y} := 1; \mathtt{x} := 1\ \mathtt{fi}$ there are no explicit, implicit or bypassing flows. Yet, if $\mathtt{y}$ was intended to be a private key (albeit a short one) and $\mathtt{x}$ is a public variable, then clearly we can learn something about $\mathtt{y}$ from knowing $\mathtt{x}$. So we say that there is a *correlation* flow from $\mathtt{x}$ to $\mathtt{y}$. For this flow to be admissible we shall assume that $\varrho(\mathtt{x}) \subseteq \varrho(\mathtt{y})$. Similarly, there is a *correlation* flow from $\mathtt{y}$ to $\mathtt{x}$ so that we shall additionally assume that $\varrho(\mathtt{y}) \subseteq \varrho(\mathtt{x})$. In summary, we shall assume that $\varrho(\mathtt{x}) = \varrho(\mathtt{y})$. (The term *correlation* flow is motivated by the fact that, viewed as random variables, the two random variables $\mathtt{x}$ and $\mathtt{y}$ are correlated rather than being independent.)

The command $\mathtt{sum}\ \mathtt{y} := 0; \mathtt{x} := 0\ [\!]\ \mathtt{y} := 1; \mathtt{x} := 1\ \mathtt{mus}$ is mostly equivalent to $\mathtt{if}\ \mathtt{true} \to \mathtt{y} := 0; \mathtt{x} := 0\ [\!]\ \mathtt{true} \to \mathtt{y} := 1; \mathtt{x} := 1\ \mathtt{fi}$ and hence also exhibits correlation flows and we shall assume that $\varrho(\mathtt{x}) = \varrho(\mathtt{y})$.

Sanitised Flows. In the command $\mathtt{y} := \mathtt{san}(\mathtt{x}, \mathtt{K})$ there is a direct and *sanitised* flow from $\mathtt{x}$ to $\mathtt{y}$. For this flow to be admissible we shall assume that $\varrho(\mathtt{x})\backslash\{\langle\mathtt{K}\rangle\} \subseteq \varrho(\mathtt{y})$. This is in line with viewing the sanitisation construct as a declassification construct that permits part of the security label to be leaked. The condition may be rewritten as $\varrho(\mathtt{x}) \subseteq \varrho(\mathtt{y}) \cup \{\langle\mathtt{K}\rangle\}$ to make the comparison to declassification rules more obvious [11].

Indirect Flows. All of the discussions above only considered direct flows taking place in one single step. Indirect flows (following the terminology of [4,5]) arise when direct flows are combined. The transitive nature of the subset relation ($\subseteq$) ensures that our treatment naturally generalises to deal also with indirect flows.

Covert Channels. Almost all considerations of information flow ignore certain computational phenomena from which information perhaps can be learned. That is, they admit covert channels. For the considerations of the present paper it is clear that we are ignoring issues related to non-termination and execution time. This is intentional is order to avoid defining an overly restrictive information flow type system.

Example 2. Let us return to the program of Fig. 3. In the command for Mux we have *direct explicit* flows from $\mathtt{in}_i$ to $\mathtt{x}$ and from $\mathtt{x}$ to $\mathtt{ch_m}$ (for $i = 1, 2$); to see this it is helpful to view an input $c\,?\,x$ as an assignment $x := c$ and an output $c\,!\,e$ as an assignment $c := e$. We have *correlation* flows between $\mathtt{x}$, $\mathtt{ch_r}$ and $\mathtt{ch_m}$ due to their appearance in the two parts of the sum-construct. Overall we have an *indirect* flow from $\mathtt{in}_i$ to $\mathtt{x}$, $\mathtt{ch_m}$ and $\mathtt{ch_r}$ (for $i = 1, 2$).

In the command for Demux we have, in addition to some explicit flows, also an *implicit* flow from $\mathtt{y_r}$ to $\mathtt{out_1}$ as well as $\mathtt{out_2}$ due to the tests on $\mathtt{y_r}$ in both branches of the conditional.

We have a *sanitised* flow from $\mathtt{u}_i$ to $\mathtt{v}_i$ in the command for $\mathtt{Consumer}_i$; the other direct flows in the process are explicit flows and so are the direct flows of $\mathtt{Producer}_i$.

The program of Fig. 3 has no bypassing flows. However, if we modify the code of Demux to be

```
var[yr : int{ }; ym : crypt{ }]
do true →
      chr ? yr ; chm ? ym ;
      if yr ≠ 2 → out1 ! ym
      ▯ yr ≠ 1 → out2 ! ym
      fi
od
```

then the implicit flows from $\mathtt{y_r}$ to $\mathtt{out_1}$ and $\mathtt{out_2}$ will also be *bypassing* flows (because of the apparent possibility that $\mathtt{y_r}$ could hold the value 3). □

$$\frac{}{\rho \vdash n : \mathtt{int}\,\{\}} \qquad \frac{}{\rho \vdash s : \mathtt{string}\,\{\}} \qquad \frac{}{\rho \vdash k : \mathtt{key}\,\{\langle k \rangle\}} \qquad \frac{}{\rho \vdash \mathtt{true} : \mathtt{bool}\,\{\}}$$

$$\frac{}{\rho \vdash x : t\,\ell} \text{ if } \rho(x) = (t, \ell) \qquad \frac{\rho \vdash e_1 : \mathtt{int}\,\ell_1 \quad \rho \vdash e_2 : \mathtt{int}\,\ell_2}{\rho \vdash e_1 + e_2 : \mathtt{int}\,(\ell_1 \cup \ell_2)}$$

$$\frac{\rho \vdash e_1 : t\,\ell_1 \quad \rho \vdash e_2 : t\,\ell_2}{\rho \vdash e_1 = e_2 : \mathtt{bool}\,(\ell_1 \cup \ell_2)} \qquad \frac{\rho \vdash e_1 : \mathtt{bool}\,\ell_1 \quad \rho \vdash e_2 : \mathtt{bool}\,\ell_2}{\rho \vdash e_1 \wedge e_2 : \mathtt{bool}\,(\ell_1 \cup \ell_2)}$$

$$\frac{\rho \vdash e_1 : t\,\ell_1 \quad \rho \vdash e_2 : \mathtt{key}\,\ell_2}{\rho \vdash \mathtt{san}(e_1, e_2) : t\,(\ell_1 \backslash \ell_2)} \qquad \frac{\rho \vdash e_1 : t\,\ell_1 \quad \rho \vdash e_2 : \mathtt{key}\,\ell_2}{\rho \vdash \mathtt{enc}(e_1, t, \ell, e_2) : \mathtt{crypt}\,\{\}} \text{ if } \ell_1 \cup \ell_2 \subseteq \ell$$

Fig. 7. Types for expressions.

5 Information Flow Type System

In this section we develop an information flow type system based on the considerations of Sect. 4. The development extends that of [14] in dealing with symmetric cryptography, communication and parallelism. It will make use of well-typing judgements for expressions, commands and processes. They are inspired by traditional approaches such as those of [18,19] but need to be extended to deal with parallelism and non-determinism. The fact that we use communication rather than shared variables reduces the complexity of the task.

Well-Typed Expressions. For expressions the judgement takes the form

$$\rho \vdash e : t\,\ell$$

It is defined by the axiom schemes and rules of Fig. 7 and will be explained below. The judgement makes use of a type environment ρ that assigns types and security labels to all variables and channels. Thus the second component of $\rho(x)$ equals $\varrho(x)$ as used in Sect. 4.

Let us first observe that every free variable x_i in an expression may give rise to an (explicit, implicit or bypassing) flow to some other variable y. As motivated in the previous section this gives rise to the condition that $\varrho(x_i) \subseteq \varrho(y)$ and taking all free variables into account we get $\bigwedge_i \varrho(x_i) \subseteq \varrho(y)$. Thanks to the lattice properties of powersets ordered by subset inclusion this is equivalent to $\bigcup_i \varrho(x_i) \subseteq \varrho(y)$.

Thus the overall idea is that $\rho \vdash e : t\,\ell$ should ensure that the type of the expression e is t and that the security label is $\ell = \bigcup_i \varrho(x_i)$ where x_i ranges over all free variables of e. This is in line with the development in [18,19] and takes care of explaining the axiom schemes and rules of Fig. 7 except for symmetric keys, sanitisation, and encryption.

For a symmetric key k we create a security label that is the singleton set of $\langle k \rangle$ being the hashed value of the key. Since we have no sub-typing rules for expressions it is easy to show that whenever $\rho \vdash e : \mathtt{key}\,\ell$ then e will contain a single key k and $\ell = \{\langle k \rangle\}$.

$$\frac{\rho \vdash e_1 : t\,\ell_1}{\rho \vdash x := e_1 : [\ell,\ell]} \text{ if } \begin{cases} \rho(x) = (t,\ell) \\ \ell_1 \subseteq \ell \\ t = \mathtt{key} \Rightarrow \ell_1 = \ell \end{cases} \qquad \frac{\rho \vdash C_1 : L_1 \quad \rho \vdash C_2 : L_2}{\rho \vdash C_1 ; C_2 : L_1 \sqcap L_2}$$

$$\frac{\bigwedge_i \rho \vdash e_i : \mathtt{bool}\,\ell_i \quad \bigwedge_i \rho \vdash C_i : L_i}{\rho \vdash \mathtt{if}\ e_1 \to C_1\ [\!]\ \cdots\ [\!]\ e_n \to C_n\ \mathtt{fi} : L_1 \sqcap \cdots \sqcap L_n} \text{ if } \begin{cases} \bigwedge_i \ell_i \sqsubseteq L_i \\ \bigwedge_{(i,j)\in\mathtt{cosat}} \ell_j \sqsubseteq L_i \\ \bigwedge_{(i,j)\in\mathtt{cosat}} \mathtt{uniq}(L_i) \end{cases}$$

$$\frac{\bigwedge_i \rho \vdash e_i : \mathtt{bool}\,\ell_i \quad \bigwedge_i \rho \vdash C_i : L_i}{\rho \vdash \mathtt{do}\ e_1 \to C_1\ [\!]\ \cdots\ [\!]\ e_n \to C_n\ \mathtt{od} : L_1 \sqcap \cdots \sqcap L_n} \text{ if } \begin{cases} \bigwedge_i \ell_i \sqsubseteq L_i \\ \bigwedge_{(i,j)\in\mathtt{cosat}} \ell_j \sqsubseteq L_i \\ \bigwedge_{(i,j)\in\mathtt{cosat}} \mathtt{uniq}(L_i) \end{cases}$$

$$\frac{\bigwedge_i \rho \vdash C_i : L_i}{\rho \vdash \mathtt{sum}\ C_1\ [\!]\ \cdots\ [\!]\ C_n\ \mathtt{mus} : L_1 \sqcap \cdots \sqcap L_n} \text{ if } \bigwedge_i \mathtt{uniq}(L_i)$$

$$\frac{\rho \vdash e_0 : t\,\ell_0}{\rho \vdash c!e_0 : [\ell,\ell]} \text{ if } \begin{cases} \rho(c) = (t,\ell) \\ \ell_0 \subseteq \ell \\ t = \mathtt{key} \Rightarrow \ell_0 = \ell \end{cases} \qquad \frac{}{\rho \vdash c?x : [\ell,\ell]} \text{ if } \begin{cases} \rho(c) = (t,\ell') \\ \rho(x) = (t,\ell) \\ \ell' \subseteq \ell \\ t = \mathtt{key} \Rightarrow \ell' = \ell \end{cases}$$

$$\frac{\rho \vdash e_1 : \mathtt{crypt}\,\ell_1 \quad \rho \vdash C : L \quad \rho \vdash e_2 : \mathtt{key}\,\ell_2}{\rho \vdash \mathtt{decrypt}\ e_1\ \mathtt{as}\ x{:}t\,\ell\ \mathtt{using}\ e_2\ \mathtt{in}\ C : [\ell,\ell] \sqcap L} \text{ if } \begin{cases} \rho(x) = (t,\ell) \\ \ell_1 \cup \ell_2 \subseteq \ell \end{cases}$$

Fig. 8. Types for commands.

For sanitisation $\mathtt{san}(e_1, e_2)$ with $\rho \vdash e_1 : t\,\ell_1$ and $\rho \vdash e_2 : \mathtt{key}\,\ell_2$ the discussion above tells us that e_2 would consist of a single key k and $\ell_2 = \{\langle k \rangle\}$. The purpose of the sanitisation construct is to express that the group of owners of the key k have agreed to consider the value of e_1 to be sufficiently sanitised to remove the hashed key $\langle k \rangle$ from the security label ℓ_1. (Clearly the nature of the group of owners is such that any one owner acts on behalf of the entire group; if this is not desired one would have to incorporate asymmetric cryptography.)

The type of an encryption $\mathtt{enc}(e_1, t, \ell, e_2)$ is $\mathtt{crypt}$ and we let the overall security label be $\{\}$ reflecting that cryptography is sufficiently powerful to let encrypted messages flow even over insecure channels. The typing rule additionally ensures that the security labels of the two subexpressions are contained in the security label of the encryption construct itself. Once we get to decryption below we shall exploit the fact that the semantics encapsulates the value with its type and security label.

Well-Typed Commands. For commands the typing judgement takes the form

$$\rho \vdash C : L$$

It is defined by the axiom schemes and rules of Fig. 8 and further explained below. The judgement makes use of a security level L being a *pair* of security labels, written as $[\ell_1, \ell_2]$ with $\ell_1 \subseteq \ell_2$. We shall allow to write $\ell \sqsubseteq [\ell_1, \ell_2]$ for $\ell \subseteq \ell_1$ and define $[\ell_1, \ell_2] \sqcap [\ell'_1, \ell'_2] = [\ell_1 \cap \ell'_1, \ell_2 \cup \ell'_2]$ (which is the greatest lower bound operation with respect to a partial order $\sqsubseteq'$ defined by $[\ell_1, \ell_2] \sqsubseteq' [\ell'_1, \ell'_2]$ whenever $\ell_1 \subseteq \ell'_1$ and $\ell_2 \supseteq \ell'_2$). We shall write $\mathtt{uniq}([\ell_1, \ell_2])$ for the condition that $\ell_1 = \ell_2$.

Let us first observe that every modified variable y_i in a command may be subject to an (explicit, implicit or bypassing) flow from some other variable x. As motivated in the previous section this gives rise to the condition that $\varrho(x) \subseteq \varrho(y_i)$ and taking all modified variables into account we get $\bigwedge_i \varrho(x) \subseteq \varrho(y_i)$. Thanks to the lattice properties of powersets ordered by subset inclusion this is equivalent to $\varrho(x) \subseteq \bigcap_i \varrho(y_i)$.

Thus the overall idea is that $\rho \vdash C : [\ell_1, \ell_2]$ should ensure that $\ell_1 = \bigcap_i \varrho(x_i)$ where x_i ranges over all modified variables of C, and this is in line with the development in [18,19]. However, we shall see that we need a bit more to deal with non-determinism and so we will additionally ensure that $\ell_2 = \bigcup_i \varrho(x_i)$ so as to record the variety of variables modified in the command.

The rule for assignment records the security label of the variable modified and checks that the explicit information flow is admissible. There is an implicit sub-typing going on because we do not require that the security labels of the expression and variable agree, and to maintain the property that the security label for a key k is the singleton $\langle k \rangle$ we forbid this sub-typing for keys.

The rule for output and the axiom scheme for input are similar to the one for assignment, essentially treating output $c\,!\,e$ as an assignment $c := e$, and input $c\,?\,x$ as an assignment $x := c$. Once again we forbid the implicit sub-typing for keys although this could be dispensed with given our current choice of modelling all channels as potentially insecure.

The rule for sequencing is straightforward given our explanation of $\rho \vdash C : L$ and the operation $L_1 \sqcap L_2$.

The rule for decryption records the security label of the variables modified, namely the security label of the variable that receives the payload, and the security labels of the variables modified in the enclosed command. We exploit that the semantics bypasses the body of the decryption, if we decrypt with an incorrect key or if the encapsulated type and security label do not match what is stated in the decryption construct. It also checks the explicit information flow from the encrypted value and the implicit flow from the key.

The rule for 'external' non-deterministic choice takes care of correlation flows. It makes use of $\mathtt{uniq}(L_i)$, i.e. $L_i = [\ell', \ell']$ for some ℓ', to ensure that all modified variables have the same security label.

The rules for conditional and iteration are essentially identical and make use of guarded commands of the form $e_1 \to C_1 \,\square\, \cdots \,\square\, e_n \to C_n$. They take care of implicit flows by checking that $\ell_i \sqsubseteq L_i$ whenever $\bigwedge_i \rho \vdash e_i : \mathtt{bool}\,\ell_i$ and $\bigwedge_i \rho \vdash C_i : L_i$. They take care of bypassing flows whenever some $e_i \wedge e_j$ is satisfiable for $i \neq j$. This is expressed using the set $\mathtt{cosat}$ that contains those *distinct* pairs (i, j) of indices such that $e_i \wedge e_j$ is satisfiable; it may be computed using a Satisfaction Modulo Theories (SMT) solver such as Z3 [1] or it may be

$$\frac{\bigwedge_i \rho_i \vdash C_i : L_i}{\vdash E \ \mathtt{par}\ D_1\, C_1 \,\Box\, \cdots \,\Box\, D_n\, C_n \ \mathtt{rap} : \checkmark} \text{ where } \rho_i = \mathsf{env}(E)\mathsf{env}(D_i)$$

Fig. 9. Types for processes.

approximated using the DAG-based heuristics described in [14]. Whenever this is the case, the condition $\ell_j \sqsubseteq L_i$ checks that the bypassing flows are admissible, and the condition $\mathtt{uniq}(L_i)$ checks the correlation flows are admissible.

Well-Typed Processes. For processes the typing judgement takes the form $\vdash P : \checkmark$. It is defined by the rule in Fig. 9 and makes use of $\mathsf{env}(\cdots)$ to construct the appropriate environments for the commands. We shall only allow the semantics to be used on well-typed programs P (i.e. satisfying $\vdash P : \checkmark$).

Example 3. The processes of Fig. 3 all type check. We shall consider a couple of the most interesting cases. For $\mathtt{Producer}_i$ we use the type environment ρ_i obtained by extending the channel environment with information about the local variable $\mathtt{m}_i$. Since $\rho_i(\mathtt{in}_i) = (\mathtt{crypt}, \{\,\})$ we get

$$\rho_i \vdash \mathtt{in}_i\,!\,\mathtt{enc}(\mathtt{m}_i, \mathtt{string}, \{\langle \mathtt{K}_0\rangle, \langle \mathtt{K}_i\rangle\}, \mathtt{K}_i) : [\{\,\}, \{\,\}]$$

because $\rho_i \vdash \mathtt{m}_i : \mathtt{string}\{\langle \mathtt{K}_0\rangle, \langle \mathtt{K}_i\rangle\}$, $\rho_i \vdash \mathtt{K}_i : \mathtt{key}\{\langle \mathtt{K}_i\rangle\}$ and $\{\langle \mathtt{K}_0\rangle, \langle \mathtt{K}_i\rangle\} \cup \{\langle \mathtt{K}_i\rangle\} \subseteq \{\langle \mathtt{K}_0\rangle, \langle \mathtt{K}_i\rangle\}$. Using this information we can show that $\mathtt{Producer}_i$ type checks with the security level $[\{\}, \{\}]$.

For $\mathtt{Consumer}_i$ we use the type environment ρ'_i obtained by extending the channel environment with information about the local variables $\mathtt{z}_i$, $\mathtt{u}_i$ and $\mathtt{v}_i$. We can establish that

$$\rho'_i \vdash \mathtt{v}_i := \mathtt{san}(\mathtt{u}_i, \mathtt{K}_i) : [\{\langle \mathtt{K}_0\rangle\}, \{\langle \mathtt{K}_0\rangle\}]$$

using that $\rho'_i(\mathtt{u}_i) = (\mathtt{string}, \{\langle \mathtt{K}_0\rangle, \langle \mathtt{K}_i\rangle\})$ and $\rho'_i(\mathtt{v}_i) = (\mathtt{string}, \{\langle \mathtt{K}_0\rangle\})$ together with $\rho'_i \vdash \mathtt{K}_i : \mathtt{key}\{\langle \mathtt{K}_i\rangle\}$. Using that $\rho'_i(\mathtt{z}_i) = (\mathtt{crypt}, \{\})$ we get

$$\rho'_i \vdash \mathtt{decrypt}\ \mathtt{z}_i\ \mathtt{as}\ \mathtt{u}_i : \mathtt{string}\{\langle \mathtt{K}_0\rangle, \langle \mathtt{K}_i\rangle\}\ \mathtt{using}\ \mathtt{K}_i\ \mathtt{in}\ \mathtt{v}_i := \mathtt{san}(\mathtt{u}_i, \mathtt{K}_i) : L_i$$

where $L_i = [\{\langle \mathtt{K}_i\rangle\}, \{\langle \mathtt{K}_i\rangle\}] \sqcap [\{\langle \mathtt{K}_0\rangle\}, \{\langle \mathtt{K}_0\rangle\}] = [\{\,\}, \{\langle \mathtt{K}_0\rangle, \langle \mathtt{K}_i\rangle\}]$. Continuing with the body of the loop and the loop itself we get that $\mathtt{Consumer}_i$ type checks with the security level L_i. □

Example 4. Let us conclude with an example that does not type check. Consider the following variant of Mux where a variable p is updated with, say, the cost charged for the service offered by the Mux-Demux system. The variable p has a security label $\{\langle \mathtt{K}\rangle\}$ where K is a new key.

```
var[x : crypt{ }, p : int{⟨K⟩}]
do true →
        sum in₁ ? x ; chᵣ ! 1 ; chₘ ! x ; p := p + 1
        □  in₂ ? x ; chᵣ ! 2 ; chₘ ! x ; p := p + 2
        mus
od
```

Now let ρ'' be the type environment obtained by extending the channel environment with information about the local variables x and p. We then get

$$\rho'' \vdash \texttt{in}_i\,?\,\texttt{x}\,;\texttt{ch}_\texttt{r}\,!\,i\,;\texttt{ch}_\texttt{m}\,!\,\texttt{x}\,;\texttt{p} := \texttt{p} + i : L_i$$

where $L_i = [\{\,\}, \{\langle\texttt{K}\rangle\}]$; here we have exploited that the security level of the three communications amounts to $[\{\,\}, \{\,\}]$ while that of the assignment is $[\{\langle\texttt{K}\rangle\}, \{\langle\texttt{K}\rangle\}]$ and their greatest lower bound is L_i. In order for the sum-construct to type check it must be the case that $\texttt{uniq}(L_i)$ holds and this clearly fails since $\{\,\} \neq \{\langle\texttt{K}\rangle\}$. Thus the revised version of Mux will be rejected by the type system, intuitively because there is a *correlation* flow between $\texttt{in}_i$ (for $i = 1, 2$) and p: controlling which channels that are used for communication with the producers will allow us to control the payment charged for the multiplexer service. To remedy this situation one might declare p as $\texttt{p} : \texttt{int}\{\,\}$. □

6 Conclusion

We developed an information flow type system tracking explicit, implicit, bypassing and correlation flows arising in Guarded Commands extended with parallelism, communication, and symmetric cryptography. By incorporating all of these constructs into the same programming language we are able to model the concurrent interaction between a number of Trusted Computing Bases. Information flow was modelled using security labels being sets of hashed symmetric keys expressing the capabilities needed for access to data. This allowed us to use symmetric cryptography to bypass the policies of information flow as well as to control the sanitisation of data.

We did not establish a non-interference result but were content with the tracking of the different kinds of information flow. Indeed, a classical non-interference result as in [18,19] will fail due to non-determinism and one would need to explore other approaches such as [15,21]. However, even these approaches would still admit the covert channels due to non-termination and timing, and many would consider our semantics unrealistic due to the absence of an explicit scheduler. Rather than complicating the type system to rectify some of these 'shortcomings' we have chosen to retain an element of simplicity.

Indeed, we think that the development is made in such a way that it remains *teachable*. We are currently using the developments on FormalMethods.dk at both Bachelor's level (ignoring non-determinism) and at Master's level (including non-determinism) and look forward to be able to incorporate the developments of the present paper.

References

1. de Moura, L., Bjørner, N.: Z3: an efficient SMT solver. In: Ramakrishnan, C.R., Rehof, J. (eds.) TACAS 2008. LNCS, vol. 4963, pp. 337–340. Springer, Heidelberg (2008). https://doi.org/10.1007/978-3-540-78800-3_24

2. De Nicola, R., Ferrari, G.L., Pugliese, R.: KLAIM: a kernel language for agents interaction and mobility. IEEE Trans. Softw. Eng. **24**(5), 315–330 (1998)
3. De Nicola, R., et al.: From flow logic to static type systems for coordination languages. Sci. Comput. Program. **75**(6), 376–397 (2010)
4. Denning, D.E.: A lattice model of secure information flow. Commun. ACM **19**(5), 236–243 (1976)
5. Denning, D.E., Denning, P.J.: Certification of programs for secure information flow. Commun. ACM **20**(7), 504–513 (1977)
6. Dijkstra, E.W.: Guarded commands, nondeterminacy and formal derivation of programs. Commun. ACM **18**(8), 453–457 (1975)
7. Fielder, A., Panaousis, E.A., Malacaria, P., Hankin, C., Smeraldi, F.: Decision support approaches for cyber security investment. Decis. Support Syst. **86**, 13–23 (2016)
8. Gollmann, D.: Computer Security, 3rd edn. Wiley, Hoboken (2011)
9. Hankin, C.: Lambda Calculi: a Guide for Computer Scientists. Oxford University Press, Oxford (1994)
10. Huth, M., Nielson, F.: Static analysis for proactive security. In: Steffen, B., Woeginger, G. (eds.) Computing and Software Science. LNCS, vol. 10000, pp. 374–392. Springer, Cham (2019). https://doi.org/10.1007/978-3-319-91908-9_19
11. Myers, A.C., Liskov, B.: Protecting privacy using the decentralized label model. ACM Trans. Softw. Eng. Methodol. **9**(4), 410–442 (2000)
12. Nielson, F., Nielson, H.R.: Atomistic Galois insertions for flow sensitive integrity. Comput. Lang. Syst. Struct. **50**, 82–107 (2017)
13. Nielson, F., Nielson, H.R.: Formal Methods: An Appetizer. Springer, Heidelberg (2019). https://doi.org/10.1007/978-3-030-05156-3. ISBN 9783030051556
14. Nielson, F., Nielson, H.R.: Lightweight information flow. In: Boreale, M., Corradini, F., Loreti, M., Pugliese, R. (eds.) Models, Languages, and Tools for Concurrent and Distributed Programming. LNCS, vol. 11665, pp. 455–470. Springer, Cham (2019). https://doi.org/10.1007/978-3-030-21485-2_25
15. Nielson, F., Nielson, H.R., Vasilikos, P.: Information flow for timed automata. In: Aceto, L., Bacci, G., Bacci, G., Ingólfsdóttir, A., Legay, A., Mardare, R. (eds.) Models, Algorithms, Logics and Tools. LNCS, vol. 10460, pp. 3–21. Springer, Cham (2017). https://doi.org/10.1007/978-3-319-63121-9_1
16. Nielson, H.R., Nielson, F.: Content dependent information flow control. J. Log. Algebr. Methods Program. **87**, 6–32 (2017)
17. Pettai, M., Laud, P.: Combining differential privacy and mutual information for analyzing leakages in workflows. In: Maffei, M., Ryan, M. (eds.) POST 2017. LNCS, vol. 10204, pp. 298–319. Springer, Heidelberg (2017). https://doi.org/10.1007/978-3-662-54455-6_14
18. Volpano, D.M., Irvine, C.E.: Secure flow typing. Comput. Secur. **16**(2), 137–144 (1997)
19. Volpano, D.M., Irvine, C.E., Smith, G.: A sound type system for secure flow analysis. J. Comput. Secur. **4**(2/3), 167–188 (1996)
20. Yang, F., Hankin, C., Nielson, F., Nielson, H.R.: Predictive access control for distributed computation. Sci. Comput. Program. **78**(9), 1264–1277 (2013)
21. Zdancewic, S., Myers, A.C.: Observational determinism for concurrent program security. In: Proceedings of Computer Security Foundations Workshop, CSFW 2003, pp. 29–43 (2003)

Modelling the Impact of Threat Intelligence on Advanced Persistent Threat Using Games

Andrew Fielder(✉)

Institute for Security Science and Technology,
Imperial College London, London, UK
a.fielder@ic.ac.uk

Abstract. System administrator time is not dedicated to just cyber security tasks. With a wide variety of activities that need to be undertaken being able to monitor and respond to cyber security incidents is not always possible. Advanced persistent threats to critical systems make this even harder to manage.

The model presented in this paper looks at the Lockheed Martin Cyber Kill Chain as a method of representing advanced persistent threats to a system. The model identifies the impact that using threat intelligence gains over multiple attacks to help better defend a system.

Presented as a game between a persistent attacker and a dedicated defender, findings are established by utilising simulations of repeated attacks. Experimental methods are used to identify the impact that threat intelligence has on the capability for the defender to reduce the likelihood of harm to the system.

1 Introduction

In cyber security there is a common expression that states that an attacker only needs to be successful in a single attack to achieve their objective, whereas those defending a system need to be successful every time. This statement is designed to demonstrate the difficulty of a defenders task in protecting a system from attack, which is particularly relevant in the case of a determined attacker.

Determined attackers and advanced persistent threats (APTs) [12] are common terms to describe attackers and attacks that seek to achieve goals through building a presence on a network. While historical examples such as Stuxnet were designed to disrupt a particular target. APTs are more frequently associated with similar value targets. There is often a particular focus on the theft of high value assets such as financial details or intellectual property.

Lockheed Martin have developed the most common representation of an APT with their Cyber Kill Chain [1]. The cyber kill chain, shown in Fig. 1 gives an abstract representation of the steps that an attacker goes through when exploiting a target. The steps are described as:

A. Di Pierro et al. (Eds.): Festschrift Hankin, LNCS 12065, pp. 216–232, 2020.
https://doi.org/10.1007/978-3-030-41103-9_8

- Reconnaissance: Investigation by the attacker into the system and potential vulnerabilities
- Weaponization: Creation of the tools that are later used to exploit the target
- Delivery: Distribution of the initial tools to try and gain access to the target
- Exploitation: Execution of the delivered tools within the target to exploit the known vulnerabilities
- Installation: Establishing a foothold within the target network in order to exploit the target network
- Command and Control: Process by which an attacker is able to directly control the actions of the target
- Actions on Objective: The attacker acts on their aim for the attack

Within this representation, there are a set of associated actions that can be taken at each stage to mitigate the threat. These are defined as, Detect, Deny, Disrupt, Degrade, Deceive, and Contain.

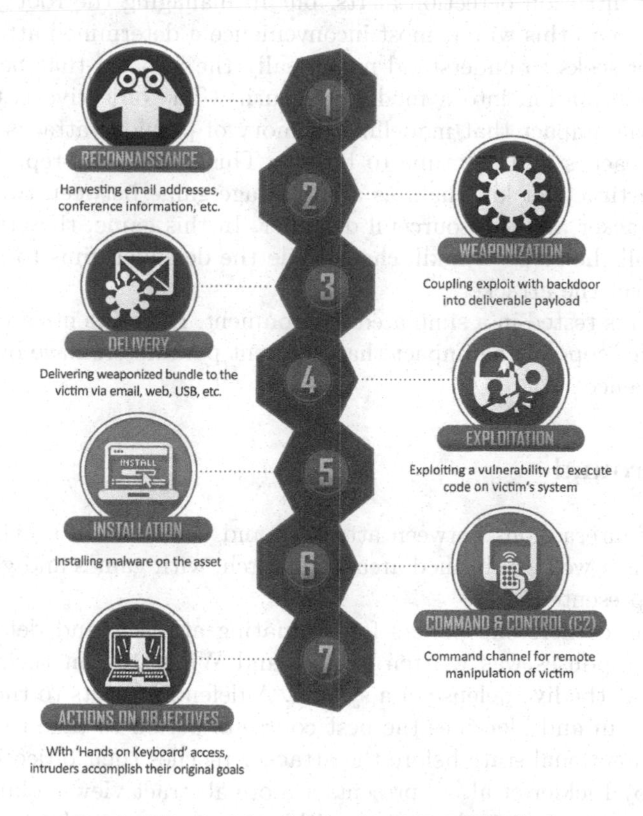

Fig. 1. The Lockheed Martin Cyber Kill Chain

Threat intelligence is given as the gathering and analysis of information for making security decisions. For APTs this is particularly important, as knowing who is trying to attack the target network, what they are trying to steal and how they are trying to achieve this can help with making more optimal security decisions.

If a system is under attack, conventional knowledge would dictate that it is always important to eject an attacker as soon as possible. However, when gathering threat intelligence, it is not always the most effective strategy. By allowing an attacker to progress further along the kill chain, they have to not only expose the intent behind their attacks, but also the tools that they use to do this. This can be especially useful, when those tools are bespoke and have taken considerable time and effort to create.

In many instances, the tools for collecting threat intelligence are not utilised effectively in defending the system. This is commonly seen when using defensive deception tools such as honeypots. To a defender, a honeypot should be producing few if any alerts from the system. This is given that those working within the environment should have no need to interact with the device. The issue is not acting on intrusion detection alerts, but in managing the root cause of an alert. Failure to do this will at most inconvenience a determined attacker.

This paper seeks to understand numerically the impacts that being able to feed threat information into a model of security. The objective is to measure if modelling the impact that modelling memory of previous attacks has on the number of breaches and the time to breach. This is done by representing the Lockheed Martin cyber kill chain as a multistage game between two players, a persistent attacker and a resourceful defender. In this game, the attacker aims to complete all the steps of a kill chain while the defender aims to identify the attack and eject the attacker.

The model is tested in a simulated environment. Experiments are performed to identify the scope of the impact that different parameters have on the use of threat intelligence.

2 Background

The study of interactions between attackers and defenders in a cyber security environment is a well established area of research, with games and game theory being well represented.

One of the cornerstone papers for evaluating attacker and defender interactions and responses was written by Lye and Wing [10]. In their paper the authors look at the live defense of a system. A defender reacts to the state that the network is in and identifies the best course of action to take to restore the system to a functional state before the attacker reaches their objective.

A paper by Fielder et al. [6] presents a more abstract view looking at where a defender should position themselves within a network in order to protect the widest range of assets. The work compares a defence-in-depth based approach to a more targeted defensive strategy. Similar work applied this to administrator time, and optimal strategies to cover multiple targets within a system where resource limitations mean full coverage is not possible [7].

A different approach using games has been provided by the Flip-It game [13]. The authors of this paper present a two person game of an attacker and a defender who with limited resources need to choose when to act in order to gain control of a specific resource. There have been many iterations on this game to build more realism or test different scenarios, of particular note is the paper by Hu et al., who have applied it to APTs [9].

A paper by Caulfield and Fielder looked at the vulnerability lifecycle [3]. The work looks at the way in which an optimal defensive strategy can be defined in an uncertain environment to counter unknown threats while still maintaining other business functionality. This is an extension of a previous work, which had more widely looked at the vulnerability lifecycle [2,8].

Work by Rass et al. [11] has looked at using game theory to protect against APT attacks. In this work, the authors present a method based on attack graphs to create an extensive form game with information sets. The authors conclude that with limited information on the attackers actions, they are able to use principles of uncertainty to generate optimal defensive solutions.

Zhang et al. [5,15] proposed a game where limited resources constrain both an attacker and a defender. They utilise an asymmetric feedback structure, where the attacker is able to hide information about its location from the defender. The paper defines algorithms for optimal solutions in two different scenarios, one where the defender announcing a strategy and one where they do not. Further examples have been studied to address the risk from APTs. In particular cloud platform security has more recently begun to utilised these approaches [4,14].

3 Kill Chain Model

Taking the foundation of the Lockheed Martin kill chain it is possible to represent an abstraction of persistent threat as an extended form game. The game represents a number of discrete states that relate to each step of the kill chain. More formally, a game (G) can be expressed as $G(K, O, S, Y, Z)$, where we define K as the players, O as the set of actions that are available to the player, and S represents the set of possible states the game can be in. Further, Y represents the payoffs for a successful attack and Z as the payoffs for successful defences.

The game is formed of two active players and a nature player. The attacker, denoted by a is the determined attacker that is trying to perform a successful APT style attack on a system. The defender, given by d, is attempting to stop the attacker from successfully attacking the system. The nature player, n, dictates a random element to the game that dictates the success and failure of actions taken by the players.

Both the attacker and defender have two actions available to them, act and learn, given as $O = c, e$.

Act, denoted by c, dictates the process of a player acting on their current understanding of the state of the attack and the environment. For the defender, this is an attempt to remove the attacker from their position in the network.

For the attacker this represents attempting to move to the next stage of the kill chain.

Learn (e) allows either player to add to their knowledge of their opponent. A player takes time to improve their capability to perform the relevant action at the current state of the game. For an attacker this represents learning about the structure of the organisation or underlying systems that they are trying to exploit. The defender acts on IDS logs and other available security mechanisms to see if they can understand what is happening in the system.

The act action success rate is defined by nature, and is defined by $p(i^d)$ and $p(i^a)$. $p(i^d)$ and $p(i^a)$ are the probability of success of the defenders and attacker action at stage i of the attack respectively. The learn action increments $p(i^d)$ or $p(i^a)$ by a value (u). The bounds of which are $0.00 <= p(i^d) <= 0.95$ and $0.00 <= p(i^a) <= 0.95$. The lower bounds assumes that there are scenarios where no information is known, and knowledge on the attack needs to be built from zero, and the upper bounds indicates that there is no certainty that an action will complete successfully.

Both actions c and e have the ability to advance the current state of the game. The learn action advance the state when $p(i^d) < 0.95$ or $p(i^a) < 0.95$, whereas the act action will only change the state if the action governed by n is successful. Each s is uniquely given by the stage of the game, i, and the probability $p(i^d)$ and $p(i^a)$. The probability of success can therefore be noted more specifically by the state s, which is seen as $p_s(i^d)$ and $p_s(i^a)$.

The mixed strategy for a player is defined as the probability that the act action is taken, and for each stage i is given as $p_d(i^c)$ for the defender and $p_a(i^c)$ for the attacker. The strategy of each player at each stage is agnostic of the associated success rate of the attack. This is based on the assumption that neither player would has certainty on the information prior to starting the attack. The strategy therefore is an approximation to the amount of time needed before they are willing to act.

The strategy is designed to highlight the emphasis on building and maintaining knowledge of the attacker and the target system for the defender and attacker respectively.

An important abstraction that needs to be considered, is that a step is not a fixed discrete time interval. APTs run some parts of the process over different time frames. The game is divided into individual discrete actions called rounds. These rounds represent a single decision made by both the attacker and defender.

The game is played over a finite horizon bound by the number of rounds (r) that the attacker has available to them to complete the objective. The game ends when the attacker either completes all seven stages of an attack or runs out of rounds. While it is possible to consider that the game could be played with an infinite horizon. Changes in technologies, priorities and staff that would occur over the course of the game would fundamentally change the environment in which the game is being played. With unlimited resources, the probability of a successful attack tends towards 1.0, as the probability of a successful defence is <1.0.

The payoffs for this game are split between the reward Y that the attacker receives for successfully breaching the system and ending the game, and payoffs Z that the defender receives for successfully stopping an attack. Each stage of the kill chain can have a different payoff $z_i \in Z$, but there is only one single reward value in Y.

The game is formed as a zero sum game, where the gain of the attacker is equal to the loss of the defender and similarly the gain for the defender is equal to the loss for an attacker. For Y this means that the value the attacker gains in the end goal is experienced as a matched direct or indirect loss for the defender. For Z, the gain of the defender through repelling an attack is measurable as a function of the time and tools wasted by the attacker in any failed attempt.

This problem can therefore be represented as a constrained optimisation problem, where the defender aims to maximise the rewards from each z_i while minimising the risk of losing Y. Similarly, the attacker is trying to achieve Y while minimising the loss from multiple failed attempts associated with each z_i.

4 Experimental Approach

Currently the state of knowledge on the effectiveness and incentives for using threat intelligence is not particularly wide. This lack of empirical data sources means that the experimentation cannot dictate with certainty what the optimal strategies for defending against APTs in a real environment would be. However, through experimentation, it is possible to highlight the impact of the different contributing factors on defending against APTs.

To approach this, a number of numerical experiments have been devised to illustrate the difference in the way in which possible constraints and parameter impact the benefit that threat intelligence provides to defending a target system.

The experiments are performed using a standard genetic algorithm (GA) based optimisation approach. Populations of strategies are generated and evolved pseudo-randomly.

Strategies are evaluated by simulating the attack and defence process. The attacker starts at the recon stage of the kill chain and attempts to successfully complete all seven steps of an attack without being removed by the defender. The attacker must complete this before the limit on number of rounds has been reached. If the attacker succeeds, then they receive the maximum reward. If the defender removes the attacker, the simulation returns them to the recon stage of the attack, and the players score the reward associated with the stage that the game was in.

The nature player is governed by the Python3.7 pseudo-random number generator. This acts only to simulate the probability of success of the act action of both players. Given the inherent uncertainty that this has, each simulation is repeated a fixed number of times, with the average payoff being used to judge fitness. In the event that there are multiple strategies that receive the same score, then the average number of steps used to achieve the goal is used. For the attacker this will be a minimisation of the number of steps used, where the defender will want to maximise the number of steps.

4.1 Experimental Configuration

The experiments presented in this section aim to identify the impact of the underlying game parameterisation on the use of threat intelligence to help secure against APTs.

All of the experiments performed in this work use the following set of parameters for the GA:

- Iterations: 1000
- Population Size: 100
- Offspring Size: 20
- Number of Evaluations: 1000
- Selection Scheme: Tournament

Here we define iterations as the maximum number of attempts that the system makes in order to reach a stable equilibrium. If convergence has not been reached after this number of iterations, then the system takes the best scoring candidates for the attacker and defender from the last generation for the evaluation.

Within the evaluation function of the system, there are a number of parameters that define how the game is played out. These parameters dictate aspects that will influence how effective the moves that the attacker and the defender undertake are. Unless otherwise stated by the configuration of the experiment, the default parameters for the game are given as:

- Rounds $= 100$
- Payoff for attacker success $(Y) = 1000$
- Initial success rate for defender $(p_0(i^d)) = 0.25$
- Initial success rate for defender $(p_0(i^a)) = 0.25$
- Learning rate $(u) = 0.05$
- Rate of learning decay $(R_u) = 0.05$
- Payoffs (Z) - Stage Incremental: $[1, 2, 3, 4, 5, 6, 7]$

The initial success rates are not set at 0 for the base configuration. While it is feasible that the players might have no knowledge at any stage, the initial probability is set to have a fundamental understanding of the situation and allow for strategies that don not require a mandatory learn action.

Much like the payoff for a successful attacker, the profile of the defence action rewards (Z) needs to be set. The values chosen for Z are significantly smaller than that of Y, this is to emphasise a distinction between failing at the core objective of protecting the network, and the incremental benefit of rewarding repelling the attacker.

The strategy utilising threat intelligence maintains the learnt information and increased probabilities of successful actions for both the attacker and defender across different attack attempts. When not using threat intelligence, these probabilities are reset to the default values for the defender between attack attempts.

4.2 Payoff for Threat Intelligence

In order to understand the potential impacts of integration of threat intelligence, there needs to be some understanding of how payoff structures impact the success of the implementation. This experiment seeks to identify if any change in the structure of the interim payoffs for the defender leads to a better defence.

For the experiments, we consider five different payoff structures for Z:

- None $[0, 0, 0, 0, 0, 0, 0]$: There is no reward gained by the defender for the successful defence. This indicates that there is no perceived benefit to the defender for ejecting the attacker beyond stopping the attacker.
- Uniform $[1, 1, 1, 1, 1, 1, 1]$: A single reward that represents that there is some perceived benefit to ejecting the attacker, no real difference is noted between the stage of the attack at which the attacker is removed.
- Stage Incremental $[1, 2, 3, 4, 5, 6, 7]$: The defender is rewarded based on the number of stages that the attacker managed to complete. This is used to denote an incremental reward for having seen more of the attacker's capabilities.
- Stage Inverse $[7, 6, 5, 4, 3, 2, 1]$: The defender is rewarded based on the concept that they optimally repel the attack at the earliest possible stage. This is demonstrated as a reduced reward the closer the attacker gets to completing their objective.
- Cumulative $[1, 3, 6, 10, 15, 21, 28]$: This positions an understanding that there is a greater reward associated with eliminating the attacker at later stages of the attack, as the defender is capable of seeing more of the tools and techniques and using that to build a greater understanding of the threat of the attacker.

4.3 Attacker Determination

A determined attacker, while having significant resources available, will not have an endless amount of them. The scenario presented is to reflect the idea that the time an attacker has dedicated to the task will still have some limits. This can be seen as either shifting requirements and objectives, or personnel replacement on a key task.

The minimum amount of rounds is set as being able to realistically attempt one complete attack, increasing to scenarios, where defence is only likely with an effective strategy. Scenarios are considered to a point where the attacker would be able to brute force the problem.

The number of rounds the attacker has available is set as $[10, 50, 100, 200, 500]$.

4.4 Initial Knowledge Levels

One of the assumptions that had previously been made, was that both the attacker and defender have a baseline understanding of how to perform the next stage of the attack or find and remove the attacker. The level of this base knowledge is not clear, and an overestimate on confidence in the defender to be able to defend the system could result in missing part of an attack.

The experiment on base level aims to address the impact that the initial level of knowledge has on the capability of the players to successfully perform their respective tasks.

Defender Variation: This experiment keeps the initial knowledge level at 0.1 for the attacker but varies the initial knowledge level for the defender in the range $[0, 0.1, 0.25, 0.5, 0.75]$. Alternative Fixed Increment: This experiment uses the same initial knowledge level for both the attacker and the defender. The range for these values is set as $[0, 0.1, 0.25, 0.5, 0.75]$.

The first variation looks to see the relative impact that the initial knowledge level has, where the last variant aims to understand how much the absolute level impacts the solution.

4.5 Effectiveness of Learning

The initial set-up dictates a fixed learning rate, which increases the probability of success for each player by 0.1. This set of experiments looks to compare the results of 4 alternative values to the base configuration. The aim of this experiment is to capture the impact of learning rates on the capability for the defender to repel attacks.

By changing the effectiveness, it is possible to capture the impact that the capability to learn has on the strategies. Higher learning rates mean that the players will need to spend less time learning, giving them more rounds to act. If the learning rate is too low, then the effort needed by either player to overcome the rate of learning decay means that in order to make the most of the threat intelligence, learning needs to be prioritised.

This experiment uses the same learning rate for both the attacker and the defender. The range for these values is set as $[0, 0.01, 0.05, 0.1, 0.2]$.

4.6 Intelligence Recall Decay

The previous experiments have defined that there is a near perfect recall of information by the defender on the capabilities of the attacker and the attacker on the defender, however this might not be strictly true in reality. Aspects of techniques and tools might change, meaning that the knowledge is not the same.

This experiment looks to vary the recall decay rate (R_u) for both players.

As R_u tends towards $1 - p_0(i^d)$ for the defender and $1 - p_0(i^a)$ for the attacker, then the system reverts to a close approximation of the scenario where no threat intelligence is being used. For this reason, the range stops at recall loss of 0.2, where the complete range of values used is $[0, 0.01, 0.05, 0.1, 0.2]$.

5 Results

The primary comparison presented in this section is the probability of the attacker successfully attacking the system, and the average number of cycles that it takes for the attacker to achieve that goal.

5.1 Payoff for Threat Intelligence

Figure 2 shows the results for using different payoff functions for a defender when they are integrating threat intel. The use of threat intelligence reduces the expected number of successful attacks by a minimum of 0.3. With the exception of the cumulative payoff method, which sees attacks succeed with a probability of 0.27, the remaining methods are grouped with a probability of approximately 0.5.

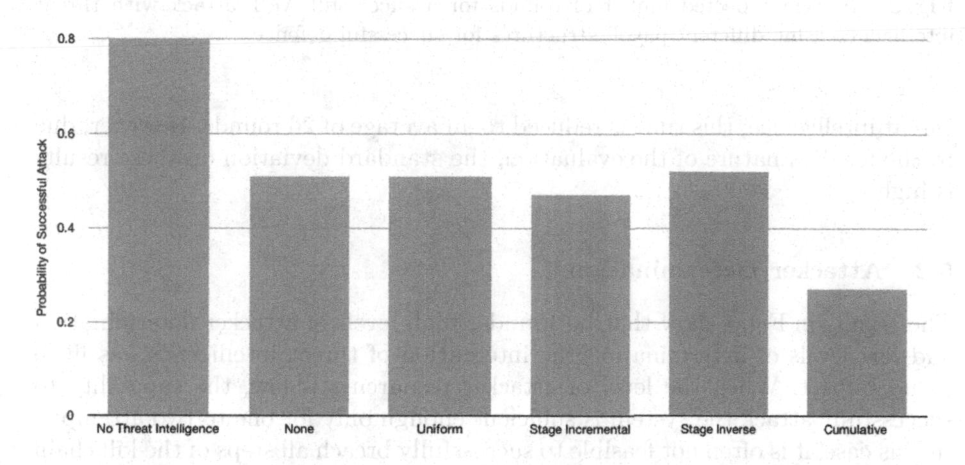

Fig. 2. Probability of a successful APT attack with threat intelligence using different payoff structures for successful defence

The improvement for the cumulative payoff, appears to be down to being able to best utilise the payoff function to provide incentives to allow attacks to penetrate quite far into the system before being remediated. While this might be a risky strategy, allowing those deeper attacks, the defender is more frequently in a position to be able to gather intelligence on more advanced stages of an attack. These stages might not be seen as frequently with other strategies, due to not being incentivised to reach the later stages of an attack.

Without this selection pressure, generated optimal solutions for other strategies will be efficient at handling early stage attacks, but in the event of a failure at the earl stages are much more vulnerable to later stage attacks.

Considering the average expected number of cycles that it takes to complete an attack, Fig. 3 shows that it is fairly consistent for those using threat intelligence, taking on average between 40 and 50 rounds to achieve. When not using

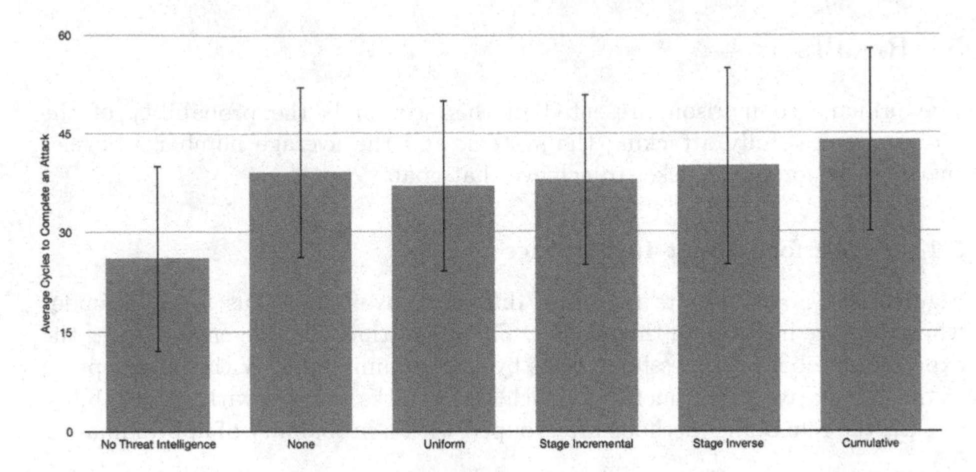

Fig. 3. Average expected umber of rounds for a successful APT attack with threat intelligence using different payoff structures for successful defence

threat intelligence, this time is reduced to an average of 26 rounds. However, due to the random nature of the evaluation, the standard deviation on these results is high.

5.2 Attacker Determination

The results in Fig. 4 show that both under high levels of attacker determination and low levels of determination the integration of threat intelligence has little to no impact. When the level of attacker resourcing is low, the capability to successfully attack the system is sufficient enough only for one to two attempts. In this case, it is often not feasible to successfully breach all steps of the kill chain before exhausting available resources. Likewise with near infinite resources, the probability of success tends to one for the attacker. This means that the use of accumulated knowledge bears little impact on the results.

However, between the extremes, there is a clear divergence. Most interesting is that the initial availability of resources that allows for attacks to be successful results in the most significant gap. However, since the attacker is able to exploit the system effectively with few resources, the growth in probability then tails off, as there are fewer cycle in which it was not able to achieve better results. This is opposed to the case where threat intelligence is used, where the initial growth is small, but the more opportunities the attacker is given, the greater the rate at which they are able to achieve successful attacks. The rate increases more rapidly until it reaches near parity with the outcomes for not using threat intelligence.

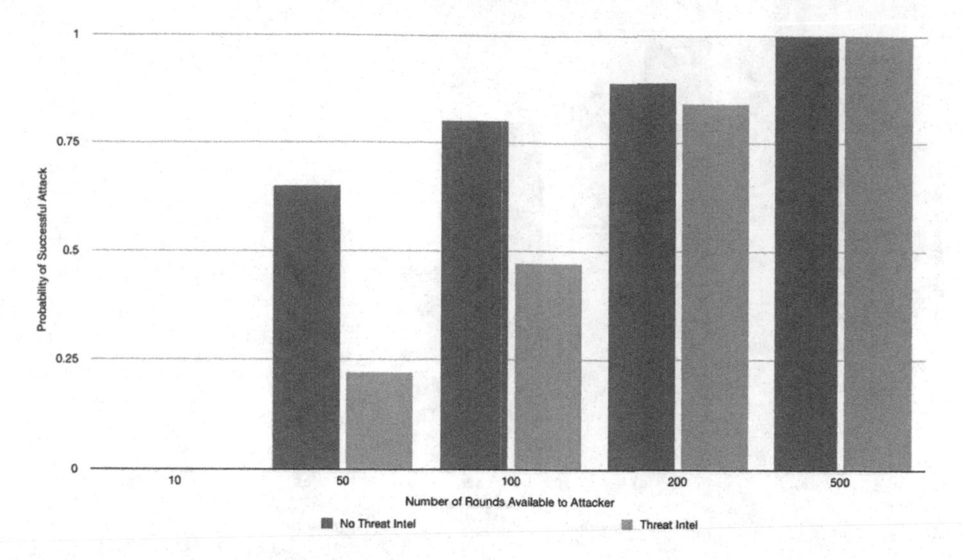

Fig. 4. Probability of a successful APT attack with threat intelligence using different levels of resource for the attacker

5.3 Initial Knowledge Levels

Figures 5 and 6 both show similar outcomes, where for high initial success rates, the defender is able to eject all attackers from the system. However, when this rate is low, the attacker is able to successfully attack more often.

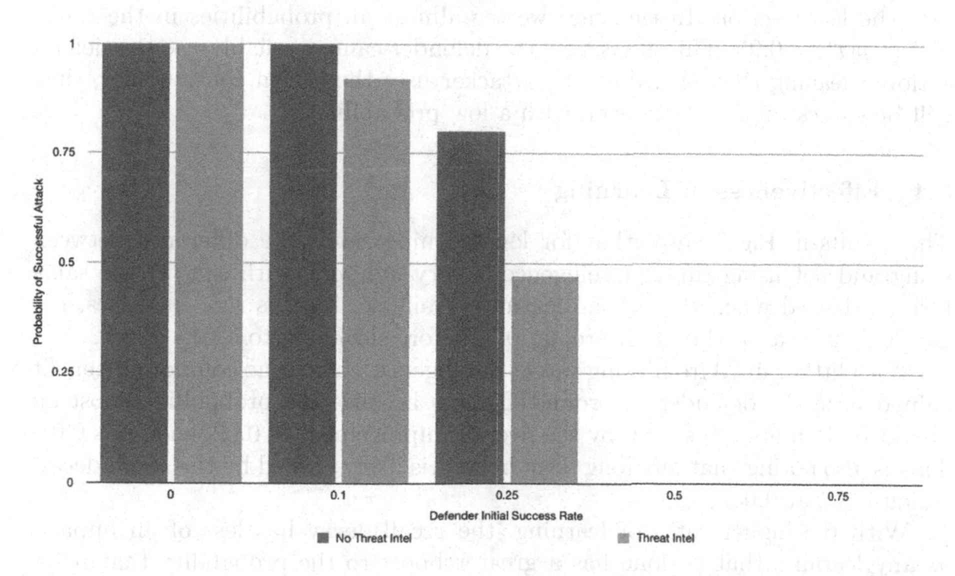

Fig. 5. Probability of a successful APT attack with threat intelligence using different levels of knowledge for the defender

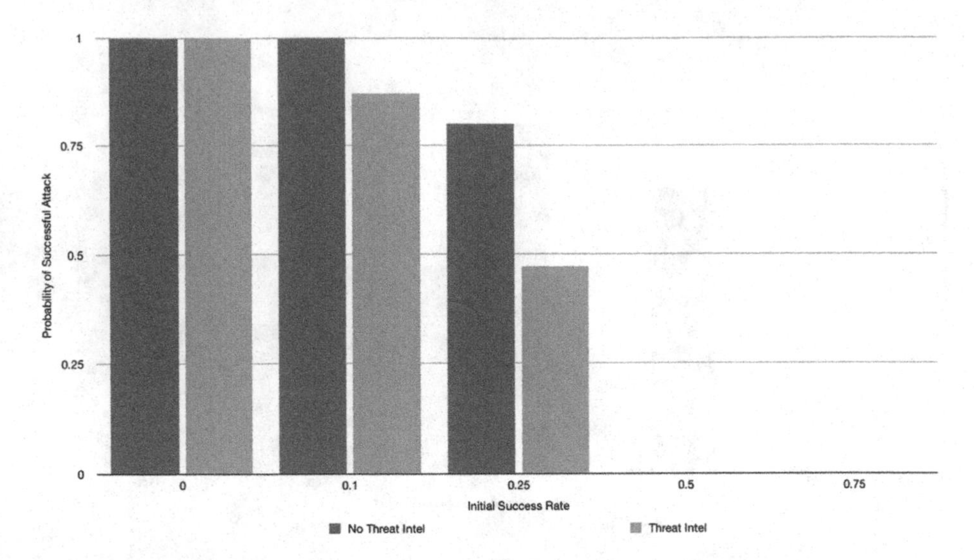

Fig. 6. Probability of a successful APT attack with threat intelligence using different initial levels of knowledge for both attacker and defender

Comparing the scenario of assigning both attacker and defender lower initial success rate to the standard case, it takes an attacker significantly longer to exploit the target. It takes on average 50 rounds to complete with threat intelligence, and 37 rounds to complete without, compared to the 38 and 26 for each reported previously.

The reason for this, is that for low rates, the players must put in more effort into the learn action. In this case we see almost all probabilities in the range $0.75 < p(i^a) < 0.95$. This is because the defender cannot reliably use the defend action, meaning that providing the attacker uses the action continuously, they will be successful over time even with a low probability.

5.4 Effectiveness of Learning

The results in Fig. 7 show that for low learning rates, the difference between using and not using threat intelligence is very minimal, with exactly the same result achieved when there is no learning available. As this rate increases, the gap widens to a maximum of around 0.3, before shrinking to a gap of 0.12.

With little gained from using threat intelligence, there is no compound benefit gained from the defender per round. This is because the probability boost on operation in minimal, shown by the overall improvement of 0.02 when u is 0.01. This is also noting that any long term benefit is likely eroded by the recall decay, which is set at 0.05.

With the higher rates of learning, the recall decay has less of an impact, as any learning that is done has a greater boost to the probability that helps the defender more quickly learn about the attacker. This accounts for the more significant drops as the rate increases. As the rate approaches 1.0, the discrepancies between using and not using threat intelligence will tend towards 0.

5.5 Intelligence Recall Decay

The rate of recall delay as shown by Fig. 8 has no impact on the optimal strategy and results of attacks against a system that does not use threat intelligence, so the main comparison is how the rate impacts the systems that do use threat intelligence to this.

Similar to the effectiveness of learning, the recall decay has a similar but inverse trajectory. When there is no decay rate, the optimal solution is able to learn effectively enough in order to prevent almost all attacks from succeeding. While it is capable of eliminating most, the inherent uncertainty and speed at which the defender can learn will prohibit it from eliminating every threat.

The effectiveness of learning about the attacker then reduces in efficiency in a predictable manner. At a rate of decay equal to 0.2, there is no difference that can be seen between using and not using threat intelligence. This means that with a rate of 0.2, the capability to learn about the attacker cannot keep up with the rate at which the attacker, defender and environment change.

6 Discussion

Given that the defender should be no worse off from a security perspective, the results should not be surprising. This is given that in the worst case scenario, the defender is in the same state of knowledge as if they were not using threat intelligence.

What is important is when and where there is justification for trying to use threat intelligence. The integration of known threats, investigation to learn techniques, and implementation of these approaches will have an associated cost.

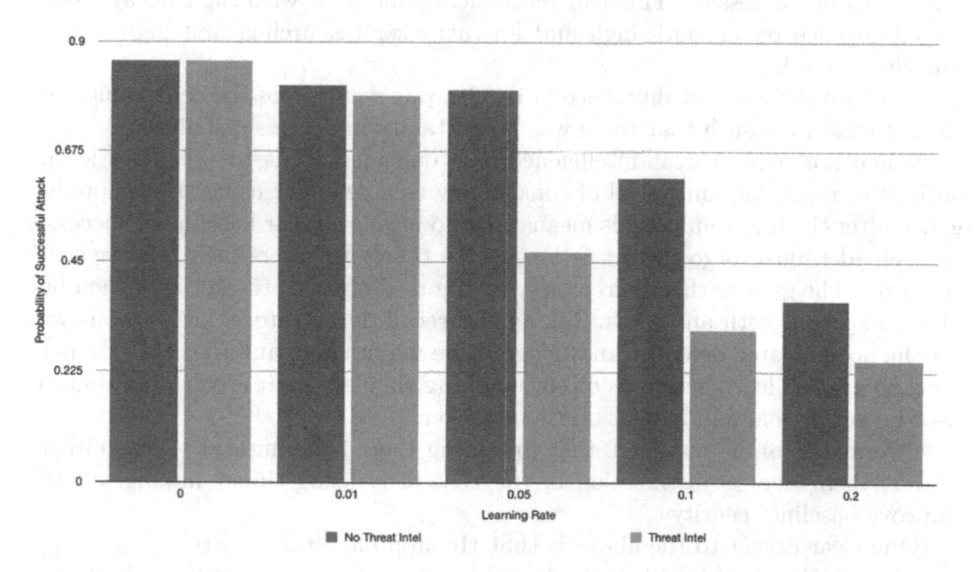

Fig. 7. Probability of a successful APT attack with threat intelligence using different rates of accumulating knowledge

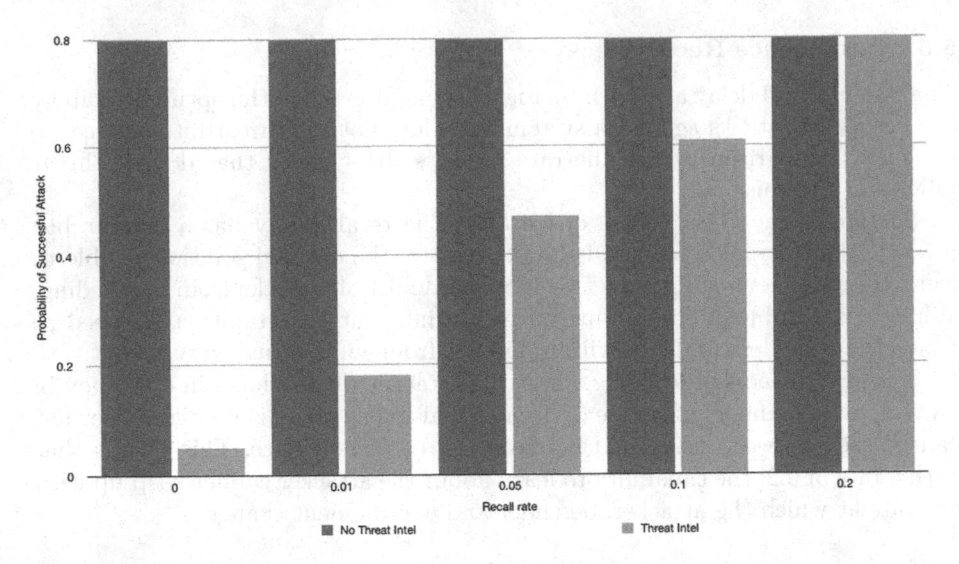

Fig. 8. Probability of a successful APT attack with threat intelligence using different intelligence recall decay rates

From the results, it is clear to see that in the extreme cases where the operational environment is similar between a defender utilising threat intelligence and not utilising threat intelligence, then the results offer no clear incentive for the utilisation of accumulated knowledge. This can be seen in cases where the incremental advantage of learning about the attacker does not impact the capability for act to be successful. This can most clearly be seen with high decay rates on intelligence recall, both high and low attacker resourcing, and high initial knowledge levels.

Likewise the greatest difference in results was seen where the constraints on the attacker were such that there was no certainty in success or failure.

When not using threat intelligence, the defender needs to use more learn actions to reach the same level of consistency as a defender using threat intelligence after the first round. This means that to have a similar baseline of success, the defender must forgo the act action, which therefore lowers their relative success rate. The greater this discrepancy, the more likely the attacker is to then be able to succeed with an attack. Taking the recall decay rate, a low rate allows for the accumulated defender knowledge to be maintained, and thus they do not need to use the learn action as often, meaning they are more frequently able to use the act action, and impact on the attacker.

Overall, it can be reasoned, that providing there is no natural bias to either player, then there is justification in the cost of utilising threat intelligence to improve baseline security.

One clear caveat to the above is that the approach taken in this paper represents a high level of uniformity in representing the stages of the kill chain. Consistent learning rates, base probabilities, and intelligence decay rates will not likely be consistent across all stages. Therefore the exact bounds at which the use of integrating threat intelligence into future rounds rounds may shift from the example cases demonstrated here. While this might impact the exact bounds, this will only impact the point at which divergence and convergence of strategies exist based on the resource and capability difference between the two players.

7 Conclusions and Further Work

This paper has presented an approach to modelling the integration of threat intelligence into the Lockheed Martin kill chain using a game formulation of an attacker trying to execute the steps of the kill chain, and a defender trying to protect the system from being exploited at each stage. This model was tested in a simulated environment, where a number of numerical examples were tested to see the impact that knowledge integration has on the capability for a defender to resist a persistent attacker.

The results presented show an expected outcome that the integration of accumulated knowledge back into the defence of the system does reduce the likelihood that an attacker will be successful, and on average elongate the amount of time and effort that an attacker has to put into . The results also begin to demonstrate the bounds under which this holds true, presenting that when capability asymmetry is too high, then the impact can be diminished to the point where there is no advantage to integrating threat intelligence.

To progress this work further, a detailed set of realistic data that can inform the parameters of the model are needed. This would allow for further investigation into the strategy space and optimal decision making of the defender against an attacker. This would allow for greater decision support as to what effective strategies are when facing a live attacker in a network, balancing the utilisation of gathering threat intelligence against removing an attacker at the earliest opportunity.

To be able to see the value of the approach, and the optimal time to invest based on the increment in security. This then adds an extra cost-benefit dimension to the problem, where the benefit of the increase in defence from integration of threat intelligence must be aligned with the cost of performing this action against utilising those resources elsewhere. This will result in further bounding the use case for the adoption of threat intelligence.

References

1. Cyber kill chain (2019). https://www.lockheedmartin.com/en-us/capabilities/cyber/cyber-kill-chain.html
2. Beres, Y., Griffin, J., Shiu, S., Heitman, M., Markle, D., Ventura, P.: Analysing the performance of security solutions to reduce vulnerability exposure window. In: 2008 Annual Computer Security Applications Conference (ACSAC), pp. 33–42. IEEE (2008)

3. Caulfield, T., Fielder, A.: Optimizing time allocation for network defence. J. Cybersecur. **1**(1), 37–51 (2015). https://doi.org/10.1093/cybsec/tyv002
4. Chen, J., Zhu, Q.: Security as a service for cloud-enabled internet of controlled things under advanced persistent threats: a contract design approach. IEEE Trans. Inf. Forensics Secur. **12**(11), 2736–2750 (2017)
5. Feng, X., Zheng, Z., Cansever, D., Swami, A., Mohapatra, P.: Stealthy attacks with insider information: a game theoretic model with asymmetric feedback. In: 2016 IEEE Military Communications Conference, MILCOM 2016, pp. 277–282. IEEE (2016)
6. Fielder, A., Li, T., Hankin, C.: Defense-in-depth vs. critical component defense for industrial control systems (2016)
7. Fielder, A., Panaousis, E., Malacaria, P., Hankin, C., Smeraldi, F.: Game theory meets information security management. In: Cuppens-Boulahia, N., Cuppens, F., Jajodia, S., Abou El Kalam, A., Sans, T. (eds.) SEC 2014. IAICT, vol. 428, pp. 15–29. Springer, Heidelberg (2014). https://doi.org/10.1007/978-3-642-55415-5_2
8. Frei, S., May, M., Fiedler, U., Plattner, B.: Large-scale vulnerability analysis. In: Proceedings of the 2006 SIGCOMM Workshop on Large-Scale Attack Defense, pp. 131–138. ACM (2006)
9. Hu, P., Li, H., Fu, H., Cansever, D., Mohapatra, P.: Dynamic defense strategy against advanced persistent threat with insiders. In: 2015 IEEE Conference on Computer Communications (INFOCOM), pp. 747–755. IEEE (2015)
10. Lye, K.W., Wing, J.M.: Game strategies in network security. Int. J. Inf. Secur. **4**(1–2), 71–86 (2005)
11. Rass, S., König, S., Schauer, S.: Defending against advanced persistent threats using game-theory. PloS One **12**(1), e0168675 (2017)
12. Tankard, C.: Advanced persistent threats and how to monitor and deter them. Netw. Secur. **2011**(8), 16–19 (2011)
13. Van Dijk, M., Juels, A., Oprea, A., Rivest, R.L.: FlipIt: the game of "stealthy takeover". J. Cryptol. **26**(4), 655–713 (2013)
14. Xiao, L., Xu, D., Xie, C., Mandayam, N.B., Poor, H.V.: Cloud storage defense against advanced persistent threats: a prospect theoretic study. IEEE J. Sel. Areas Commun. **35**(3), 534–544 (2017)
15. Zhang, M., Zheng, Z., Shroff, N.B.: A game theoretic model for defending against stealthy attacks with limited resources. In: Khouzani, M.H.R., Panaousis, E., Theodorakopoulos, G. (eds.) GameSec 2015. LNCS, vol. 9406, pp. 93–112. Springer, Cham (2015). https://doi.org/10.1007/978-3-319-25594-1_6

Security Metrics at Work on the Things in IoT Systems

Chiara Bodei[1], Pierpaolo Degano[1], Gian-Luigi Ferrari[1], and Letterio Galletta[2](✉)

[1] Dipartimento di Informatica, Università di Pisa, Pisa, Italy
{chiara.bodei,pierpaolo.degano,gian-luigi.ferrari}@unipi.it
[2] IMT School for Advanced Studies, Lucca, Italy
letterio.galletta@imtlucca.it

Abstract. The Internet of Things (IoT) is deeply changing our society. Daily we use smart devices that automatically collect, aggregate and exchange data about our lives. These data are often pivotal when they are used e.g. to train learning algorithms, to control cyber-physical systems, and to guide administrators to take crucial decisions. As a consequence, security attacks on devices can cause severe damages on IoT systems that take care of essential services, such as delivering power, water, transport, and so on. The difficulty of preventing intrusions or attacks is magnified by the big amount of devices and components IoT systems are composed of. Therefore, it is crucial to identify the most critical components in a network of devices and to understand their level of vulnerability, so as to detect where it is better to intervene for improving security. In this paper, we start from the modelling language IoT-LySa and from the results of Control Flow Analysis that statically predict the manipulation of data and their possible trajectories. On this basis, we aim at deriving possible graphs of how data move and which are their dependencies. These graphs can be analysed, by exploiting some security metrics - among which those introduced by Barrere, Hankin et al. - offering system administrators different estimates of the security level of their systems.

1 Introduction

Security issues can jeopardise the growth of the Internet of Things (IoT). The risk of intrusions or attacks is magnified by the big amount of devices and components these systems are composed of. Things are less and less material, while the effects of cyber attacks are far from being only virtual. The presence of actuators that act on the physical world makes security particularly crucial. Nevertheless there is some reluctance in adopting security countermeasures because they can impact on costs and performance. As a consequence, techniques able to evaluate where

Partially supported by Università di Pisa PRA_2018_66 *DECLWARE: Metodologie dichiarative per la progettazione e il deployment di applicazioni* and by MIUR project PRIN 2017FTXR7S *IT MATTERS* (Methods and Tools for Trustworthy Smart Systems).

A. Di Pierro et al. (Eds.): Festschrift Hankin, LNCS 12065, pp. 233–255, 2020.
https://doi.org/10.1007/978-3-030-41103-9_9

it is more important to intervene for improving security can be useful for IoT designers in choosing which security mechanisms to use.

We intend here to exploit the Security by Design development methodology, introduced in [3,4,9] by taking inspiration from the work by Barrère, Hankin, et al. [1,20], where the authors propose some security metrics for identifying critical components in ICS (Industrial Control System) networks.

Our methodology is based on a kernel of a modelling language IoT-LySa to describe the structure of an IoT system and its interactive capabilities. This language is endowed with a suitable Control Flow Analysis (CFA) that approximates the system by providing an abstract model of behaviour. These abstractions allow predicting how (abstract) data may flow inside the system. In [7] the CFA also includes a data path analysis that supports tracking of the propagation of data, by identifying their possible trajectories among the computing nodes.

In [1], the security level of a system is obtained as the minimum set of CPS (Cyber-Physical System) components that an attacker must compromise in order to impair the functionalities of the system. Moreover, each component is associated to a score, i.e. a value that quantitatively represents the effort (in terms of cost) required by an attacker to compromise it. Scores allow designers to compute the set of components with the minimal cost for the attacker. The metric relies on AND/OR graphs that represent the complex dependencies among network components, and the problem consists in finding a minimal weighted vertex cut in the graph.

Inspired by [1,20], we use a graph model of the CFA results, in particular, we model the data dependency and the logical conditions on which the trigger of a given actuator depends. More precisely, the obtained graph encodes the dependencies of the condition on the raw data collected by sensors and on the logical and aggregation functions applied to the sensor data in processing the decision (a full formalisation of how to derive the graph from the analysis results is left as a future work). For each sub-term of the abstract value, the analysis gives the location where it is introduced and computed and its possible trajectories across the network. By using this information, designers can investigate the security of their system, by reasoning on which is the minimal set of data that can be altered (by tampering the corresponding nodes) in order to affect the capacity of the system to correctly trigger the actuators. In the presence of more choices, by associating score to each node, it is possible to compare the different solutions and establish which are the cheaper ones. The composition of data in the graph can drive the reasoning in a fashion similar to that introduced in [1].

Structure of the Paper. The paper is organised as follows. Our approach is introduced in Sect. 2, with an illustrative example that serves as case study. We briefly recall the process algebra IoT-LySa in Sect. 3 and the adopted CFA in Sect. 4. Conclusions are in Sect. 5. The Appendix completes the formal treatment of the semantics and of the analysis.

2 Motivating Example

In this section we illustrate our methodology through a simple scenario concerning the fire alarm and suppression system of an academic department building.

The Scenario. Suppose that the building has three floors: (0) at the ground floor there is the library; (1) at the first floor there is the office area; (2) at the second floor there is the IT center and, inside it, the control room of the whole building.

In every room of each floor there are a smoke detector and a heat detector together with a pump that may be activated in case of fire. We assume that these sensors and the actuator are controlled by a smart device associated to the room, called *Room controller*. For example, each of the k rooms of the Library has its own controller R_{L_i} (with $i \in [1, k]$). Furthermore, at each floor there is a device working as *Floor controller* (e.g. at the ground floor there is the Library controller L) that receives the data sensed in each room, elaborates them to detect possible fires and sends them to the control room.

Also, each Floor controller is equipped with a sensor that checks the water level of the suppression system at that floor.

The Control Room receives data from each floor and, in case of fire, raises the alarm, by calling the Fire Station, and triggers the evacuation plan.

The IoT-LySa *Model.* The IoT-LySa model of the network N of our building is in Table 1. It has a finite number of nodes running in parallel (the operator $|$ denotes the parallel composition of nodes). A node represents a smart device as a Room controller, e.g. R_{L_1} for the first room of the Library, a Floor controller, e.g. L for the Library, and the Control Room CR. Furthermore, each node is uniquely identified by a label ℓ and it consists of control processes (the software) and, possibly, of sensors and actuators (the cyber-physical components).

In IoT-LySa communication is multi-party: each node sends information (in the form of tuples of terms) to a set of nodes, provided that they are within the same transmission range. Outputs and inputs must match to allow communication. In more detail, the output $\langle\langle E_1, \cdots, E_k \rangle\rangle \triangleright L$ represents that the tuple $E_1, \cdots, E_k$ is sent to the nodes with labels in L. The input is instead modelled as $(E_1, \cdots, E_j; x_{j+1}, \cdots, x_k)$ and embeds pattern matching. In receiving an output tuple $E'_1, \cdots, E'_k$ of the same size (arity), the communication succeeds provided that the first j elements of the output match the corresponding first elements of the input (i.e. $E_1 = E'_1, \cdots, E_j = E'_j$), and then the variables occurring in the input are bound to the corresponding terms in the output. In other words, our primitive for input tests the components of the received message. Suppose e.g. to have a process P waiting a message that P knows to include the value v, together with a datum that is not known from P. The input pattern tuple would be: $(v; x)$. If P receives the matching tuple $\langle v, d \rangle$, the variable x is bound to v, since the first component of the tuple matches with the corresponding value. Finally, note that terms are uniquely annotated to support the Control Flow Analysis (see the next sections).

We first examine the nodes R_{Li}, located in the rooms of the Library:

$$R_{Li} = \ell_{Li} : [P_{Li} \parallel S_{Li1} \parallel S_{Li2} \parallel B_{Li}]$$

Table 1. Fire Detection System: the controllers of Office area and IT Center are omitted because similar to those of the Library.

The Network of Smart Devices

$$\begin{aligned}
N &= CR \mid L \mid O \mid IT \mid \\
&\quad R_{L1} \mid \ldots \mid R_{Lk} \mid R_{O1} \mid \ldots \mid R_{Ok} \mid R_{I1} \mid \ldots \mid R_{Ik}
\end{aligned}$$

Room controllers of the Library with $i \in [1, k]$

$$\begin{aligned}
R_{Li} &= \ell_{Li} : [P_{Li} \parallel S_{Li1} \parallel S_{Li2} \parallel B_{Li}] \\
P_{Li} &= \mu h.(z_{Li1}^{r_{Li1}} := 1_{Li}^{s_{Li1}}).(z_{Li2}^{r_{Li2}} := 2_{Li}^{s_{Li2}}). \\
&\quad \langle\langle i, z_{Li1}^{r_{Li1}}, z_{Li2}^{r_{Li2}} \rangle\rangle \triangleright \{\ell_{Library}\}.(z_{check_{Li}}^{r_{ch_{Li}}}).\langle z_{check_{Li}}^{r_{ch_{Li}}}, i_{Li}, Go \rangle.h \\
S_{Li1} &= \mu h.(\tau.1_{Li} := s_{Li}).\tau.h \\
S_{Li2} &= \mu h.(\tau.2_{Li} := h_{Li}).\tau.h \\
A_{Li} &= \mu h.(\!| i_{Li}, \{\mathsf{Go}\} |\!).\tau.h
\end{aligned}$$

Floor controller of the Library

$$\begin{aligned}
L &= \ell_L : [P_L \parallel S_L \parallel B_L] \\
P_L &= \mu h.(w_L^{f_L} := 1_L^{s_L}).check_L^{f_{ch_L}} := w_L^{f_L} \geq^{g_L} th_L.\langle check_L^{f_{ch_L}}, i_L, Refill \rangle. \\
&\quad (1; w_{L11}^{f_{L11}}, w_{L12}^{f_{L12}}). \\
&\quad check_{L1}^{f_{L1}} := and_{L1}^{a_{L1}}(check_L^{f_{ch_L}}, or_{L1}^{o_{L1}}(w_{L11}^{f_{L11}} \geq^{g_{L11}} th_{L11}, w_{L12}^{f_{L12}} \geq^{g_{L12}} th_{L12})). \\
&\quad \langle\langle check_{L1}^{f_{L1}} \rangle\rangle \triangleright \{\ell_{L_1}\}. \\
&\quad \ldots \\
&\quad (k; w_{Lk1}^{f_{Lk1}}, w_{Lk2}^{f_{Lk1}}). \\
&\quad check_{Lk}^{f_{Lk}} := and_{Lk}^{a_{Lk}}(check_L^{f_{ch_L}}, or_{Lk}^{o_{Lk}}(w_{Lk1}^{f_{Lk1}} \geq^{g_{Lk1}} th_{Lk1}, w_{Lk2}^{f_{Lk2}} \geq^{g_{Lk2}} th_{Lk2})). \\
&\quad \langle\langle check_{Lk}^{f_{Lk}} \rangle\rangle \triangleright \{\ell_{L_k}\}. \\
&\quad \langle\langle L, p^{f_P}(check_{L1}^{f_{L1}}, \ldots, check_{Lk}^{f_{Lk}}) \rangle\rangle \triangleright \{\ell_{L_{CR}}\}.h \\
S_L &= \mu h.(\tau.1_L := s_L).\tau.h \\
A_L &= \mu h.(\!| i_L, \{\mathsf{Refill}\} |\!).\tau.h
\end{aligned}$$

Room controller and Floor controller of the Office area with $i \in [1, k]$

$$\begin{aligned}
R_{Oi} &= \ldots \\
O &= \ldots
\end{aligned}$$

Room controller and Floor controller of the IT Center with $i \in [1, k]$

$$\begin{aligned}
R_{Ii} &= \ldots \\
IT &= \ldots
\end{aligned}$$

Control Room

$$\begin{aligned}
CR &= \ell_{CR} : [P_{CR} \parallel A_{CR} \parallel B_{CR}] \\
P_{CR} &= \mu h.(L, x_{check_L}^{c_{ch_L}}).alarm_L^{c_{al_L}} := (x_{check_L}^{c_{ch_L}} \geq^{cr_L} th_L).\langle alarm_L{}^{c_{al_L}}, 1_{CR}, ring \rangle \parallel \\
&\qquad (O, x_{check_O}^{c_{ch_O}}).alarm_O^{c_{al_O}} := (x_{check_O}^{c_{ch_O}} \geq^{cr_O} th_O).\langle alarm_O{}^{c_{al_O}}, 1_{CR}, ring \rangle \parallel \\
&\qquad (I, x_{check_I}^{c_{ch_I}}).alarm_I^{c_{al_I}} := (x_{check_I}^{c_{ch_I}} \geq^{cr_I} th_I).\langle alarm_I{}^{c_{al_I}}, 1_{CR}, ring \rangle.h \\
A_{CR} &= \mu h.(1_{CR}, \{\mathsf{Ring}\}).\tau.h
\end{aligned}$$

where the label ℓ_{Li} uniquely identifies the node. The control process P_{Li} represents the logic of the device and manages the smoking sensor S_{Li1}, the heating sensor S_{Li1} and the suppression actuator A_{Li}. Whereas B_{Li} represents those node components we are not interested in here, such as the device store Σ_{Li}. All these components run in parallel (this is the meaning of the parallel composition operator $\|$ for components inside nodes).

Intuitively, each node R_{Li} collects the data from its sensors, transmits them to the Library controller L, and waits for its decision. In IoT-LySA, sensors communicate the sensed data to the control processes of the node by storing them in their reserved location $1_{L11}^{s_{L11}}$ and $1_{L12}^{s_{L12}}$ of the shared store. The action τ denotes internal actions of the sensor we are not interested in modelling. The construct $\mu h.$ denotes the iterative behaviour of processes, of sensors and actuators, where h is the iteration variable.

More in detail, the control process P_{Li}: (i) stores in the variables $z_{Li1}^{r_{Li1}}$ and $z_{Li2}^{r_{Li2}}$ (where r_{Li1} and r_{Li2} are the variable annotations) the data collected by the smoking and heating sensors, through the two assignments: $z_{Li1}^{r_{Li1}} := 1_{L11}^{s_{L11}}$ and $z_{Li2}^{r_{Li2}} := 1_{L12}^{s_{L12}}$; (ii) transmits the collected data to the Library controller L, with the output $\langle\!\langle i, z_{Li1}^{r_{Li1}}, z_{Li2}^{r_{Li2}} \rangle\!\rangle \triangleright \{\ell_{Library}\}$; (iii) waits for the decision of the Library controller on the input $(z_{check_{Li}}^{r_{ch_{Li}}})$; (iv) forwards the decision to the suppression actuator A_{Li}, with the command $\langle z_{check_{Li}}^{r_{ch_{Li}}}, i_{Li}, Go\rangle$: if $z_{check_{Li}}^{r_{ch_{Li}}}$ is true, the water suppression (action Go) is activated.

In the Library controller L, the process P_L: (i) stores in the variables $w_L^{f_L}$ the data provided by the water level sensor; (ii) checks whether the water level is above a given threshold th_L and if *Refill* is necessary, gives the command $\langle checkL^{f_{ch_L}}, i_L, Refill\rangle$ to the actuator A_L; (iii) waits for the data of each room R_{Li}, with the input $(i; w_{Li1}^{f_{L11}}, w_{Li2}^{f_{Li2}})$; (iv) elaborates the water level and the values received in a variable $check_{Li}^{f_{Li}}$ and provides a decision on the suppression or not in the room R_{Li}; (v) sends to the Control Room Node CR the aggregation p of all the variables $check_{Li}^{f_{Li}}$ previously collected.

Finally, the Control Room node CR: (i) collects the result of aggregation from each floor and puts it in the variable $x_{check_B}^{c_{ch_B}}$ (with $B = L, O, I$); (ii) according to some threshold values, decide to activate an alarm call to the Fire Station.

Again B_L and B_{CR} abstract other components we are not interested in, among which, their stores Σ_L and Σ_{CR}.

Security Analysis. Once having specified a system, a designer can use the CFA to statically predict its behaviour. The CFA predicts the "shape" of the data that nodes may exchange and elaborate across the network.

In particular, the CFA can predict, for each piece of data, from which sensors that piece can be derived and which manipulations it may be subjected to. Consider, e.g. the first sensor S_{Li1} of the i-th Library Room controller (the second sensor S_{Li2} is treated analogously). We abstractly represent data produced by this sensor with the term $1_{Li1}^{s_{Li1}}$, encoding both the identity of the sensor S_{Li1} and of the corresponding node L_i. The term $g_{Li1}(1_{Li1}^{s_{Li1}}, th_{L11})$ denotes instead a piece of data that is obtained by applying the logical function *greater than* (abbreviated with g) to the data from the sensor $1^{s_{Li1}}_{Li1}$ and the constant th_{L11}.

Furthermore, the analysis approximates the trajectories across the network of each piece of data. For instance, we can discover that the data $1^{s_{Lij}}_{Lij}$ collected by the sensors S_{Lij} (with $j \in [1,2]]$) may traverse the path $S_{Lij}, \ell_{Li}, \ell_L, \ell_{Li}$ meaning that they are generated in the sensor S_{Lij}, go to the node ℓ_L and then come back to ℓ_{Li}. Similarly, data produced by the sensor S_L in the node labelled ℓ_L are abstractly represented by the term $1^{s_L}_L$ and may traverse the path S_L, ℓ_L.

Finally, the analysis can also predict on which data the trigger of an actuator may depend on.

By using the CFA results, system designers can investigate the security and robustness of their system through a "what if" reasoning.

For example, in our scenario assume that some nodes can be attacked and that data passing through them tampered with. We would like to study how this may impact on the capacity of the system to correctly trigger the actuators. Consider the suppression system of the generic room controller R_{L_i} of the Library floor. The question we would like to answer is: which are the data that once altered can affect the trigger of the command *Go*? We know that this command is given when the value of the variable $z_{check_{Li}}$ is true (the condition of the trigger), thus, knowing how this value is computed and from which nodes its components come is critical. The analysis predicts (see Sect. 4 for details) that a possible way to compute the relevant value is the following

$$and^{a_{Li}}_{Li}(g_L(1^{s_L}_L, th_L), or^{o_{Li}}_{Li}(g_{Li1}(1^{s_{Li1}}_{Li1}, th_{Li1}), g_{Li2}(1^{s_{Li2}}_{Li2}, th_{Li1}))).$$

This "abstract" term encodes the fact that the condition of the trigger depends on the values of the sensors in the node R_{L_i} and L ($1^{s_L}_L$, $1^{s_{Li1}}_{Li1}$ and $1^{s_{Li2}}_{Li2}$), and on a suitable combination of the results of their comparisons with the corresponding thresholds (e.g. g_L, and_{Li}, or_{Li}) carried out by the node L. This information can be easily retrieved by looking at the labels on each sub-term, which indicate the location where that term is introduced and computed. In a sense the "abstract" term encodes the supply chain of the fire detection system.

As mentioned above, for each piece of data the analysis computes its possible trajectories inside the system: the data collected by the sensors S_{Lij} (with $j \in [1,2]]$) may traverse the path $S_{Lij}, \ell_{Li}, \ell_L, \ell_{Li}$ and those produced by the sensor S_L in the node labelled ℓ_L may traverse the path S_L, ℓ_L.

To simplify the reasoning on the analysis results, one can build a graph representation as the one of Fig. 1. The root represents the actuator whose activation we want to reason about; the node under the root is the trigger condition; the other nodes represent the aggregation functions and the data sensors that contribute to the evaluation of the relevant condition. Moreover, the nodes are labelled so as to record in which node of the network the corresponding computations occur. Intuitively, the actuation *Go* depends on the evaluation of the abstract value: $and^{a_{Li}}_{Li}(g_L(1^{s_L}_L, th_L), or^{o_{Li}}_{Li}(g_{Li1}(1^{s_{Li1}}_{Li1}, th_{Li1}), g_{Li2}(1^{s_{Li2}}_{Li2}, th_{Li1})))$, that, in turn, depends on the evaluation of the logic conjunction of the results of the two branches: $g_L(1^{s_L}_L, th_L)$ and $or^{o_{Li}}_{Li}(g_{Li1}(1^{s_{Li1}}_{Li1}, th_{Li1}), g_{Li2}(1^{s_{Li2}}_{Li2}, th_{Li1}))$.

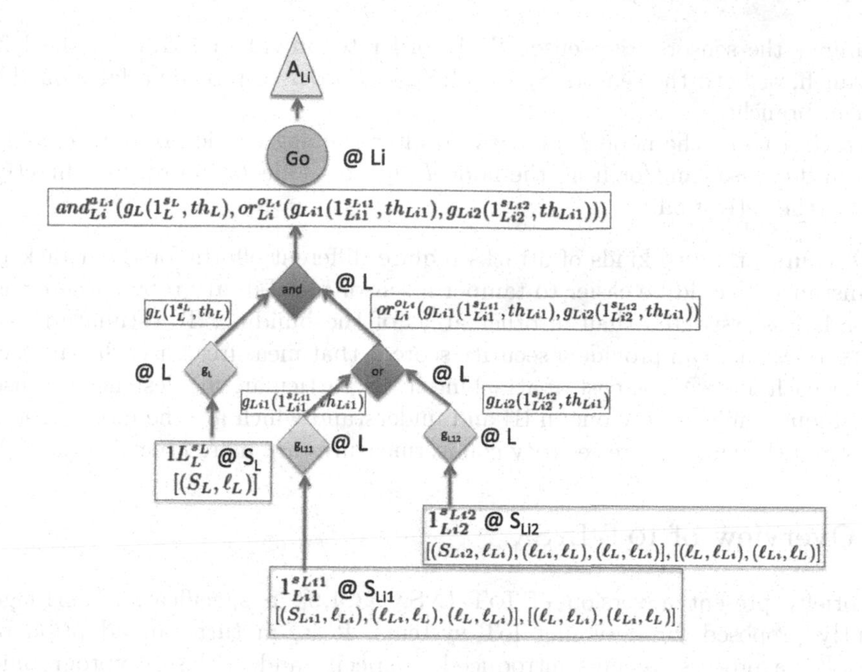

Fig. 1. Composition data graph

The first depends on the evaluation of the function g_L applied to the data coming from the sensor S_L, while the second depends from the logic disjunction of the outputs of the two further branches $g_{Li1}(1^{s_{Li1}}_{Li1}, th_{Li1})$ and $g_{Li2}(1^{s_{Li2}}_{Li2}, th_{Li2})$, resulting from the application of the functions g_{Li1} and g_{Li2} to the data coming from the sensors S_{li1} and S_{Li2}, respectively.

Using this graph one can study how an attacker can operate to force a specific behaviour of the actuator, and which nodes are involved. E.g. in case of fire, one can prevent activating of the fire suppression or, on the contrary, force it in order to flood the rooms. Note that although similar in shape to Attack Trees [23], our graphs have a different goal: while the first represents the steps of an attack (the goal being the root node and different ways of achieving it as leaf nodes), we represent in the nodes the components of a system to attack and their dependencies.

For the sake of simplicity, hereafter we focus on the structure induced by the AND and OR functions in the graph, and neglect the comparison functions g_L, g_{Li1}, g_{Li2}. Suppose the attacker wants to impair the trigger of Go. This can be done by complementing the boolean value $and^{a_{Li}}_{Li}(g_L(1^{s_L}_L, th_L), or^{o_{Li}}_{Li}(g_{Li1}(1^{s_{Li1}}_{Li1}, th_{Li1}), g_{Li2}(1^{s_{Li2}}_{Li2}, th_{Li1})))$. To turn the value `true` to `false`, it suffices that the attacker forces just one AND branch to `false`. As a consequence, an attacker can:

- tamper the sensors: the sensor S_L in order to provide a `false` on the left branch, or both the sensors S_{Li1} and S_{Li2} in order to provide a false on the `true` branch;
- attack a node: the node L in order to alter the aggregation of data coming from its sensor and/or from the node L_i or the node L_i in order to directly alter the activation.

Of course, the two kinds of attacks require different efforts for the attacker. For instance it could be easier to tamper a sensor in a Library room, where the access is less restricted than in other areas of the building. To estimate these efforts a designer can provide a security score ϕ that measures for each sensor S and for each node N the cost of attacking it. In particular, the designer can use a technique similar to the one in [1] and understand which are the more critical nodes and therefore where security countermeasures are more crucial.

3 Overview of IoT-LySa

We briefly present a version of IoT-LySa [3,4,9], a specification language recently proposed for designing IoT systems. It is, in turn, an adaption of LySa [2], a process calculus introduced to specify and analyse cryptographic protocols and checking their security properties (see e.g. [13,14]).

The calculus IoT-LySa differs from other process algebraic approaches introduced to model IoT systems, e.g. [16–18], because it provides a design framework that includes a static semantics to support verification techniques and tools for checking properties of IoT applications.

Syntax. Systems in IoT-LySa are pools of nodes (things), each hosting a shared store for internal communication, sensors and actuators, and a finite number of control processes that detail how data are to be processed and exchanged. We assume that each sensor (actuator) in a node with label ℓ is uniquely identified by an index $i \in \mathcal{I}_\ell$ ($j \in \mathcal{J}_\ell$, resp). A sensor is an active entity that reads data from the physical environment at its own fixed rate, and deposits them in the local store of the node. Actuators are instead passive: they just wait for a command to become active and operate on the environment. Data are represented by terms, which carry annotations $a, a', a_i, ... \in \mathcal{A}$ identifying their occurrences. Annotations are used in the analysis and do not affect the dynamic semantics in Table 5. We assume the existence of a finite set $\mathcal{K}$ of secret keys owned by nodes, exchanged at deployment time in a secure way, as it is often the case [24]. The encryption function $\{E_1, \cdots, E_r\}_{k_0}$ returns the result of encrypting values E_i for $i \in [1, r]$ under the shared key k_0. We assume to have perfect cryptography. The term $f(E_1, \cdots, E_r)$ is the application of function f to r arguments; we assume as given a set of primitive functions, typically for aggregating or comparing values. We assume the sets $\mathcal{V}$, $\mathcal{I}_\ell$, $\mathcal{J}_\ell$, $\mathcal{K}$ be pairwise disjoint.

The syntax of systems of nodes and of its components is as follows.

$\mathcal{N} \ni N ::=$ *systems of nodes*		$\mathcal{B} \ni B ::=$ *node components*	
0	empty system	Σ_ℓ	node store
$\ell : [B]$	single node ($\ell \in \mathcal{L}$)	P	process
$N_1 \mid N_2$	par. composition	S	sensor (label $i \in \mathcal{I}_\ell$)
		A	actuator (label $j \in \mathcal{J}_\ell$)
		$B \parallel B$	par. composition

$\mathcal{N} \ni N ::=$ *systems of nodes*

0	empty system
$\ell : [B]$	single node ($\ell \in \mathcal{L}$, the set of labels)
$N_1 \mid N_2$	parallel composition of nodes

$\mathcal{B} \ni B ::=$ *node components*

Σ_ℓ	node store
P	process
S	sensor, with a unique identifier $i \in \mathcal{I}_\ell$
A	actuator, with a unique identifier $j \in \mathcal{J}_\ell$
$B \parallel B$	parallel composition of node components

Each node $\ell : [B]$ is uniquely identified by a label $\ell \in \mathcal{L}$ that may represent further information on the node (e.g. node location). Sets of nodes are described through the (associative and commutative) operator $\mid$ for parallel composition. The system 0 has no nodes. Inside a node $\ell : [B]$ there is a finite set of components combined by means of the parallel operator $\parallel$. We impose that there is a *single* store $\Sigma_\ell : \mathcal{X} \cup \mathcal{I}_\ell \rightarrow \mathcal{V}$, where $\mathcal{X}, \mathcal{V}$ are the sets of variables and of values (integers, booleans, ...), respectively.

The store is essentially an array whose indexes are variables and sensors identifiers $i \in \mathcal{I}_\ell$ (no need of α-conversions). We assume that store accesses are atomic, e.g. through CAS instructions [15]. The other node components are control processes P, and sensors S (less than $\#(\mathcal{I}_\ell)$), and actuators A (less than $\#(\mathcal{J}_\ell)$) the actions of which are in *Act*. The syntax of processes is as follows.

$P ::= 0$	inactive process
$\langle\langle E_1, \cdots, E_r \rangle\rangle \triangleright L.\, P$	asynchronous multi-output $L \subseteq \mathcal{L}$
$(E_1, \cdots, E_j; x_{j+1}, \cdots, x_r)^X.\, P$	input (with matching and tag)
$\mathsf{decrypt}\ E\ \mathsf{as}\ \{E_1, \cdots, E_j; x_{j+1}, \cdots, x_r\}_{k_0}\ \mathsf{in}\ P$	decryption with key k_0 (with match.)
$E?P : Q$	conditional statement
h	iteration variable
$\mu h.\ P$	tail iteration
$x^a := E.\, P$	assignment to $x \in \mathcal{X}$
$\langle E, j, \gamma \rangle.\, P$	output of action γ to actuator j on condition E

The prefix $\langle\langle E_1, \cdots, E_r \rangle\rangle \triangleright L$ represents a simple form of multi-party communication: the tuple obtained by evaluating $E_1, \ldots, E_r$ is asynchronously sent to the nodes with labels in L that are "compatible" (according to, among other attributes, a proximity-based notion). The input prefix $(E_1, \cdots, E_j; x_{j+1}, \cdots, x_r)$ receives a r-tuple, provided that its first j elements match the corresponding input ones, and then assigns the variables (after ";") to the received values. Otherwise, the r-tuple is not accepted. A process repeats its behaviour, when defined through the tail iteration construct $\mu h.P$ (where h is the iteration variable).

The process decrypt E as $\{E_1, \cdots, E_j; x_{j+1}, \cdots, x_r\}_{k_0}$ in P tries to decrypt the result of the expression E with the shared key $k_0 \in \mathcal{K}$. Also in this case, if the pattern matching succeeds, the process continues as P and the variables $x_{j+1}, \ldots, x_r$ are suitably assigned.
Sensors and actuators have the form:

S ::= *sensors*		A ::= *actuators*	
0	inactive sensor	0	inactive actuator
$\tau.S$	internal action	$\tau.A$	internal action
$i := v.\, S$	store of $v \in \mathcal{V}$ by the ith sensor	$(\!| j, \Gamma |\!).\, A$	command for actuator j ($\Gamma \subseteq Act$)
		$\gamma.A$	triggered action ($\gamma \in Act$)
h	iteration var.	h	iteration var.
$\mu h\,.\,S$	tail iteration	$\mu h\,.\,A$	tail iteration

A sensor can perform an internal action τ or put the value v, gathered from the environment, into its store location i. An actuator can perform an internal action τ or execute one of its actions γ, received from its controlling process. Note that in the present version of IoT-LySA the process triggers the action γ of the actuator, provided the related condition E is satisfied. This construct emphasises that actuation is due to a decision based on the aggregation and elaboration of collected data. Sensors and actuators can iterate. For simplicity, here we neither provide an explicit operation to read data from the environment, nor we describe the impact of actuator actions on the environment (see [9]).

Finally, the syntax of terms is as follows:

$\mathcal{E} \ni E$ ::= *annotated terms*		$\mathcal{M} \ni M$ ::= *terms*	
M^a	with $a \in \mathcal{A}$	v	value ($v \in \mathcal{V}$)
		i	sensor location ($i \in \mathcal{I}_\ell$)
		x	
		$\{E_1, \cdots, E_r\}_{k_0}$	encryption with key $k_0 \in \mathcal{K}$
		$f(E_1, \cdots, E_r)$	function on data

The encryption function $\{E_1, \cdots, E_k\}_{k_0}$ returns the result of encrypting values E_i for $i \in [1, k]$ under k_0 representing a shared key in $\mathcal{K}$. The term $f(E_1, \cdots, E_n)$ is the application of function f to n arguments; we assume given a set of primitive functions, typically for aggregating or comparing values, be them computed or representing data in the environment.

Operational Semantics. The semantics is based on a standard structural congruence and a two-level *reduction relation* $\rightarrow$ defined as the least relation on nodes and its components, where we assume the standard denotational interpretation $[\![E]\!]_\Sigma$ for evaluating terms. As examples of semantic rules, we show the rules (Ev-out) and (Multi-com) in Table 2, that drive asynchronous IoT-LySA multi-communications, and the rules (A-com1) and (A-com2) used to communicate with actuators. The complete semantics is in Appendix, where however we omit the rules handling errors, e.g. when a node fails receiving a message.

In the (Ev-out) rule, to send a message $\langle\!\langle v_1, ..., v_r \rangle\!\rangle$ obtained by evaluating $\langle\!\langle E_1, ..., E_r \rangle\!\rangle$, a node with label ℓ spawns a new process, running in parallel with

Table 2. Communication semantic rules.

(Ev-out)

$$\frac{\bigwedge_{i=1}^{r} v_i = [\![E_i]\!]_{\Sigma}}{\Sigma \parallel \langle\!\langle E_1, \cdots, E_r \rangle\!\rangle \triangleright L.\, P \parallel B \;\rightarrow\; \Sigma \parallel \langle\!\langle v_1, \cdots, v_r \rangle\!\rangle \triangleright L.0 \parallel P \parallel B}$$

(Multi-com)

$$\frac{\ell_2 \in L \wedge\ Comp(\ell_1, \ell_2) \wedge\ \bigwedge_{i=1}^{j} v_i = [\![E_i]\!]_{\Sigma_2}}{\begin{array}{c}\ell_1 : [\langle\!\langle v_1, \cdots, v_r \rangle\!\rangle \triangleright L.\, 0 \parallel B_1] \ \mid\ \ell_2 : [\Sigma_2 \parallel (E_1, \cdots, E_j; x_{j+1}^{a_{j+1}}, \cdots, x_r^{a_r}).Q \parallel B_2] \;\rightarrow \\ \ell_1 : [\langle\!\langle v_1, \cdots, v_r \rangle\!\rangle \triangleright L \setminus \{\ell_2\}.\, 0 \parallel B_1] \ \mid\ \ell_2 : [\Sigma_2\{v_{j+1}/x_{j+1}, \cdots, v_r/x_r\} \parallel Q \parallel B_2]\end{array}}$$

(A-com1)

$$\frac{\gamma \in \Gamma \wedge [\![E]\!]_{\Sigma} = \mathtt{true}}{\langle E, j, \gamma \rangle.\, P \parallel (\!| j, \Gamma |\!).\, A \parallel B \;\rightarrow\; P \parallel \gamma.\, A \parallel B}$$

(A-com2)

$$\frac{\gamma \in \Gamma \wedge [\![E]\!]_{\Sigma} = \mathtt{false}}{\langle E, j, \gamma \rangle.\, P \parallel (\!| j, \Gamma |\!).\, A \parallel B \;\rightarrow\; P \parallel (\!| j, \Gamma |\!).\, A \parallel B}$$

the continuation P; this new process offers the evaluated tuple to all the receivers with labels in L. In the (Multi-com) rule, the message coming from ℓ_1 is received by a node labelled ℓ_2, provided that: (i) ℓ_2 belongs to the set L of possible receivers, (ii) the two nodes satisfy a compatibility predicate *Comp* (e.g. when they are in the same transmission range), and (iii) that the first j values match with the evaluations of the first j terms in the input. When this match succeeds the variables after the semicolon ";" are assigned to corresponding values of the received tuple. Moreover, the label ℓ_2 is removed by the set of receivers L of the tuple. The spawned process terminates when all the receivers have received the message ($L = \emptyset$).

In the rules (A-com1) and (A-com2) a process with prefix $\langle E, j, \gamma \rangle$ commands the jth actuator to perform the action γ, if it is one of its actions and it the condition E is true, according to the standard denotational interpretation $[\![E]\!]_{\Sigma}$.

4 Control Flow Analysis

Here we present a variant of the CFA of [7], following the same schema of the ones in [4,8]. It approximates the abstract behaviour of a system of nodes and tracks the trajectories of data. Intuitively, abstract values "symbolically" represent runtime data so as to encode where they have been introduced and elaborated. Here the analysis has a new component to track the conditions that may trigger actuators. Finally, we show how to use the CFA results to check which are the possible trajectories of the data and which of these data affect actuations.

Abstract Values. Abstract values represent data from sensors and resulting from aggregation functions and encryptions. Furthermore, for analysis reasons we also record into abstract values the annotations of the expression where they are generated. During the execution nodes may generate encrypted terms with an arbitrarily nesting level, due to recursion. To deal with them, in the analysis we also introduce the special abstract values $(\top, a)$ denoting terms with a depth greater than a threshold d. In the analysis specification we maintain the depth of abstract values smaller than d using the cut function $\lfloor - \rfloor_d$. It is inductively defined on the structure of the abstract value and cut it when the relevant depth is reached. Formally, abstract values are defined as follows, where $a \in \mathcal{A}$.

$$\begin{array}{ll}
\hat{\mathcal{V}} \ni \hat{v} ::= \textit{abstract terms} & \\
(\top, a) & \text{value denoting cut} \\
(v, a) & \text{value for clear data} \\
(f(\hat{v}_1, \cdots, \hat{v}_n), a) & \text{value for aggregated data} \\
(\{\hat{v}_1, \cdots, \hat{v}_n\}_{k_0}, a) & \text{value for encrypted data}
\end{array}$$

Abstract values are pairs where the first component is a value and the second one records the annotation of the expression where the value was computed. In particular, v abstracts the concrete value from sensors or computed by a function in the concrete semantics; $f(\hat{v}_1, \cdots, \hat{v}_n)$ represents a value obtained by applying the aggregation function f to $\hat{v}_1, \cdots, \hat{v}_n$; $\{\hat{v}_1, \cdots, \hat{v}_n\}^a_{k_0}$ abstracts encrypted data.

To simplify the notation, hereafter we write abstract values (v, a) as v^a and indicate with $\downarrow_i$ the projection function on the ith component of the pair. We naturally extend the projection to sets, i.e. $\hat{V}_{\downarrow_i} = \{\hat{v}_{\downarrow_i} | \hat{v} \in \hat{V}\}$, where $\hat{V} \subseteq \hat{\mathcal{V}}$.

We denote with $\mathbf{A}$ the function that recursively extracts all the annotations from an abstract value, e.g. $\mathbf{A}(f(v_1^{a_1}, v_2^{a_2}), a) = \{a_1, a_2, a\}$.

Definition 1. *Give an abstract value $\hat{v} \in \hat{\mathcal{V}}$, we define the set of labels $\mathbf{A}(\hat{v})$ inductively as follows.*

- $\mathbf{A}(\top, a) = \mathbf{A}(v, a) = \{a\}$
- $\mathbf{A}(f(\hat{v}_1, \cdots, \hat{v}_n), a) = \{a\} \cup \bigcup_{i=1}^{n} \mathbf{A}(\hat{v}_i)$
- $\mathbf{A}(\{\hat{v}_1, \cdots, \hat{v}_n\}_{k_0}, a) = \{a\} \cup \bigcup_{i=1}^{n} \mathbf{A}(\hat{v}_i)$

CFA Validation and Correctness. Here, we define our CFA that approximates system behaviours, e.g. communications and exchanged data, and in particular, micro-trajectories of data. As usual with the CFA, the analysis is specified through a set of inference rules expressing when the analysis results are valid. The analysis result is a tuple $(\widehat{\Sigma}, \kappa, \Theta, T, \alpha)$ (a pair $(\widehat{\Sigma}, \Theta)$ when analysing a term), called *estimate* for N (for E), where $\widehat{\Sigma}$, κ, Θ, T, and α are the following *abstract domains*:

- the disjoint union $\widehat{\Sigma} = \bigcup_{\ell \in \mathcal{L}} \hat{\Sigma}_\ell$ where each function $\hat{\Sigma}_\ell : \mathcal{X} \cup \mathcal{I}_\ell \rightarrow 2^{\hat{\mathcal{V}}}$ maps a given sensor in $\mathcal{I}_\ell$ or a given variable in $\mathcal{X}$ to a set of abstract values;
- a set $\kappa : \mathcal{L} \rightarrow \mathcal{L} \times \bigcup_{i=1}^{k} \hat{\mathcal{V}}^i$ of the messages that may be received by the node ℓ;
- a set $\Theta : \mathcal{L} \rightarrow \mathcal{A} \rightarrow 2^{\hat{\mathcal{V}}}$ of the information of the actual values computed by each labelled term M^a in a given node ℓ, at run time;
- a set $T = \mathcal{A} \rightarrow (\mathcal{L} \times \mathcal{L})$ of possible micro-trajectories related to the abstract values;
- a set $\alpha : \mathcal{L} \rightarrow \Gamma \rightarrow 2^{\hat{\mathcal{V}}}$ that connects the abstract values that can reach the condition related to an action γ.

With respect to the previous analyses of IoT-LySA, such as the ones in [4,7], the component α is new, and also the combined use of the above five components is new and allows us to potentially integrate the present CFA with the previous ones. For simplicity, we do not have here the component that tracks which output tuples that may be accepted in input as in [7]. Note that the component α allows us to determine the trajectories and the provenance of the values reaching the annotated condition. Using such information, we can therefore reason on how critical decisions may depend on them.

An available estimate has to be validated correct. This requires that it satisfies the judgments defined according to the syntax of nodes, node components and terms, fully presented in Appendix (see Tables 6 and 7). They are defined by a set of clauses. Here, we just show some examples.

The judgements for labelled terms have the form $(\widehat{\Sigma}, \Theta) \models_{\ell} M^a$. Intuitively, they require $\Theta(\ell)(a)$ includes all the abstract values $\hat{v}$ associated to M^a, e.g. if the term is a sensor identifier i^a, $\Theta(\ell)(a)$ includes (i, a) and the micro-trajectory (S_i, ℓ) belongs to $T(a)$, if the term is a variable x^a, $\Theta(\ell)(a)$ includes the abstract values bound to x collected in $\widehat{\Sigma}_{\ell}$. The judgements for nodes have the form $(\widehat{\Sigma}, \kappa, \Theta, T, \alpha) \models N$. The rule for a single node $\ell : [B]$ requires that B is analysed with judgements $(\widehat{\Sigma}, \kappa, \Theta, T, \alpha) \models_{\ell} B$. As examples of clauses, we consider the clauses for communication in Table 3. An estimate is valid for *multi-output*, if it is valid for the continuation of P and the set of messages communicated by the node ℓ to each node ℓ' in L, includes all the messages obtained by the evaluation of the r-tuple $\langle\langle M_1^{a_1}, \cdots, M_r^{a_r} \rangle\rangle$. More precisely, the rule (i) finds the sets $\Theta(\ell)(a_i)$ for each term $M_i^{a_i}$, and (ii) for all tuples of values $(\hat{v}_1, \cdots, \hat{v}_r)$ in $\Theta(\ell)(a_1) \times \cdots \times \Theta(\ell)(a_r)$ it checks whether they belong to $\kappa(\ell')$ for each $\ell' \in L$. Symmetrically, the rule for *input* requires that the values inside messages that

Table 3. Communication CFA rules.

$$\frac{\bigwedge_{i=1}^{k} (\widehat{\Sigma}, \Theta) \models_{\ell} M_i^{a_i} \;\wedge\; (\widehat{\Sigma}, \kappa, \Theta, T, \alpha) \models_{\ell} P \;\wedge \quad \forall \hat{v}_1, \cdots, \hat{v}_r : \bigwedge_{i=1}^{r} \hat{v}_i \in \Theta(\ell)(a_i) \;\Rightarrow \forall \ell' \in L : (\ell, \langle\langle \hat{v}_1, \cdots, \hat{v}_r \rangle\rangle) \in \kappa(\ell')}{(\widehat{\Sigma}, \kappa, \Theta, T, \alpha) \models_{\ell} \langle\langle M_1^{a_1}, \cdots, M_r^{a_r} \rangle\rangle \triangleright L.\, P}$$

$$\frac{\begin{array}{c}\bigwedge_{i=1}^{j} (\widehat{\Sigma}, \Theta) \models_{\ell} M_i^{a_i} \;\wedge \\ \forall (\ell', \langle\langle \hat{v}_1, \cdots, \hat{v}_r \rangle\rangle) \in \kappa(\ell) : \; Comp(\ell', \ell) \Rightarrow \\ (\bigwedge_{i=j+1}^{r} \hat{v}_i \in \hat{\Sigma}_{\ell}(x_i) \;\wedge \\ (\ell', \langle\langle \hat{v}_1, \cdots, \hat{v}_r \rangle\rangle) \in \rho(X) \;\wedge\; \forall a \in \mathbf{A}(\hat{v}_i).(\ell, \ell') \in T(a) \\ \wedge (\widehat{\Sigma}, \kappa, \Theta, T, \alpha) \models_{\ell} P)\end{array}}{(\widehat{\Sigma}, \kappa, \Theta, T, \alpha) \models_{\ell} (M_1^{a_1}, \cdots, M_j^{a_j}; x_{j+1}^{a_{j+1}}, \cdots, x_r^{a_r}).\, P}$$

$$\frac{\begin{array}{c}(\widehat{\Sigma}, \Theta) \models_{\ell} M^a \wedge \\ \Theta(\ell)(a) \subseteq \alpha(\ell)(\gamma) \wedge (\widehat{\Sigma}, \kappa, \Theta, T, \alpha) \models_{\ell} P\end{array}}{(\widehat{\Sigma}, \kappa, \Theta, T, \alpha) \models_{\ell} \langle M^a, j, \gamma \rangle.\, P}$$

can be sent to the node ℓ, passing the pattern matching, are included in the estimates of the variables $x_{j+1}, \cdots, x_r$. More in detail, the rule analyses each term $M_i^{a_i}$, and requires that for any message that the node with label ℓ can receive, i.e. $(\ell', \langle\langle \hat{v}_1, \cdots, \hat{v}_j, \hat{v}_{j+1}, \ldots, \hat{v}_r \rangle\rangle)$ in $\kappa(\ell)$, provided that the two nodes can communicate (i.e. $Comp(\ell', \ell)$), the abstract values $\hat{v}_{j+1}, \ldots, \hat{v}_r$ are included in the estimates of $x_{j+1}, \cdots, x_r$. Furthermore, the micro-trajectory (ℓ, ℓ') is recorded in the T component for each annotation related (via $\mathbf{A}$) to the abstract value $\hat{v}_i$, to record that the abstract value $\hat{v}_i$ coming from the node ℓ can reach the node labelled ℓ', e.g. if $\hat{v}_i = (f((v_{i1}, a_{i1}), (v_{i2}, a_{i2})), a_i)$, then the micro-trajectory is recorded in $T(a_i)$, $T(a_{i1})$ and $T(a_{i2})$.

The rule for actuator trigger predicts in the component α that a process at node ℓ may trigger the action γ for the actuator j, based on the abstract values included in the analysis of the term M^a.

Example 1. To better understand how our analysis works, we apply it to the following simple system of three nodes $N_1 \mid N_2 \mid N_3$, where S_1 is a sensor of the first node and P_i' and B_i (with $i = 1, 2, 3$) abstract other components we are not interested in.

$$\begin{array}{l}
N_1 = \ell_1 : [P_1 \parallel S_1 \parallel B_1] \mid \\
N_2 = \ell_2 : [P_2 \parallel B_2] \mid \\
N_3 = \ell_3 : [P_3 \parallel B_3] \\
P_1 \ = x^{a1} = 1^{s1}.\langle\langle x^{a_1} \rangle\rangle \triangleright \ell_2.\, P_1' \\
P_2 \ = (; x_2^{b_2}).\langle\langle f^m(x_2^{b_x}) \rangle\rangle \triangleright \ell_3.P_2' \\
P_3 \ = (; y_3^{c_3}).P_3'
\end{array}$$

Every valid estimate $(\widehat{\Sigma}, \kappa, \Theta, T, \alpha)$ must include at least the following entries, with $d = 4$.

$$\begin{array}{lll}
\widehat{\Sigma}_{\ell_1}(x^{a_1}) \supseteq \{1^{s_1}\} & \widehat{\Sigma}_{\ell_2}(x^{b_2}) \supseteq \{1^{s_1}\} & \widehat{\Sigma}_{\ell_3}(y^{c_3}) \supseteq \{f^m(1^{s_1})\} \\
T(s_1) \supseteq \{(S_1, \ell_1)\} & T(a_1) \supseteq \{(\ell_1, \ell_2)\} & T(s_1) \supseteq \{(\ell_2, \ell_3)\} \\
\Theta(\ell_1)(s_1) \supseteq \{1^{s_1}\} & \Theta(\ell_2)(b_2) \supseteq \{f^m(1^{s_1})\} & T(m) \supseteq \{(\ell_2, \ell_3)\} \\
\kappa(\ell_2) \supseteq \{(\ell_1, \langle\langle 1^{s_1} \rangle\rangle)\} & \kappa(\ell_3) \supseteq \{(\ell_2, \langle\langle f^m(1^{s_1}) \rangle\rangle\} & \\
T(s_1) \supseteq \{(\ell_1, \ell_2)\} & &
\end{array}$$

Indeed, an estimate must satisfy the checks of the CFA rules. The validation of the system requires the validation of each node, i.e. $(\widehat{\Sigma}, \kappa, \Theta, T, \alpha) \models N_i$ and of the processes there included, i.e. $(\widehat{\Sigma}, \kappa, \Theta, T, \alpha) \models_{\ell_i} P_i$, with $i = 1, 2, 3$. In particular, the validation of the process P_1, i.e. $x^{a_1} := 1^{s_1}.P_1'$ holds because $1^{s_1} \in \Theta(\ell_1)(s_1)$ and $(S_1, \ell_1) \in T(s_1)$, $1^{s_1} \in \widehat{\Sigma}_{\ell_1}(x^{a_1})$ and the continuation holds as well. In particular, $\langle\langle x^{a_1} \rangle\rangle \triangleright \{\ell_2\}$ holds because the checks required by CFA clause for output succeed. We can indeed verify that $(\widehat{\Sigma}, \Theta) \models_\ell x^{a_1}$ holds because $1^{s_1} \in \widehat{\Sigma}_{\ell_1}(x^{a_1})$, according to the CFA clause for variables. Furthermore $(\ell_1, \langle\langle 1^{s_1} \rangle\rangle) \in \kappa(\ell_2)$. This suffices to validate the output, by assuming that the continuation P_1' is validated as well. We have the following instantiation of the clause for output.

$$\frac{\dfrac{1^{s_1} \in \Theta(\ell_1)(s_1) \ \wedge\ (S_1,\ell_1) \in T(s_1)}{(\widehat{\Sigma},\Theta) \models_{\ell_1} 1^{s_1}} \ \wedge\ (\widehat{\Sigma},\kappa,\Theta,T,\alpha) \models_{\ell_1} P_1' \ \wedge \quad 1^{s_1} \in \Theta(\ell_1)(s_1) \Rightarrow (\ell_1, \langle\!\langle 1^{s_1} \rangle\!\rangle) \in \kappa(\ell_2)}{(\widehat{\Sigma},\kappa,\Theta,T,\alpha) \models_{\ell_1} \langle\!\langle 1^{s_1} \rangle\!\rangle \triangleright \{\ell_2\}.\, P_1'}$$

Instead $(\widehat{\Sigma},\kappa,\Theta,T,\alpha) \models_{\ell_1} (;x_2^{b_2}).\langle\!\langle f(x_2^{b_x})^m \rangle\!\rangle \triangleright \ell_3.P_2'$ holds because the checks for the CFA clause for input succeed. From $(\ell_1, \langle\!\langle 1^{s_1} \rangle\!\rangle) \in \kappa(\ell_2)$, we can indeed obtain that $\widehat{\Sigma}_{\ell_2}(x^{b_2}) \supseteq \{1^{s_1}\}$, and that $T(s_1) \supseteq \{(\ell_1,\ell_2)\}$. The other entries can be similarly validated as well.

The following theorem establishes the correctness of our CFA w.r.t. the dynamic semantics. The statement relies on the agreement relation $\bowtie$ between the concrete and the abstract stores. Its definition is immediate, since the analysis only considers the second component of the extended store, i.e. the abstract one: $\Sigma_\ell^i \bowtie \widehat{\Sigma}_\ell$ iff $w \in \mathcal{X} \cup \mathcal{I}_\ell$ such that $\Sigma_\ell^i(w) \neq \bot$ implies $(\Sigma_\ell^i(w))_{\downarrow_2} \in \widehat{\Sigma}_\ell(w)$.

It is also possible to prove the existence of a (minimal) estimate, as in [4], since estimates form a Moore family with a minimal element.

Theorem 1 (Subject reduction). *If $(\widehat{\Sigma},\kappa,\Theta,T,\alpha) \models N$ and $N \to N'$ and $\forall \Sigma_\ell^i$ in N it is $\Sigma_\ell^i \bowtie \widehat{\Sigma}_\ell$, then $(\widehat{\Sigma},\kappa,\Theta,T,\alpha) \models N'$ and $\forall \Sigma_\ell^{i\prime}$ in N' it is $\Sigma_\ell^{i\prime} \bowtie \widehat{\Sigma}_\ell$.*

The proofs follow the same schema of [4]. In particular, we use an instrumented denotational semantics for expressions, the values of which are pairs $\langle v, \hat{v} \rangle$, where v is a concrete value and $\hat{v}$ is the corresponding abstract value. The store (Σ_ℓ^i with an undefined $\bot$ value) is accordingly extended. Our semantics (see Table 5 in Appendix) just uses the projection on the first component.

Checking Trajectories. We now recall the notion of trajectories of data, in turn composed by micro-trajectories representing a single hop in the communication. It is a slight simplification of the notion presented in [8].

Definition 2. *Given a set of labels $\mathcal{L}$, we define a* micro-trajectory *μ as a pair $(\ell,\ell') \in (\mathcal{L} \times \mathcal{L})$. A* trajectory *$\tau$ is a list of micro-trajectories $[\mu_1, ..., \mu_n]$, such that $\forall \mu_i, \mu_{i+1}$ with $\mu_i = (\ell_i, \ell_i')$ and $\mu_{i+1} = (\ell_{i+1}, \ell_{i+1}')$, $\ell_i' = \ell_{i+1}$.*

A k-trajectory *is a list of micro-trajectories $[\mu_1, ..., \mu_k]$ of length k.*

In our analysis, one starts from a set of micro-trajectories and suitably compose them to obtain longer trajectories, in turn composed. As expected, for composing trajectories the head of the second must be equal to tail of the first. Technically, we use a closure of a set of micro-trajectories, the inductive definition of which follows.

Definition 3. *Given a set of micro-trajectories $S \in (\mathcal{L} \times \mathcal{L})$, its closure $Close(S)$ is defined as*

- $\forall (\ell,\ell') \in S.\ [(\ell,\ell')] \in Close(S)$;
- $\forall [L,(\ell,\ell')],\ [(\ell',\ell''),L''] \in S.\ [L,(\ell,\ell'),(\ell',\ell''),L''] \in Close(S)$.

The set $Close(S)$ contains trajectories of any size. To obtain only the set of k-trajectories it suffices to the subset consisting only of those of length k.

Given a term E annotated by a, the over-approximation of its possible trajectories is obtained by computing the trajectory closure of the set composed by all the micro-trajectories (ℓ, ℓ') in $T(a)$.

$$Trajectories(E^a) = Close(T(a))$$

Therefore, our analysis enables traceability of data. For every exchanged message $\langle\!\langle v_1, \ldots, v_r \rangle\!\rangle$, the CFA keeps track of the possible composition of each of its components and of the paths of each of its components v_i and, in turn, for each v_i it keeps recursively track of the composition and of the paths of the possible data used to compose it. As a corollary of Theorem 1, it follows that κ predicts all the possible inter-node communications, and that our analysis records the micro-trajectory in the T component of each abstract value possibly involved in the communication.

Example 2. Back to our previous example, note that from $T(s_1) \supseteq \{(\ell_1, \ell_2)\}$ and $T(s_1) \supseteq \{(\ell_2, \ell_3)\}$, we can obtain the trajectory $[(\ell_1, \ell_2), (\ell_2, \ell_3)]$, by applying *Close* to $T(s_1)$.

Example 3. Consider now our running example on the fire system network in Sect. 2. Every valid estimate $(\widehat{\Sigma}, \kappa, \Theta, T, \alpha)$ must include at least the entries in Table 4, assuming $d = 4$, and where we overload the symbols g_L, g_{Li1}, g_{Li2}, by meaning both the labels and the comparison functions to check whether the argument are above the given thresholds (they are constants and their labels are omitted for simplicity).

Since we are interested in understanding which data may affect the action go, consider $\alpha_{Li}(Go)$ that includes, according to the CFA results, the following abstract value

$$and_{Li}^{a_{Li}}(g_L(1_L^{s_L}, th_L), or_{Li}^{o_{Li}}(g_{Li1}(1_{Li1}^{s_{Li1}}, th_{Li1}), g_{Li2}(1_{Li2}^{s_{Li2}}, th_{Li1}))).$$

Furthermore, for each sub-term of the abstract value, by using the component T of the analysis, we can retrieve the possible trajectories, in particular for the sensor values with $j = 1, 2$ (the subscribed numbers recall their length):

$$\begin{array}{l} Trajectories_3(1_{Lij}^{s_{Lij}}) = Close(T(s_{Lij})) \supseteq \{[(S_{Lij}, \ell_{Li}), (\ell_{Li}, \ell_L), (\ell_L, \ell_{Li})]\} \\ Trajectories_1(s_L) = Close(T(s_L)) \supseteq \{[(S_L, \ell_L)]\}\} \end{array}$$

With an approach and a technique similar to those in [1], we aim at identifying the possible minimal sets of nodes that must be tampered in order to alter the result of the value the actuation of *Go* depends on and at exploiting the score metric to compare them in order to determine the ones with minimal cost for the attacker.

In the case the attacker would like to tamper the sensors in order to force the result to be `false`, as mentioned above, there are two possible minimal choices, i.e. tampering:

Table 4. Fire Detection System Analysis: some entries, where $j = 1, 2$.

$$\widehat{\Sigma}_{\ell_{Li}}(z_{Lij}^{r_{Lij}}) \supseteq \{1_{Lij}^{s_{Lij}}\},\ \Theta(\ell_{Li})(r_{Lij}) \supseteq \{1_{Li1}^{s_{Lij}}\},\ T(s_{Lij}) \supseteq \{(S_{Lij}, \ell_{Li})\},$$
$$\kappa(\ell_{L_i}) \supseteq \{(\ell_L, \langle\langle and_{Li}^{a_{Li}}(g_L(1_L^{s_L}, th_L), or_{Li}^{o_{Li}}(g_{Li1}(1_{Li1}^{s_{Li1}}, th_{Li1}), g_{Li2}(1_{Li2}^{s_{Li2}}, th_{Li1})))\rangle\rangle)\}$$
$$\alpha_{Li}(Go) \supseteq \Theta(\ell_{Li})(r_{ch_{Li}}) \supseteq and_{Li}^{a_{Li}}(g_L(1_L^{s_L}, th_L), or_{Li}^{o_{Li}}(g_{Li1}(1_{Li1}^{s_{Li1}}, th_{Li1}), g_{Li2}(1_{Li2}^{s_{Li2}}, th_{Li1})))\}$$
$$\widehat{\Sigma}_{\ell_{Li}}(z_{check_{Li}}^{r_{ch_{Li}}}) \supseteq \{and_{Li}^{a_{Li}}(g_L(1_L^{s_L}, th_L), or_{Li}^{o_{Li}}(g_{Li1}(1_{Li1}^{s_{Li1}}, th_{Li1}), g_{Li2}(1_{Li2}^{s_{Li2}}, th_{Li1})))\}$$
$$\Theta(\ell_{Li})(r_{ch_{Li}}) \supseteq \{and_{Li}^{a_{Li}}(g_L(1_L^{s_L}, th_L), or_{Li}^{o_{Li}}(g_{Li1}(1_{Li1}^{s_{Li1}}, th_{Li1}), g_{Li2}(1_{Li2}^{s_{Li2}}, th_{Li1})))\}$$
$$T(s_{Lij}) \supseteq \{(\ell_L, \ell_{Li})\},$$
$$\widehat{\Sigma}_{\ell_L}(w_L^{f_L}) \supseteq \{1_L^{s_L}\},\ \Theta(\ell_L)(f_L) \supseteq \{1_L^{s_L}\},\ T(s_L) \supseteq \{(S_L, \ell_L)\}$$
$$\widehat{\Sigma}_L(check_L^{f_{ch_L}}) \supseteq \{g_L(1_L^{s_L}, th_L)\}, \Theta(\ell_L)(f_{ch_L}) \supseteq \{g_L(1_L^{s_L}, th_L)\}$$
$$\alpha_L(Refill) \supseteq \Theta(\ell_L)(f_{ch_L}) \supseteq \{g_L(1_L^{s_L}, th_L)\}$$
$$\widehat{\Sigma}_{\ell_L}(w_{Lij}^{f_{Lij}}) \supseteq \{1_{Lij}^{s_{Lij}}\},\ \Theta(\ell_L)(f_{Lij}) \supseteq \{1_{Lij}^{s_{Lij}}\},\ T(s_{Lij}) \supseteq \{(\ell_{Li}, \ell_L)\},$$
$$\widehat{\Sigma}_L(check_{Li}^{f_{ch_{Li}}}) \supseteq \{and_{Li}^{a_{Li}}(g_L(1_L^{s_L}, th_L), or_{Li}^{o_{Li}}(g_{Li1}(1_{Li1}^{s_{Li1}}, th_{Li1}), g_{Li2}(1_{Li2}^{s_{Li2}}, th_{Li1})))\},$$
$$\Theta(\ell_L)(f_{ch_{Li}}) \supseteq \{and_{Li}^{a_{Li}}(g_L(1_L^{s_L}, th_L), or_{Li}^{o_{Li}}(g_{Li1}(1_{Li1}^{s_{Li1}}, th_{Li1}), g_{Li2}(1_{Li2}^{s_{Li2}}, th_{Li1})))\}$$
$$\kappa(\ell_L) \supseteq \{(\ell_{L_i}, \langle\langle i, 1_{Li1}^{s_{Li1}}, 1_{Li2}^{s_{Li2}}\rangle\rangle)\}$$
$$Trajectories_3(1_{Lij}^{s_{Lij}}) = Close(T(s_{Lij})) \supseteq \{[(S_{Lij}, \ell_{Li}), (\ell_{Li}, \ell_L), (\ell_L, \ell_{Li})]\}$$
$$Trajectories_1(s_L) = Close(T(s_L)) \supseteq \{[(S_L, \ell_L)]\}\}$$
$$T(f_p) \supseteq \{[(\ell_L, \ell_{CR})]\}$$
$$\kappa(\ell_{CR}) \supseteq \{(\ell_L, \langle\langle p^{f_p}(and_{L1}^{a_{L1}}(g_L(1_L^{s_L}, th_L), or_{L1}^{o_{L1}}(g_{L11}(1_{L11}^{s_{L11}}, th_{L11}), g_{L12}(1_{L12}^{s_{L12}}, th_{L12}))), \ldots,$$
$$and_{Lk}^{a_{Lk}}(g_L(1_L^{s_L}, th_L), or_{Lk}^{o_{Lk}}(g_{Lk1}(1_{Lk1}^{s_{Lk1}}, th_{Lk1}), g_{Lk2}(1_{Lk2}^{s_{Lk2}}, th_{Lk2}))))\rangle\rangle)\}$$

- the sensor S_L in order to alter thc sensed data and force $g_L(1_L^{s_L}, th_L)$ to provide a `false` on the left branch, or
- both the sensors S_{Li1} and S_{Li2} in order to provide a `false` on the right branch.

To estimate the cost of the different strategies of attack, we assume that a table of scores is known that associates a score $\phi(S)$, $\phi(\ell)$ to each sensor S and to each node with label ℓ, respectively. Suppose that the scores are:

$$\phi(S_L) = 3 \qquad \phi(S_{Li1}) = 1 \qquad \phi(S_{Li2}) = 1.5$$

Under these hypotheses, for the attacker is more convenient to attack the two sensors in the Room Controller than attacking the sensor of the Library controller. The overall cost of the first option is indeed 2.5, while the cost of the second one is 3.

From this kind of reasoning designers can guide their choices on how to reduce the risk of impairing some critical actuations, e.g. by introducing some redundancy in the sources of data, such as adding new hardware and software components.

5 Conclusions

We started from the modelling language IoT-LySA and from the results of its CFA, and showed how administrators can exploit a graph, built from the analysis results, that represents the data dependencies and the data trajectories. In particular, the graph encodes how the condition that drives a critical actuation depends on the raw data collected by sensors and on the logical and aggregation functions applied to the sensor data. This information allows designers to reason, along the lines of [1], on which data can be altered (by tampering the corresponding nodes) to impact on the capacity of the system to correctly trigger the actuators. Since each node is associated with a score that measures its compromise cost, it is also possible to compare the different solutions and establish the cheapest. Other metrics such as the ones suggested still in [1] can be exploited as well. Then, we discussed how this graph could be used as input to compute different security metrics, e.g. those of [1], for estimating the cost of attacks and devise suitable countermeasures.

Actually, the analysis is quite general and its results can be exploited as a starting point for many other different investigations on the behaviour of a given system. For example, the graph built from the analysis results can be used: to check whether a system respects policies that rule information flows among nodes, by allowing some flows and forbidding others; to carry out a taint analysis for detecting whether critical decisions may depend on tainted data. To this aim, we could integrate our present analysis with the one in [3] in the first case and with the taint analysis of [8] in the second one. Answering to these questions can help designers to detect the potential vulnerabilities related to the presence of dangerous nodes, and can determine possible solutions and mitigation strategies.

A similar Control Flow Analysis is presented in [10]: it is there used to over-approximate the behaviour of KLAIM processes and to track how tuple data can move in the network.

As a future work we plan to fully formalise how to derive the graph from the analysis results. We would like also to compare our approach with that of [19], looking for possible synergies. Its authors also start from a hybrid process calculus for modelling both CPSs and their potential attacks, and propose a threat model that can be used to assess attack tolerance and to estimate the impact of a successful attack.

Another line of future work is linking our approach to that used in [21,22], for ensuring a certain level of quality service of a system even when in the presence of not completely reliable data. In the cited paper, the authors introduce the Quality Calculus that allows defining and reasoning on software components that have a sort of backup plan in case the ideal behaviour fails due to unreliable communication or data.

Finally, since in many IoT system the behaviour of node depends on the computational context they are immersed in, we plan to extend IoT-LySa with constructs for representing contexts along the lines of [11,12], and to study their security along the lines of [5,6].

Appendix

Operational Semantics of IoT-LySa

Our reduction semantics is based on the following *Structural congruence* $\equiv$ on nodes and node components. It is standard except for rule (4) that equates a multi-output with no receivers and the inactive process, and for the fact that inactive components of a node are all coalesced.

(1) $(\mathcal{N}/_\equiv, |, 0)$ is a commutative monoid
(2) $(\mathcal{B}/_\equiv, \|, 0)$ is a commutative monoid
(3) $\mu\, h\,.\, X \equiv X\{\mu\, h\,.\, X/h\}$ for $X \in \{P, A, S\}$
(4) $\langle\langle E_1, \cdots, E_r\rangle\rangle : \emptyset.\, 0 \equiv 0$

The two-level *reduction relation* $\rightarrow$ is defined as the least relation on nodes and its components satisfying the set of inference rules in Tables 2 and 5. For the sake of simplicity, we use one relation. We assume the standard denotational interpretation $[\![E]\!]_\Sigma$ for evaluating terms.

The first two semantic rules implement the (atomic) asynchronous update of shared variables inside nodes, by using the standard notation $\Sigma\{-/-\}$. According to (S-store), the ith sensor uploads the value v, gathered from the environment, into its store location i. According to (Asgm), a control process updates

Table 5. Reduction semantics (the upper part on node components, the lower one on nodes), where $X \in \{S, A\}$ and $Y \in \{N, B\}$, without the rules (Ev-out), (Multi-com), (A-com1) and (A-com2), discussed in Table 2.

(S-store)
$$\frac{}{\Sigma \parallel i^a := v^{a'}.\, S_i \parallel B \rightarrow \Sigma\{v/i\} \parallel S_i \parallel B}$$

(Asgm)
$$\frac{[\![E]\!]_\Sigma = v}{\Sigma \parallel x^a := E.\, P \parallel B \rightarrow \Sigma\{v/x\} \parallel P \parallel B}$$

(Cond1)
$$\frac{[\![E]\!]_\Sigma = \texttt{true}}{\Sigma \parallel E?\, P_1 : P_2 \parallel B \;\rightarrow\; \Sigma \parallel P_1 \parallel B}$$

(Cond2)
$$\frac{[\![E]\!]_\Sigma = \texttt{false}}{\Sigma \parallel E?\, P_1 : P_2 \parallel B \;\rightarrow\; \Sigma \parallel P_2 \parallel B}$$

(Act)
$$\frac{}{\gamma.A \;\rightarrow\; A}$$

(Int)
$$\frac{}{\tau.\, X \;\rightarrow\; X}$$

(Decr)
$$\frac{[\![E]\!]_\Sigma = \{v_1, \cdots, v_r\}_{k_0} \;\wedge\; \bigwedge_{i=1}^{j} v_i = [\![E'_i]\!]_\Sigma}{\Sigma \parallel \mathsf{decrypt}\ E\ \mathsf{as}\ \{E'_1, \cdots, E'_j; x_{j+1}^{a_{j+1}}, \cdots, x_r^{a_r}\}_{k_0}\ \mathsf{in}\ P \parallel B \rightarrow \Sigma\{v_{j+1}/x_{j+1}, \cdots, v_r/x_r\} \parallel P \parallel B}$$

(Ev-out)
$$\frac{\bigwedge_{i=1}^{r} v_i = [\![E_i]\!]_\Sigma}{\Sigma \parallel \langle\langle E_1, \cdots, E_r\rangle\rangle \triangleright L.\, P \parallel B \;\rightarrow\; \Sigma \parallel \langle\langle v_1, \cdots, v_r\rangle\rangle \triangleright L.0 \parallel P \parallel B}$$

(Node)
$$\frac{B \;\rightarrow\; B'}{\ell : [B] \;\rightarrow\; \ell : [B']}$$

(ParN)
$$\frac{N_1 \rightarrow N'_1}{N_1 | N_2 \rightarrow N'_1 | N_2}$$

(ParB)
$$\frac{B_1 \rightarrow B'_1}{B_1 \| B_2 \rightarrow B'_1 \| B_2}$$

(CongrY)
$$\frac{Y'_1 \equiv Y_1 \rightarrow Y_2 \equiv Y'_2}{Y'_1 \rightarrow Y'_2}$$

the variable x with the value of E. The rules for conditional (Cond1) and (Cond2) are as expected. The rule (Act) says that the actuator performs the action γ. Similarly, for the rules (Int) for internal actions for representing activities we are not interested in. The communication rules (Ev-out), (Multi-com), (A-com1) and (A-com2) that drive asynchronous multi-communications and communication with actuators are discussed in Sect. 3. The rule (Decr) tries to decrypt the result $\{v_1, \cdots, v_r\}_k$ of the evaluation of E with the key k_0, and matches it against the pattern $\{E'_1, \cdots, E'_j; x_{j+1}, \cdots, x_r\}_{k_0}$. As for communication, when this match succeeds the variables after the semicolon ";" are assigned to values resulting from the decryption. The last rules propagate reductions across parallel composition ((ParN) and (ParB)) and nodes (Node), while (CongrY) is the standard reduction rule for congruence for nodes and node components.

Control Flow Analysis of IoT-LySa

Our CFA is specified in a logical form through a set of inference rules expressing the validity of the analysis results, where the function $\lfloor - \rfloor_d$ to cut all the terms with a depth greater than a given threshold d, with the special abstract values $\top^b$, is defined as follows.

$$
\begin{aligned}
&\lfloor \top^b \rfloor_d = \top^b \\
&\lfloor v^b \rfloor_d = v^b \\
&\lfloor \{\hat{v}_1, \cdots, \hat{v}_r\}^b_{k_0} \rfloor_0 = \top^b \\
&\lfloor \{\hat{v}_1, \cdots, \hat{v}_r\}^b_{k_0} \rfloor_d = \{\lfloor \hat{v}_1 \rfloor_{d-1}, \cdots, \lfloor \hat{v}_r \rfloor_{d-1}\}^b_{k_0} \\
&\lfloor f(\hat{v}_1, \cdots, \hat{v}_r) \rfloor_d = f(\lfloor \hat{v}_1 \rfloor_{d-1}, \cdots, \lfloor \hat{v}_r \rfloor_{d-1})^b
\end{aligned}
$$

Table 6. Analysis of labelled terms $(\widehat{\Sigma}, \Theta) \models_\ell M^a$.

$$
\frac{(i,a) \in \Theta(\ell)(a) \ \wedge \ (S_i, \ell) \in T(a)}{(\widehat{\Sigma}, \Theta) \models_\ell i^a} \qquad \frac{(v,a) \in \Theta(\ell)(a)}{(\widehat{\Sigma}, \Theta) \models_\ell v^a} \qquad \frac{\widehat{\Sigma}_\ell(x) \subseteq \Theta(\ell)(a)}{(\widehat{\Sigma}, \Theta) \models_\ell x^a}
$$

$$
\frac{\begin{array}{c} \bigwedge_{i=1}^k (\widehat{\Sigma}, \Theta) \models_\ell M_i^{a_i} \ \wedge \\ \forall\, \hat{v}_1, .., \hat{v}_r : \bigwedge_{i=1}^r \hat{v}_i \in \Theta(\ell)(a_i) \ \Rightarrow \ (\lfloor \{\hat{v}_1, .., \hat{v}_r\}_{k_0} \rfloor_d, a) \in \Theta(\ell)(a) \end{array}}{(\widehat{\Sigma}, \Theta) \models_\ell \{M_1^{a_1}, .., M_r^{a_r}\}^a_{k_0}}
$$

$$
\frac{\begin{array}{c} \bigwedge_{i=1}^k (\widehat{\Sigma}, \Theta) \models_\ell M_i \ \wedge \\ \forall\, \hat{v}_1, .., \hat{v}_r : \bigwedge_{i=1}^r \hat{v}_i \in \Theta(\ell)(a_i) \ \Rightarrow \ (f(\hat{v}_1, .., \hat{v}_r), a) \in \Theta(\ell)(a) \end{array}}{(\widehat{\Sigma}, \Theta) \models_\ell f(M_1^{a_1}, .., M_r^{a_r})^a}
$$

The result or *estimate* of our CFA is a tuple $(\widehat{\Sigma}, \kappa, \Theta, T, \alpha)$ (a pair $(\widehat{\Sigma}, \Theta)$ when analysing a term) that satisfies the judgements defined by the axioms and rules of Tables 6, 3 and 7.

Table 7. Analysis of nodes $(\widehat{\Sigma}, \kappa, \Theta, T, \alpha) \models N$, and of node components $(\widehat{\Sigma}, \kappa, \Theta, T, \alpha) \models_\ell B$, without the rules introduced in Table 3.

$$\frac{}{(\widehat{\Sigma}, \kappa, \Theta, T, \alpha) \models 0} \qquad \frac{(\widehat{\Sigma}, \kappa, \Theta, T, \alpha) \models_\ell B}{(\widehat{\Sigma}, \kappa, \Theta, T, \alpha) \models \ell : [B]} \qquad \frac{(\widehat{\Sigma}, \kappa, \Theta, T, \alpha) \models N_1 \wedge (\widehat{\Sigma}, \kappa, \Theta, T, \alpha) \models N_2}{(\widehat{\Sigma}, \kappa, \Theta, T, \alpha) \models N_1 \mid N_2}$$

$$\frac{\forall\, i \in \mathcal{I}_\ell . i^\ell \in \hat{\Sigma}_\ell(i)}{(\widehat{\Sigma}, \kappa, \Theta, T, \alpha) \models_\ell \Sigma} \qquad \frac{}{(\widehat{\Sigma}, \kappa, \Theta, T, \alpha) \models_\ell S} \qquad \frac{}{(\widehat{\Sigma}, \kappa, \Theta, T, \alpha) \models_\ell A}$$

$$\frac{\begin{array}{c}(\widehat{\Sigma}, \Theta) \models_\ell M^a \wedge \bigwedge_{i=1}^{j} (\widehat{\Sigma}, \Theta) \models_\ell M_i^{a_i} \wedge \\ \forall \{\hat{v}_1, \cdots, \hat{v}_r\}_{k_0}^b \in \Theta(\ell)(a) \Rightarrow \left(\bigwedge_{i=j+1}^{r} \hat{v}_i \in \hat{\Sigma}_\ell(x_i) \wedge (\widehat{\Sigma}, \kappa, \Theta, T, \alpha) \models_\ell P\right)\end{array}}{(\widehat{\Sigma}, \kappa, \Theta, T, \alpha) \models_\ell \mathsf{decrypt}\ M^a\ \mathsf{as}\ \{M_1^{a_1}, \cdots, M_j^{a_j}; x_{j+1}^{a_{j+1}}, \cdots, x_r^{a_r}\}_{k_0}\ \mathsf{in}\ P}$$

$$\frac{\begin{array}{c}(\widehat{\Sigma}, \Theta) \models_\ell M^a \wedge \\ \forall\, \hat{v} \in \Theta(\ell)(a) \Rightarrow \hat{v} \in \hat{\Sigma}_\ell(x) \wedge (\widehat{\Sigma}, \kappa, \Theta, T, \alpha) \models_\ell P\end{array}}{(\widehat{\Sigma}, \kappa, \Theta, T, \alpha) \models_\ell x^{a_x} := M^a . P}$$

$$\frac{\begin{array}{c}(\widehat{\Sigma}, \Theta) \models_\ell M^a \wedge \\ (\widehat{\Sigma}, \kappa, \Theta, T, \alpha) \models_\ell P_1 \wedge (\widehat{\Sigma}, \kappa, \Theta, T, \alpha) \models_\ell P_2\end{array}}{(\widehat{\Sigma}, \kappa, \Theta, T, \alpha) \models_\ell M^a ? P_1 : P_2} \qquad \frac{(\widehat{\Sigma}, \kappa, \Theta, T, \alpha) \models_\ell B_1 \wedge (\widehat{\Sigma}, \kappa, \Theta, T, \alpha) \models_\ell B_2}{(\widehat{\Sigma}, \kappa, \Theta, T, \alpha) \models_\ell B_1 \| B_2}$$

$$\frac{}{(\widehat{\Sigma}, \kappa, \Theta, T, \alpha) \models_\ell 0} \qquad \frac{(\widehat{\Sigma}, \kappa, \Theta, T, \alpha) \models_\ell P}{(\widehat{\Sigma}, \kappa, \Theta, T, \alpha) \models_\ell \mu h .\, P} \qquad \frac{}{(\widehat{\Sigma}, \kappa, \Theta, T, \alpha) \models_\ell h}$$

We do not comment the clauses discussed in Sect. 4. The judgement $(\widehat{\Sigma}, \Theta) \models_\ell M^a$, defined by the rules in Table 6, requires that $\Theta(\ell)(a)$ includes all the abstract values $\hat{v}$ associated to M^a. In the case of sensor identifiers, i^a and values v^a must be included in $\Theta(\ell)(a)$. In the case of sensor identifier also the micro-trajectory (S_i, ℓ) must be included in $T(a)$. The rule for analysing compound terms requires that the components are in turn analysed. The penultimate rule deals with the application of an r-ary encryption. To do that (i) it analyses each term $M_i^{a_i}$, and (ii) for each r-tuple of values $(\hat{v}_1, \cdots, \hat{v}_r)$ in $\Theta(\ell)(a_1) \times \cdots \times \Theta(\ell)(a_r)$, it requires that the abstract structured value $\{\hat{v}_1, \cdots, \hat{v}_r\}_{k_0}^a$, cut at depth d, belongs to $\Theta(\ell)(a)$. The special abstract value $\top^a$ will end up in $\Theta(\ell)(a)$ if the depth of the term exceeds d. The last rule is for the application of an r-ary function f. Also in this case, (i) it analyses each term $M_i^{a_i}$, and (ii) for all r-tuples of values $(\hat{v}_1, \cdots, \hat{v}_r)$ in $\Theta(\ell)(a_1) \times \cdots \times \Theta(\ell)(a_r)$, it requires that the composed abstract value $f(\hat{v}_1, \cdots, \hat{v}_r)^a$ belongs to $\Theta(\ell)(a)$.

The judgements for nodes with the form $(\widehat{\Sigma}, \kappa, \Theta, T, \alpha) \models N$ are defined by the rules in Table 7. The rules for the *inactive node* and for *parallel composition* are standard. The rule for a single node $\ell : [B]$ requires that its internal components B are in turn analysed; in this case we the use rules with judgements $(\widehat{\Sigma}, \kappa, \Theta, T, \alpha) \models_\ell B$, where ℓ is the label of the enclosing node. The rule connecting actual stores Σ with abstract ones $\widehat{\Sigma}$ requires the locations of sensors to contain the corresponding abstract values. The rule for sensors is trivial, because we are only interested in the users of their values.

The rules for processes require to analyse the immediate sub-processes. The rule for *decryption* is similar to the one for communication: it also requires that the keys coincide. The rule for *assignment* requires that all the values $\hat{v}$ in the estimate $\Theta(\ell)(a)$ for M^a belong to $\widehat{\Sigma}_\ell(x)$. The rules for the *inactive process*, for *parallel composition*, and for *iteration* are standard (we assume that each iteration variable h is uniquely bound to the body P).

References

1. Barrère, M., Hankin, C., Nicolaou, N., Eliades, D.G., Parisini, T.: Identifying security-critical cyber-physical components in industrial control systems. CoRR abs/1905.04796 (2019). http://arxiv.org/abs/1905.04796
2. Bodei, C., Buchholtz, M., Degano, P., Nielson, F., Nielson, H.R.: Static validation of security protocols. J. Comput. Secur. **13**(3), 347–390 (2005)
3. Bodei, C., Degano, P., Ferrari, G.L., Galletta, L.: A step towards checking security in IoT. In: Proceedings of ICE 2016. EPTCS, vol. 223, pp. 128–142 (2016)
4. Bodei, C., Degano, P., Ferrari, G.-L., Galletta, L.: Where do your IoT ingredients come from? In: Lluch Lafuente, A., Proença, J. (eds.) COORDINATION 2016. LNCS, vol. 9686, pp. 35–50. Springer, Cham (2016). https://doi.org/10.1007/978-3-319-39519-7_3
5. Bodei, C., Degano, P., Galletta, L., Salvatori, F.: Linguistic mechanisms for context-aware security. In: Ciobanu, G., Méry, D. (eds.) ICTAC 2014. LNCS, vol. 8687, pp. 61–79. Springer, Cham (2014). https://doi.org/10.1007/978-3-319-10882-7_5
6. Bodei, C., Degano, P., Galletta, L., Salvatori, F.: Context-aware security: linguistic mechanisms and static analysis. J. Comput. Secur. **24**(4), 427–477 (2016)
7. Bodei, C., Galletta, L.: Tracking data trajectories in IoT. In: Mori, P., Furnell, S., Camp, O. (eds.) Proceedings of the 5th International Conference on Information Systems Security and Privacy (ICISSP2019). ScitePress (2019)
8. Bodei, C., Galletta, L.: Tracking sensitive and untrustworthy data in IoT. In: Proceedings of the First Italian Conference on Cybersecurity (ITASEC 2017), vol. 1816, pp. 38–52. CEUR (2017)
9. Bodei, C., Degano, P., Ferrari, G.L., Galletta, L.: Tracing where IoT data are collected and aggregated. Log. Methods Comput. Sci. **13**(3), 1–38 (2017)
10. Bodei, C., Degano, P., Ferrari, G.-L., Galletta, L.: Revealing the trajectories of KLAIM tuples, statically. In: Boreale, M., Corradini, F., Loreti, M., Pugliese, R. (eds.) Models, Languages, and Tools for Concurrent and Distributed Programming. LNCS, vol. 11665, pp. 437–454. Springer, Cham (2019). https://doi.org/10.1007/978-3-030-21485-2_24
11. Degano, P., Ferrari, G.L., Galletta, L.: A two-component language for COP. In: Proceedings of 6th International Workshop on Context-Oriented Programming, COP@ECOOP 2014, pp. 6:1–6:7. ACM (2014)
12. Degano, P., Ferrari, G.L., Galletta, L.: A two-component language for adaptation: design, semantics, and program analysis. IEEE Trans. Softw. Eng. **42**(6), 505–529 (2016)
13. Gao, H., Bodei, C., Degano, P.: A formal analysis of complex type flaw attacks on security protocols. In: Meseguer, J., Roşu, G. (eds.) AMAST 2008. LNCS, vol. 5140, pp. 167–183. Springer, Heidelberg (2008). https://doi.org/10.1007/978-3-540-79980-1_14
14. Gao, H., Bodei, C., Degano, P., Riis Nielson, H.: A formal analysis for capturing replay attacks in cryptographic protocols. In: Cervesato, I. (ed.) ASIAN 2007. LNCS, vol. 4846, pp. 150–165. Springer, Heidelberg (2007). https://doi.org/10.1007/978-3-540-76929-3_15

15. Herlihy, M.: Wait-free synchronization. ACM Trans. Program. Lang. Syst. **13**(1), 124–149 (1991)
16. Lanese, I., Bedogni, L., Felice, M.D.: Internet of Things: a process calculus approach. In: Proceedings of the 28th Annual ACM Symposium on Applied Computing. SAC 2013, pp. 1339–1346. ACM (2013)
17. Lanotte, R., Merro, M.: A semantic theory of the Internet of Things. In: Lluch Lafuente, A., Proença, J. (eds.) COORDINATION 2016. LNCS, vol. 9686, pp. 157–174. Springer, Cham (2016). https://doi.org/10.1007/978-3-319-39519-7_10
18. Lanotte, R., Merro, M.: A semantic theory of the Internet of Things. Inf. Comput. **259**(1), 72–101 (2018)
19. Lanotte, R., Merro, M., Muradore, R., Viganò, L.: A formal approach to cyber-physical attacks. In: 30th IEEE Computer Security Foundations Symposium, pp. 436–450. IEEE Computer Society (2017)
20. Nicolaou, N., Eliades, D.G., Panayiotou, C.G., Polycarpou, M.M.: Reducing vulnerability to cyber-physical attacks in water distribution networks. In: 2018 International Workshop on Cyber-Physical Systems for Smart Water Networks, CySWater@CPSWeek, pp. 16–19. IEEE Computer Society (2018)
21. Nielson, H.R., Nielson, F., Vigo, R.: A calculus for quality. In: Păsăreanu, C.S., Salaün, G. (eds.) FACS 2012. LNCS, vol. 7684, pp. 188–204. Springer, Heidelberg (2013). https://doi.org/10.1007/978-3-642-35861-6_12
22. Nielson, H.R., Nielson, F., Vigo, R.: A calculus of quality for robustness against unreliable communication. J. Log. Algebraic Methods Program. **84**(5), 611–639 (2015)
23. Schneier, B.: Attack trees. Dr Dobb's J. **24**(12), 436–450 (1999)
24. Zillner, T.: ZigBee Exploited (2015). https://www.blackhat.com/docs/us-15/materials/us-15-Zillner-ZigBee-Exploited-The-Good-The-Bad-And-The-Ugly-wp.pdf

New Program Abstractions for Privacy

Sebastian Hunt[1](✉) and David Sands[2]

[1] City, University of London, London, UK
s.hunt@city.ac.uk
[2] Chalmers University of Technology, Gothenburg, Sweden

Abstract. Static program analysis, once seen primarily as a tool for optimising programs, is now increasingly important as a means to provide quality guarantees about programs. One measure of quality is the extent to which programs respect the privacy of user data. Differential privacy is a rigorous quantified definition of privacy which guarantees a bound on the loss of privacy due to the release of statistical queries. Among the benefits enjoyed by the definition of differential privacy are compositionality properties that allow differentially private analyses to be built from pieces and combined in various ways. This has led to the development of frameworks for the construction of differentially private program analyses which are private-by-construction. Past frameworks assume that the sensitive data is collected centrally, and processed by a trusted curator. However, the main examples of differential privacy applied in practice - for example in the use of differential privacy in Google Chrome's collection of browsing statistics, or Apple's training of predictive messaging in iOS 10 -use a purely local mechanism applied at the data source, thus avoiding the collection of sensitive data altogether. While this is a benefit of the local approach, with systems like Apple's, users are required to completely trust that the analysis running on their system has the claimed privacy properties.

In this position paper we outline some key challenges in developing static analyses for analysing differential privacy, and propose novel abstractions for describing the behaviour of probabilistic programs not previously used in static analyses.

1 Purpose and Aims

Differential privacy [6] perhaps represents the most rigorous and robust approach to privacy today. Unlike anonymisation methods which focus on properties of the data such as ensuring that there are several records with a given attribute, or that certain fields have been deleted, it is a property of the general mechanism (algorithm) used to release the data (and thus independent of the data itself); an algorithm which inputs sensitive data and outputs public data (typically used to compute some statistical property if the data) satisfies differential privacy if, for any input, adding or removing the data for any one individual makes very little observable difference to the overall result. For the purpose of this work we will not delve into the specific technical details of the definition.

A. Di Pierro et al. (Eds.): Festschrift Hankin, LNCS 12065, pp. 256–267, 2020.
https://doi.org/10.1007/978-3-030-41103-9_10

Example: Randomized Response. As an example, suppose that a data analyst wishes to answer the question: *what percentage of browser users have visited websites commonly used to facilitate the download of copyrighted material?* Suppose this information can be determined from the browsing history stored in your browser. One way to give a useful but necessarily approximate answer to this question, at the same time as limiting the privacy risk for the individual, is to use the following procedure: each respondent flips two coins; if both are heads then answer "Yes", if both are tails then answer "No", and otherwise answer the query truthfully. The data analyst is able to make a statistical estimate of the true percentage: if there are y "Yes" answers from 10000 respondents then we expect 2500 random "Yes" answers, 2500 random "No" answers, so the answer to the question can be estimated as $(y - 2500)/5000$. At the same time, anyone intercepting a "Yes" answer from any one individual cannot know whether it was generated by honesty or randomness – even if the response becomes public data the respondent can plausibly deny having visited such websites. In this specific example we can informally think of the increase in privacy risk as a multiplicative factor of $0.75/0.25 = 3$, so if there was already a 0.1% chance of, say, someone launching an investigation into whether a given IP address has been used to share copyrighted data, the risk in participation would at most increase to $3 \times 0.1\%$. By adjusting the probabilities of the coin flips one can increase the degree of privacy at the expense of either having lower accuracy in the reported result, or of requiring more data points to compensate for the increased noise.

This algorithm is differentially private [5]. Differential privacy is a parameterised definition, and the parameter, referred to as "epsilon", bounds (the natural logarithm of) how much multiplicative difference there is between an analysis using my data or someone else's. In the case of this particular algorithm we would say that it is ϵ-differentially private, with $\epsilon = \ln 3$. Every nontrivial differentially private algorithm operates by the addition of noise in some form. The particular algorithm described in the example is based on a 50-year-old survey technique called *randomised response*, and was designed to persuade respondents to tell truthful answers to potentially embarrassing or incriminating questions.

Frameworks for Differential Privacy. Differential privacy enjoys a number of useful properties that make it, in theory, an excellent foundation for robust privacy-aware information release. In particular it satisfies a number of useful compositional properties that allow the construction of differentially private algorithms from well-behaved data transformations and differentially private components. Making use of these, a number of differential privacy frameworks have emerged which support the construction of differentially private mechanisms. These leverage general compositional properties of differential privacy to simplify the static verification or dynamic enforcement of a desired amount of privacy. Examples of systems of this ilk are,

- PINQ [22], wPINQ [23], ProPer [11], and EKTELO [25] which dynamically monitor how data is used, and ensure that the computation never exceeds a given privacy budget, or

- Fuzz [13,15] and LightDP [26] where programs are written in a language with a custom type system or special verification annotations, and where static type checking/verification provides differential privacy guarantees.

All of these frameworks focus on verification of mechanisms which are assumed to have access to the whole data set. This implies the existence of a trusted database curator who holds the sensitive data, and who has the responsibility to apply the mechanism to the data and to keep track of a global privacy risk (Fig. 1). We refer to this as the *centralised model*.

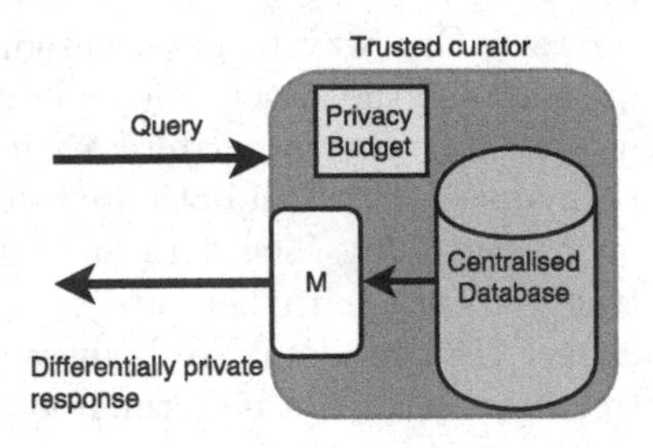

Fig. 1. Centralised Model (figure from [10])

Local Differential Privacy. The randomised response mechanism described above has a different trust model, and is called *local differential privacy* [4,19] or simply "the local model" [7]. In the local model the privacy mechanism is applied locally, at each respondent (Fig. 2).

The local model benefits from the fact that there is no longer a need to centrally store sensitive data – privacy is managed at the source (your cell-phone, car, web browser...). This removes the need for a trusted curator, and lowers the security risk of data breach. Perhaps for these reasons, the local model is the flavour of differential privacy which was first to be used in the actual "real-world" instances of differential privacy, by Google (in the Chrome browser) [12] and Apple (in iOS 10 and MacOS) [1].

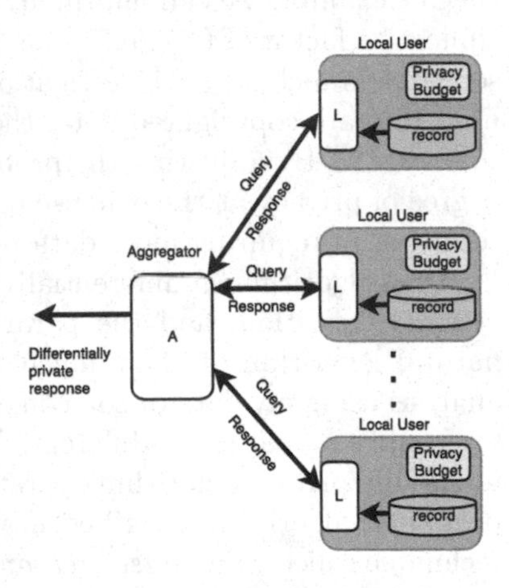

Fig. 2. Local Model (fig. from [10])

Research Goals

The local analyses by Apple and Google require a great deal of trust on the part of the user: you have to trust that they implemented their algorithms correctly on your device, and that the algorithms are indeed differentially private, not just for a single round of communication, but even when statistics are reported over time. Apple, in particular, did not initially report on the intended quantity of privacy ("epsilon") secret, and a recent reverse-engineering study of their algorithms [24] suggests that this trust is not well founded, and concludes

> "We call for Apple to make its implementation of privacy-preserving algorithms public and to make the rate of privacy loss fully transparent and tuneable by the user"

This comment aligns well with what we view as key design criteria for a **framework for local differential privacy**: the differential privacy properties of the mechanisms which deliver results based on sensitive data should be statically verifiable, and the verification should be simple enough to be done on the fly, for example in the respondent's device.

In rest of this paper we outline some key problems and possibilities in working towards statically verifiable local differential privacy. Most prior general frameworks for enforcing or verifying differential privacy are focused on the centralised model of a trusted curator providing access to a raw database. In Sect. 2 we outline one approach to verifiable local differential privacy, PreTPost, due to Ebadi and Sands [10], and discuss its limitations.

The limitations of PreTPost motivate a more general static analysis approach. We follow the philosophy outlined by Malacaria in factoring a quantitative information flow (QIF) analysis via a dependency analysis: [21]:

> "⟨the lattice of information⟩ allows for an elegant analysis decomposition of QIF into two steps, the first being an algebraic interpretation, the second being a numerical evaluation"

The "algebraic interpretation" referred to here is the use of equivalence relations, dubbed "the lattice of information" [20], but also known (in a more general form) as the lattice of *partial equivalence relations* where it was first used by the authors to express static analysis of dependency, referred to as a "binding time analysis" in [17], and as a "constancy analysis" in [16].

The gist of the approach articulated by Malacaria is to first determine which public outputs depend on which sensitive inputs, and then to instantiate that dependency numerically as a quantity. In our setting the quantity we want to measure is the epsilon of differential privacy: a bound on the largest proportional change in probability of obtaining any particular output when the user's sensitive input data is changed.

To this end we outline some specific challenges and opportunities in realising this programme for (local) differential privacy.

In Sect. 3 we discuss a shortcoming in dependency analyses that leads to imprecision in the quantitative step, namely the inability to describe a disjunctive dependency (to depend on one thing in some executions, or another thing in others, but never both in the same execution). This turns out to be a crucial distinction for properties such as differential privacy, because it allows us to use max rather than sum when we perform the quantitative instantiation step.

In Sect. 4 we propose a new way to instantiate privacy cost. Rather than directly instantiating with differential privacy costs (or their logarithm), our proposal is to abstract the behaviour of probabilistic programs (i.e. a new abstract domain) in a way which will provide greater precision and versatility; it is based on the idea of a geometric interpretation of differential privacy, a *privacy region* introduced by Kairouz, Oh, and Viswanath [18].

2 Verifiable Local Differential Privacy

PreTPost is a framework for implementing verifiable local differential privacy [10]. PreTPost leverages the simple observation that local differential privacy is preserved by arbitrary data pre-processing. The PreTPost framework requires a data analysis to be decomposed into a pre-processor (Pre), a simple core probabilistic transformation (T), and a post-processor (Post) (Fig. 3).

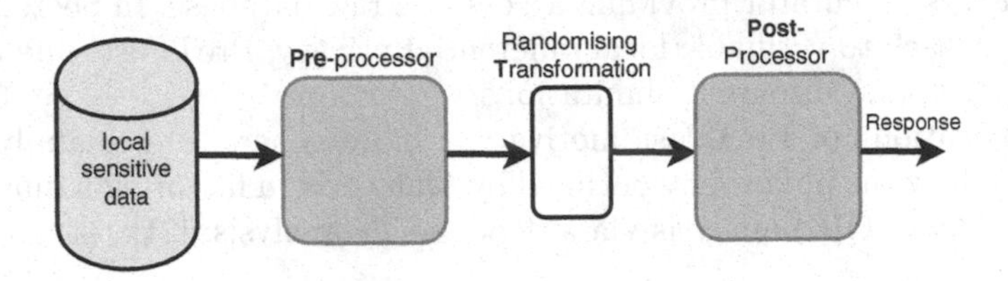

Fig. 3. PreTPost schema

The point of this schema is that an analysis delivered to a respondent in the form of a $\langle \text{Pre}, T, \text{Post} \rangle$ triple can be easily analysed: the quantity of differential privacy, often just referred to as "epsilon" but what we will refer to as the *privacy cost*, can be bounded by the cost of T alone since, unlike in the case of centralised differential privacy, the pre-processing cannot inflate the privacy cost of a subsequent differentially private operation. It is shown that this decomposition is possible and straightforward for a range of local analysis algorithms from the literature. Ebadi has implemented a prototype implementation of the PreTPost framework using sandboxed execution to prevent a malicious pre-processor from communicating the sensitive data directly, from bypassing the randomising transformation, or communicating via a covert timing channel [9].

Limitations of PreTPost. While the PreTPost schema provides a simple route to analysis of a proposed differentially private data processing algorithm, there are some limitations:

- The proposed algorithm must be refactored into the PreTPost format, which might not be the most natural way to express the algorithm.
- In practice (including in the PreTPost implementation) there are not only sensitive inputs, but also public inputs from other sources (data which is not considered sensitive) such as local data, as well as local public outputs (which may be used to modify subsequent public inputs). Although we don't anticipate major problems with these generalisations, they are still outside the simple schema of PreTPost.
- It cannot account for algorithms which reduce the sensitivity of the input data. For example, suppose an algorithm works by bitwise randomisation of a bit vector generated by a pre-processing of the sensitive input. If pre-processing yields an n-bit vector in which only a fixed number of bits k depend

on the sensitive data, then the privacy cost is k times the cost associated with the randomisation, whereas PreTPost necessarily assumes all n bits are sensitive; this example is reminiscent of the Bloom filter used in the full Rappor system [12];
- it cannot account for repeated randomisation (also a feature of the full Rappor system), i.e., where some randomised data is further randomised (not a common operation in differentially private algorithms, but used by the Rappor system).

While it is not clear the extent to which these limitations are show stoppers for PreTPost, by developing a more general and expressive static analysis that is not based on a fixed program schema we aim to gain a more fundamental understanding of the problem of abstracting and verifying differentially private algorithms. The remaining sections discuss some of the building blocks for such an analysis.

3 Dependency Analysis: The Need for Disjunction

Our aim is to address the limitations of PreTPost described above, in part by using dependency analysis to deal with the complex and subtle interdependencies that arise in realistic implementations and which prevent a simple, clean separation into three phases. However, it turns out that to get good results we need a semantic notion of dependency that is more expressive than the one used in standard dependency analyses. Moreover, this richer notion of dependency is relevant and potentially useful even when a mechanism *can* be cleanly separated in PreTPost style.

In this section we outline *why* we need a more general notion (a notion of *disjunction*), and *how* this might be represented in an analysis. We do not, however, go into the details of the semantic model for this generalised form, which is work in progress.

As it stands, PreTPost has nothing to say about how the post-processing phase uses the data supplied by the probabilistic transformation T. While post-processing the outputs of T can never increase their privacy cost, it can *decrease* it. Consider the code in Figs. 4, 5 and 6. Taken together these pieces of code define three alternative versions of a mechanism we call Three-Bits, which processes a 3-bit private value (an integer between 0 and 7, stored in `x`) and outputs a randomised result in `y`. All three versions share the same pre-processing and probabilistic phases but differ in their post-processing. The Pre-phase projects out the three bits of `x` into `a`, `b`, `c` and the T-phase independently randomises the bits. Suppose that the privacy cost of `Ran` is ϵ.

Post-processing (A) simply reassembles the randomised bits into a new 3-bit integer. Version (B) does the same but neglects to include bit `b`. Version (C) includes bit `a` in the result and, depending on `a`, includes either `b` or `c`, but never both together.

A standard dependency analysis will infer that, after (A), `y` depends on all three of `a`, `b`, `c` whereas after (B) it depends on `a` and `c` but not `b`. This allows

```
a = x % 2;
b = (x/2) % 2;
c = (x/4) % 2;
```

Fig. 4. Three-Bits: Pre

```
a = Ran(a);
b = Ran(b);
c = Ran(c);
```

Fig. 5. Three-Bits: T

```
y = a + 2*b + 4*c;
```

(A)

```
y = a + 4*c;
```

(B)

```
if (a == 0)
    y = a + 2*b;
else
    y = a + 4*c;
```

(C)

Fig. 6. Three-Bits: Post

us to infer that the privacy cost for Three-Bits-A is the sum of the costs for a, b, c (a total of 3ϵ) but the privacy cost for Three-Bits-B is the sum only of the costs for a and c (a total of 2ϵ).

Now consider (C). A standard dependency analysis will infer that, as for (A), y depends on a, b, c and an analysis using this dependency information would therefore also assign a privacy cost of 3ϵ to Three-Bits-C. But this is overly conservative. On any given run of (C), the value of y reveals *either* the values of a and c, *or* the values of a andc, but never all three together. This disjunctive dependency behaviour is reflected in the true privacy cost of Three-Bits-C, which is only 2ϵ. More generally, for a disjunctive dependency it turns out to be sound for differential privacy to take the *maximum* cost across the disjuncts, rather than the sum. Hence the cost for (C) may be calculated as $\max(\text{cost}(\mathtt{a}) + \text{cost}(\mathtt{b}), \text{cost}(\mathtt{a}) + \text{cost}(\mathtt{c})) = 2\epsilon$, in contrast with the cost for (A) which is $\text{cost}(\mathtt{a}) + \text{cost}(\mathtt{b}) + \text{cost}(\mathtt{c}) = 3\epsilon$.

A standard dependency analysis assigns to each output variable y a dependency set of input variables on which it may depend, ie a set X such that any choice of initial values for the variables in X completely determines the resulting value of y. Example (C) suggests that for our purposes it might be natural to lift such an analysis to represent a disjunctive dependency by a *set of sets* of variables. The dependency for y could then be represented as

$$\{\{\mathtt{a}, \mathtt{b}\}, \{\mathtt{a}, \mathtt{c}\}\}$$

Adapting standard approaches from abstract interpretation ([3,14]) it is relatively straightforward to lift an existing dependency analysis in this way (though the resulting algorithmic complexity may present practical challenges). However, it is not immediately obvious how to give a satisfactory *semantics* to such a set of sets. In particular, the required semantics is *not* simple logical disjunction of dependency properties: in example (C) it is *neither* true that the value of y is determined solely by $\{\mathtt{a}, \mathtt{b}\}$ *nor* that it is determined solely by $\{\mathtt{a}, \mathtt{c}\}$.

Our work in progress suggests that a satisfactory semantics for disjunctive dependency is obtainable by generalising the usual non-interference condition in an appropriate way. We conclude this section with some hints at the direction in which we are aiming.

The standard non-interference condition can give meaning to e.g., "output x depends on inputs y and z" by using equivalence relations over states (mappings from variables to values). For this example we need two such relations, $=_{\{x\}}$, which relates two stores if they agree on the value of variable x, and $=_{\{y,z\}}$, which relates two stores if they agree on the values of both y and z. Then the semantics of "output x depends on inputs y and z" is: when the program is run on two stores related by $=_{\{y,z\}}$, you end up with two stores related by $=_{\{x\}}$.

While there are *specific* relations within the full lattice of equivalence relations on stores that exhibit disjunctive dependencies such as "output x depends on y and z *or* y and w", they can only express such a disjunction by saying exactly *how* it arises. For example we could build an equivalence relation on input stores that allows us to express the more specific disjunction "when $p(y)$ then x depends on y and z but otherwise it depends on y and w". What we are aiming for is a method that will allow us to give a semantics for such a disjunction more abstractly, directly from the relations $=_{\{y,z\}}$ and $=_{\{y,w\}}$, by working with suitable *sets* of relations that take into account all the ways that such a disjunction might arise.

4 The Abstract Domain of Privacy Regions

As mentioned in the introduction, for the quantitative phase of our analysis, we propose a novel abstract domain for analysis of probabilistic programs based on the idea of privacy regions. A key design goal is to allow us to leverage the framework of abstract interpretation [2]. Recall that abstract interpretation is a framework for semantics-based analysis which works by interpreting programs over an abstract domain in place of the concrete (standard) semantic domain, where each abstract value a denotes a concrete property γa. Each program construct is then given an abstract interpretation which soundly approximates its standard semantics: if a is mapped to a', the standard semantics transforms the property γa to some property $P \subseteq \gamma a'$. The abstract semantics of a program is then computed by a fixed-point iteration. The abstract domain is typically required to be a complete lattice and the design of the framework ensures that it is always sound to over-approximate by computing abstract values higher in the lattice. This freedom may be used (at the cost of reduced precision) to force the fixed-point iteration to converge within an acceptable time limit. Here, we give a lightweight, simplified account of the idea of privacy regions and explain why they have the appropriate structure to serve as an abstract domain.

To motivate the definitions we need to recall the generalisation of differential privacy known as *approximate* or (ϵ, δ)-differential privacy. In this variant (equivalent to ϵ-DP when $\delta = 0$) one weakens the requirement so that the differential privacy property may fail with some probability (typically very small) given by δ. This weakening can be used to obtain much better accuracy when composing analyses.

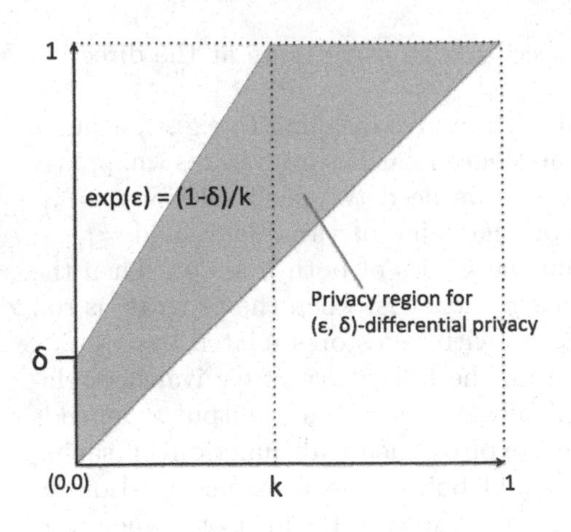

Fig. 7. (ϵ, δ) privacy region

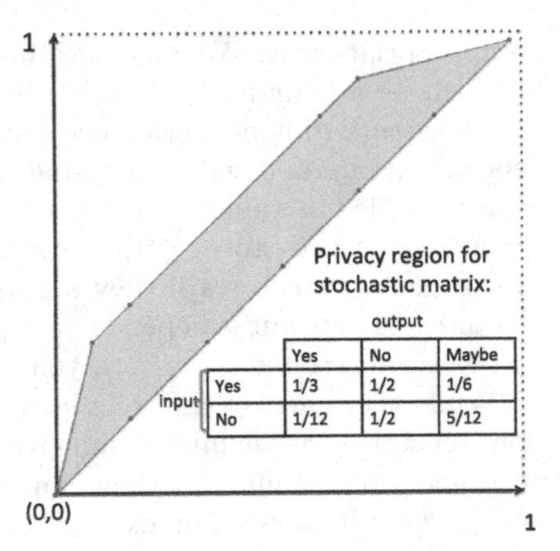

Fig. 8. Example privacy region

A key observation of [18] is that the (ϵ, δ)-differential privacy properties enjoyed by a mechanism can be characterised geometrically. For each choice of (ϵ, δ) define its *privacy region*[1] to be the closed region of the unit square above the line $x = y$ and below the line $y = e^{\epsilon}x + \delta$ as pictured in Fig. 7. Suppose that M has domain A and range B (ie M maps each value in A to a distribution over B). Then M is (ϵ, δ)-DP iff the following set is contained in the (ϵ, δ) privacy region:

$$\{(x, y) \mid a, a' \in A, S \subseteq B, x = \Pr[M(a) \in S], y = \Pr[M(a') \in S], y \geq x\} \quad (*)$$

(Note that any such set is symmetrical about $y = 1 - x$, because $\Pr[M(a) \in S] = 1 - \Pr[M(a) \in \overline{S}]$).

The privacy region for M, denoted $\mathcal{R}(M)$, is then defined as the intersection of *all* the (ϵ, δ) privacy regions such that M is (ϵ, δ)-DP, ie all the (ϵ, δ) privacy regions which contain the set (*). Equivalently, and more constructively, we can define $\mathcal{R}(M)$ to be the convex closure of (*): an example of a privacy region for a mechanism (specified as a stochastic matrix) is given in Fig. 8, where the points generated by this construction are marked. Note that each upper edge of a privacy region $\mathcal{R}(M)$ witnesses a distinct (ϵ, δ)-DP property, where ϵ is the log of the gradient and δ is the y-intercept; all of these properties hold simultaneously for M (and, indeed, for any mechanism whose privacy region is contained in $\mathcal{R}(M)$). In this example, the three upper edges witness the properties ($\epsilon = \ln 4, \delta = 0$), ($\epsilon = 0, \delta = 1/4$), and ($\epsilon = \ln(1/4), \delta = 3/4$). (The two vertices of the middle edge – $(1/12, 4/12)$ and $(8/12, 11/12)$ – are generated by the output events $\{\text{Yes}\}$ and $\{\text{No}, \text{Maybe}\}$, respectively.)

[1] For convenience, our definition is a rotation by 90° in the unit square of the region defined by [18] and we restrict to the region above $y = x$ (their definition, after rotation, is symmetric about $y = x$).

Our key proposal is that the set of privacy regions forms a suitable abstract domain for static program analysis based on the principles of abstract interpretation (as outlined at the start of this section):

1. Privacy regions form a *complete lattice* ordered by subset inclusion, with the bottom element represented by the line $x = y$ ($(0, 0)$-differential privacy) and top element being the upper left half of the unit square (no differential privacy), and the meet and join given by intersection and convex closure of the union, respectively.
2. The semantic content of a privacy region is all the (ϵ, δ)-regions in which it is contained, so any $R \supseteq \mathcal{R}(M)$ which can be inferred by analysis is a safe approximation for M.
3. Privacy regions subsume the notion of distance between distributions (as used in pure ϵ-DP): the "leading edge" of a privacy region (rising from $(0, 0)$) defines an (ϵ, δ)-DP property where $\delta = 0$ and, by convexity, where ϵ is maximal.
4. Even when ϵ is the only property of interest, the extra information carried by privacy regions provides a better abstraction, yielding a more precise bound on ϵ than can be obtained using distance alone.
5. [18] provides a variety of useful results that characterise privacy regions, in particular describing the privacy region of a mechanism built from the composition of multiple mechanisms – very natural operations in a static program analysis.
6. We have been able to construct novel abstractions of function composition and pairing, which are essential ingredients of a static analysis.
7. There are many natural ways to safely coarsen a privacy region (to make it bigger, and thus less precise) – something which is a prerequisite to approximating fixed-point computations necessary to abstract the behaviour of iterative or recursive computation.

5 Conclusions

We have identified two key steps in our goal of constructing a static analysis of local differential privacy: the ability to abstract disjunctive dependency properties, and the use of privacy regions as an abstract domain for privacy cost. We have a promising approach for the semantics of disjunctive dependency, and believe this could be of independent interest. It remains, of course, to show that this can be combined with privacy regions to perform a useful static analysis.

Acknowledgements. This work was partly funded by the Swedish Foundation for Strategic Research (SSF) under the projects WebSec and by the Swedish Research Council (VR).

References

1. Apple Press Release: Apple previews iOS 10, the biggest iOS release ever (2016). https://www.apple.com/newsroom/2016/06/apple-previews-ios-10-biggest-ios-release-ever. Accessed 22 July 2017

2. Cousot, P., Cousot, R.: Abstract interpretation: a unified lattice model for static analysis of programs by construction or approximation of fixpoints. In: Proceedings 4th Annual ACM Symposium on Principles of Programming Languages, pp. 238–252 (1977)
3. Cousot, P., Cousot, R.: Systematic design of program analysis frameworks. In: Proceedings of the 6th ACM SIGACT-SIGPLAN Symposium on Principles of Programming Languages. POPL 1979, pp. 269–282. ACM, New York (1979). https://doi.org/10.1145/567752.567778
4. Duchi, J.C., Jordan, M.I., Wainwright, M.J.: Local privacy and statistical minimax rates. In: 2013 51st Annual Allerton Conference on Communication, Control, and Computing (Allerton), pp. 1592–1592, October 2013. https://doi.org/10.1109/Allerton.2013.6736718
5. Dwork, C.: Differential privacy. In: Bugliesi, M., Preneel, B., Sassone, V., Wegener, I. (eds.) ICALP 2006. LNCS, vol. 4052, pp. 1–12. Springer, Heidelberg (2006). https://doi.org/10.1007/11787006_1
6. Dwork, C., McSherry, F., Nissim, K., Smith, A.: Calibrating noise to sensitivity in private data analysis. In: Halevi, S., Rabin, T. (eds.) TCC 2006. LNCS, vol. 3876, pp. 265–284. Springer, Heidelberg (2006). https://doi.org/10.1007/11681878_14
7. Dwork, C., Roth, A.: The algorithmic foundations of differential privacy. Found. Trends Theoret. Comput. Sci. **9**, 211–407 (2014). https://doi.org/10.1561/0400000042
8. Ebadi, H.: Dynamic Enforcement of Differential Privacy. Ph.D. thesis, Chalmers University of Technology, March 2018
9. Ebadi, H.: The PreTPost Framework (2018). https://github.com/ebadi/preTpost
10. Ebadi, H., Sands, D.: PreTPost: a transparent, user verifiable, local differential privacy framework (2018). https://github.com/ebadi/preTpost. Also appears in [8]
11. Ebadi, H., Sands, D., Schneider, G.: Differential privacy: now it's getting personal. In: Proceedings of the 42nd Annual ACM SIGPLAN-SIGACT Symposium on Principles of Programming Languages. POPL 2015, pp. 69–81. ACM (2015). https://doi.org/10.1145/2676726.2677005
12. Erlingsson, Ú., Pihur, V., Korolova, A.: RAPPOR: randomized aggregatable privacy-preserving ordinal response. In: CCS. ACM (2014)
13. Gaboardi, M., Haeberlen, A., Hsu, J., Narayan, A., Pierce, B.C.: Linear dependent types for differential privacy. In: Proceedings of the 40th Annual ACM SIGPLAN-SIGACT Symposium on Principles of Programming Languages. POPL 2013, pp. 357–370. ACM, New York (2013). https://doi.org/10.1145/2429069.2429113
14. Giacobazzi, R., Ranzato, F.: Optimal domains for disjunctive abstract interpretation. Sci. Comput. Program. **32**(1), 177–210 (1998). https://doi.org/10.1016/S0167-6423(97)00034-8,http://www.sciencedirect.com/science/article/pii/S0167642397000348. 6th European Symposium on Programming
15. Haeberlen, A., Pierce, B.C., Narayan, A.: Differential privacy under fire. In: Proceedings of the 20th USENIX Conference on Security. SEC 2011, pp. 33–33. USENIX Association, Berkeley (2011). http://dl.acm.org/citation.cfm?id=2028067.2028100
16. Hunt, S.: Abstract interpretation of functional languages: from theory to practice. Ph.D. thesis, Imperial College London, UK (1991)
17. Hunt, S., Sands, D.: Binding time analysis: a new perspective. In: Proceedings of the ACM Symposium on Partial Evaluation and Semantics-Based Program Manipulation (PEPM 1991), pp. 154–164. ACM Press (1991)

18. Kairouz, P., Oh, S., Viswanath, P.: The composition theorem for differential privacy. IEEE Trans. Inf. Theory **63**(6), 4037–4049 (2017)
19. Kairouz, P., Oh, S., Viswanath, P.: Extremal mechanisms for local differential privacy. J. Mach. Learn. Res. **17**(17), 1–51 (2016). http://jmlr.org/papers/v17/15-135.html
20. Landauer, J., Redmond, T.: A lattice of information. In: CSFW (1993)
21. Malacaria, P.: Algebraic foundations for information theoretical, probabilistic an guessability measures of information flow. CoRR abs/1101.3453 (2011). http://arxiv.org/abs/1101.3453
22. McSherry, F.: Privacy integrated queries. In: Proceedings of the 2009 ACM SIGMOD International Conference on Management of Data (SIGMOD). Association for Computing Machinery, Inc., June 2009
23. Proserpio, D., Goldberg, S., McSherry, F.: Calibrating data to sensitivity in private data analysis: a platform for differentially-private analysis of weighted datasets. Proc. VLDB Endow. **7**(8), 637–648 (2014). https://doi.org/10.14778/2732296.2732300
24. Tang, J., Korolova, A., Bai, X., Wang, X., Wang, X.: Privacy loss in Apple's implementation of differential privacy on MacOS 10.12. CoRR abs/1709.02753 (2017). http://arxiv.org/abs/1709.02753
25. Zhang, D., McKenna, R., Kotsogiannis, I., Hay, M., Machanavajjhala, A., Miklau, G.: EKTELO: a framework for defining differentially-private computations. In: Proceedings of the 2018 International Conference on Management of Data, SIGMOD Conference 2018, Houston, TX, USA, 10–15 June 2018, pp. 115–130 (2018). https://doi.org/10.1145/3183713.3196921
26. Zhang, D., Kifer, D.: LightDP: towards automating differential privacy proofs. In: POPL (2017)

Optimizing Investments in Cyber Hygiene for Protecting Healthcare Users

Sakshyam Panda[1(✉)], Emmanouil Panaousis[2], George Loukas[2], and Christos Laoudias[3]

[1] University of Surrey, Guildford, UK
s.panda@surrey.ac.uk
[2] University of Greenwich, London, UK
{e.panaousis,g.loukas}@greenwich.ac.uk
[3] University of Cyprus, Nicosia, Cyprus
laoudias.christos@ucy.ac.cy

Abstract. Cyber hygiene measures are often recommended for strengthening an organization's security posture, especially for protecting against social engineering attacks that target the human element. However, the related recommendations are typically the same for all organizations and their employees, regardless of the nature and the level of risk for different groups of users. Building upon an existing cybersecurity investment model, this paper presents a tool for optimal selection of cyber hygiene safeguards, which we refer as the Optimal Safeguards Tool (OST). The model combines game theory and combinatorial optimization (0-1 Knapsack) taking into account the probability of each user group to being attacked, the value of assets accessible by each group, and the efficacy of each control for a particular group. The model considers indirect cost as the time employees could require for learning and trainning against an implemented control. Utilizing a game-theoretic framework to support the Knapsack optimization problem permits us to optimally select safeguards' application levels minimizing the aggregated expected damage within a security investment budget.

We evaluate OST in a healthcare domain use case. In particular, on the Critical Internet Security (CIS) Control group 17 for implementing security awareness and training programs for employees belonging to the ICT, clinical and administration personnel of a hospital. We compare the strategies implemented by OST against alternative common-sense defending approaches for three different types of attackers: Nash, Weighted and Opportunistic. Our results show that Nash defending strategies are consistently better than the competing strategies for all attacker types with a minor exception where the Nash defending strategy, for a specific game, performs at least as good as other common-sense approaches. Finally, we illustrate the alternative investment strategies on different Nash equilibria (called plans) and discuss the optimal choice using the framework of 0-1 Knapsack optimization.

Keywords: Cybersecurity · Cyber hygiene · Healthcare · Optimization · Training and awareness · CIS control · Game theory

A. Di Pierro et al. (Eds.): Festschrift Hankin, LNCS 12065, pp. 268–291, 2020.
https://doi.org/10.1007/978-3-030-41103-9_11

1 Introduction

In the last few years, several cybersecurity incidents have taken place in the healthcare sector, including the WannaCry ransomware, which influenced globally the cybersecurity landscape[1]. The 2018 Ponemon Cost of a Data Breach study[2] shows that the healthcare industry has the highest cost per record breached in a cyber incident, at $408. This is almost twice the equivalent cost per record breached in the financial sector. This calls for the effective preparation of healthcare organizations in an ever-evolving cyber attack landscape. An example project that is addressing this from the perspective of training the users in the sector is H2020 CUREX project[3], which allows a healthcare provider to assess the realistic cybersecurity and privacy risks they are exposed to [1].

Yet, a recent report from Mckinsey[4] states that almost all companies systematically over-invest in the protection of assets that have no risk while at the same time they under-fund the protection of high-risk assets. Furthermore, regarding bearing costs of cybersecurity controls, in a survey from KPMG[5], 43% of correspondents stated that they did not increase their cybersecurity budget even though high profile security breaches have been widely known. So, effective risk management is not only about assessing the risk correctly but also about selecting the controls that are optimal given the cost constraints of adopting them. To address the challenge of optimal control selection, in this paper, we formulate a model and tool for suggesting mathematically optimal *cyber hygiene* strategies minimising the cyber risk.

Regarding cyber hygiene, we adopt the recent definition proposed by [2], which relates it to "*the cyber security practices that online consumers should engage in to protect the safety and integrity of their personal information on their Internet enabled devices from being compromised in a cyber-attack.*"

Towards the goal of optimizing cyber hygiene, we extend the model presented in [3] so that:

- the Attacker's target is a user group (focusing on social engineering attacks) instead of (asset, vulnerability) pair of the system;
- the Indirect cost of a safeguards' application depends not only on the safeguard itself but also on the size of the user group (i.e., number of users) and more specifically it increases with the group size;
- we adopt an aggregated risk model, as the objective function of Knapsack optimization problem, rather than the weakest-link defending against a variety of attacks that can cause, in total, highest aggregated damage and;

[1] https://www.telegraph.co.uk/technology/2018/10/11/wannacry-cyber-attack-cost-nhs-92m-19000-appointments-cancelled.

[2] https://securityintelligence.com/series/ponemon-institute-cost-of-a-data-breach-2018.

[3] https://cordis.europa.eu/project/rcn/220350/factsheet/en.

[4] https://www.mckinsey.com/business-functions/risk/our-insights/cyber-risk-measurement-and-the-holistic-cybersecurity-approach.

[5] https://advisory.kpmg.us/content/dam/advisory/en/pdfs/cyber-report-healthcare.pdf.

– we use a "small" healthcare case study as a preliminary example to evaluate the OST against other common-sense approaches for a number of attacking strategies that have not been simulated in [3].

Our analysis results show that the game-theoretic approach increases risk control efficacy, by selecting an optimal combination of safeguard application levels, compared with alternative common-sense approaches. In addition, our use case designed for the healthcare domain exhibits a number of interchangeably optimal investment strategies subject to a budget constraint under the framework of 0-1 Knapsack optimization.

The remainder of this paper is organized as follows. Section 2 presents the related work in both the fields of (i) user-oriented cybersecurity safeguards and (ii) optimization of cybersecurity countermeasures including security investments. Section 3 presents both the game-theoretic model used to determine optimal cybersecurity safeguard plans as well as the optimization problem modeled and solved to derive the best ways to invest in these safeguards given a limited available budget. In Sect. 4, we undertake comparisons of the game-theoretic defending strategies against alternative common-sense approaches as well as we plot the results of the Knapsack optimization to illustrate the optimal investment solutions. Finally, Sect. 5 concludes this paper by summarizing its main contributions and highlighting future work to be undertaken to further improve the performance and the usability of our model.

2 Related Work

This work has been inspired by a previous work of Fielder et al. [3] where the authors have proposed decision support methodologies for the optimal choice of cybersecurity controls within an investment budget. They have addressed cybersecurity investment decisions by proposing different approaches; a game-theoretic approach, a combinatorial optimization approach and a mix of both called *hybrid*. This paper utilizes the latter method to recommend the optimal choice of safeguards for healthcare organizations. In this section, we discuss two classes of work relevant to this paper: literature on *cyber hygiene in healthcare* - more specifically on the *user-oriented cybersecurity safeguards*, and literature on *optimal selection of cybersecurity safeguards.* Note that the literature covered on optimal selection of cybersecurity safeguards mainly highlight work beyond the literature covered in [3].

2.1 Cyber Hygiene in Healthcare

There have been growing concerns that the existing cybersecurity posture of healthcare organizations are insufficient and this has already impacted the confidentiality [4] and integrity of medical data [5]. Further, many healthcare organizations are still using legacy systems such as Windows XP and Windows NT 3.1 which Microsoft has long stopped supporting[6], allowing adversaries to easily

[6] https://www.itpro.co.uk/public-sector/27740/nine-in-10-nhs-trusts-still-use-windows-xp.

breach the defenses (e.g., WannaCry attacks on NHS[7]). In general, healthcare organizations being rich sources of valuable data and relatively weaker security postures have become attractive targets for cybercrime [6]. The weaker security posture that they exhibit is primarily due to lack of adequate cybersecurity budget resulting in limited access to technology and expertise [7].

Besides, investment in cybersecurity has not been traditionally considered essential for healthcare systems as emphasis has predominantly been upon providing patient care and people believed that there would be no motivation to attack them. On the other hand, the increasing use of Internet of Things (IoT) technologies in healthcare has widened the attack surface beyond electronic health record databases and privacy issues to physical safety [8]. Alongside technical aspects, the role of the user in cybersecurity is paramount, as a significant proportion of attacks target the users directly through deceptive means such as application masquerading and spear-phishing. This is particularly the case in healthcare as deceiving a nurse, doctor, healthcare IT professional or administrator can impact the privacy and physical safety of patients [9].

With the increasing usage of technology, the role that humans play in underlying security processes will continually expand. Heartfield and Loukas [10] have developed a framework involving humans to effectively detect and report semantic social engineering attacks against them. Their results illustrate that involving users significantly improves the cyber threat detection rate affirming the importance of the human in cybersecurity. This further depicts that humans can no longer be seen as a threat and/or vulnerability in cybersecurity.

Acknowledging the importance of human in cybersecurity along with the increase in the severity of breaches, security experts, policymakers and governments are urging to improve cyber hygiene. Such et al. [11] have demonstrated that Cyber Essentials[8] have worked well for SMEs in mitigating threats exploiting vulnerabilities remotely using commodity-level exploitation tools. From a human-cyber interaction perspective, Vishwanath et al. [2] have demonstrated that cyber hygiene practices positively impact individuals' cyber attitude which is pivotal to cyber safety. These studies have actively exhibited that even general concepts of basic cyber hygiene work in different organizational contexts and can convincingly reduce cyber risk.

Security training in healthcare has been studied for over 20 years. It ranges from an exploratory analysis of the factors that healthcare professionals need to focus on, up to highly targeted digital applications (e.g., [12]) and platforms for raising awareness of healthcare data privacy and security risks. Furnell et al. [13] discussed the necessity to promote information security issues and the need for appropriate training and awareness initiatives in healthcare institutions. They have highlighted factors to consider while designing training and awareness programmes to familiarize healthcare personnel with basic security concepts and procedures.

---

[7] https://www.nao.org.uk/wp-content/uploads/2017/10/Investigation-WannaCry-cyber-attack-and-the-NHS-Summary.pdf.

[8] https://www.gov.uk/government/publications/cyber-essentials-scheme-overview.

The effect to which security training and awareness programmes work for different users has been studied from multiple angles. The authors have shown that specifically for deception-based attacks, such as semantic social engineering [14], where self-study and work-based training are considerably more effective than formal education in cybersecurity [15]. Besides, the perceived origin of training materials i.e., from security experts, third party agencies, or peers can have large impacts on security outcomes [16].

2.2 Optimal Selection of Cybersecurity Controls

Cybersecurity has become a key factor in determining the growth of organizations relying on information systems as it is not only a defensive measure but also has become a strategic decision providing a competitive advantage over rivalry firms. Further, the potential loss due to cyber incidents has encouraged organizations to imperatively consider cybersecurity investment decisions, especially in deriving the optimum level of investments between risk treatment options. The objective of cybersecurity investment methodologies is to compute an optimal distribution of cybersecurity budget and one of the initial work studying this was performed by Gordon and Loeb [17].

Beyond previous works such as [3,18,19] and the related work investigated there, Nagurney et al. [20] have proposed a game-theoretic supply chain network model with retailers competing to maximize their expected profits. This maximization is based on determining optimal product transactions and cybersecurity investments under budget constraints. Along the direction of optimal cybersecurity investments, Wang [21] investigated the cybersecurity investment balance between acquiring knowledge and expertise, and deploying mitigation techniques. On the other hand, Chronopoulos et al. [22] have opted a real options approach to analyze the performance of optimal cybersecurity controls on organizations. In particular, the authors have analyzed the effects of the cost of cyber attacks and the time of arrival of cybersecurity controls on the organization's optimal strategy. Similar to these papers, our work also considers the choice of the optimal strategy based on the efficacy of the control towards mitigating cyber risks.

Most closely, in terms of methodology, related recent work on optimal cybersecurity investment is [23] where the authors have investigated the balance between investing in self-protection and cyber insurance. The key difference is that their optimization minimizes expected risk and cyber insurance premium, while our model optimizes considering the efficacy of control in mitigating the aggregated residual risk and the security investment budget. Besides this, our work uses a unique combination of game theory and combinatorial optimization inspired by [3].

3 Optimal Cyber Hygiene Safeguards Model

3.1 System Model

Our model assists in acquiring an optimal selection of safeguards using game theory and combinatorial optimization. We assume $\mathcal{U}$ be the set of potential user groups consisting of employees of a healthcare organization. Any employee of a user group being susceptible to malicious activities can use any of the safeguards from the set of available safeguards $\mathcal{S}$ to improve their defense posture. However, each safeguard has a set of implementation levels $\mathcal{L}$ with each level having different efficacies in improving the security posture of user groups.

Each user group i is associated with an impact value which expresses the level of expected damage to the healthcare organization, given a successful attack against a user of a group i. This impact is equivalent to the overall asset value in association with user group i and may relate to *confidentiality*, *integrity*, and *availability*. We further consider A_i to be a random variable that expresses the overall value of the assets that the user group i has access to. For simplicity, we let the users of a group have the same *access privileges*, thus having access to assets of the same value. Users of different groups have different *access privileges* due to their different roles (e.g., IT personnel, healthcare practitioners, and administration) and access to different assets. The vulnerability of a user group i, i.e., the probability of being compromised by an attack, is captured by the security level S_i exhibited by the user group i. We assume that S_i increases with the number of safeguards applied as well as their application level.

Furthermore, we denote R_i as the threat occurrence, i.e., the probability of a threat to attack the i user group, and L_i as the expected loss associated with a user group i. Using the well-known risk assessment formula, risk = (likelihood of being attacked) x (probability of success of this attack) x probable loss [24], we compute the risk as

$$L_i = R_i \, S_i \, A_i. \tag{1}$$

An attack against a user group i is partially mitigated by the efficacy value of the implemented cybersecurity safeguard p_{cj}. The efficacy parameter, modeled as a random variable, depends on the selected application level and can be represented as $E(j,i)\colon \mathcal{L} \times \mathcal{U} \to [0,1)$. It is evident from real-world practices that different implementation levels work differently on different users and this has motivated us in considering $E(j,i)$ rather than a single efficacy value for the level j against all user groups i. Note that $E(j,i)$ is determined by the application level j and the user group i. Due to the existence of 0-day vulnerabilities, we assume that $E(j,i) \neq 1$.

Remark 1. Different users have different likelihood of adopting a measure. A cyber hygiene measure works only when it is adopted, and this adaption rate distinguishes human users from systems. For example, a user may decide not to implement a cyber hygiene measure due to unfitting usability (e.g, hard to remember complex passwords) even if the optimization framework recommends otherwise.

Let $S(j,i)$ be the security level of a user group i when level j is implemented and can be expressed as $S(j,i) = 1 - E(j,i)$. Replacing L_i and S_i as $L(j,i)$ and $S(j,i)$, respectively, in formula 1, we compute the *cybersecurity loss* for a safeguard application level j and target i as

$$L(j,i) = R_i\, A_i\, [1 - E(j,i)]. \tag{2}$$

Equation 2 implies the expected damage of the Defender when a user group i is successfully compromised given the investigated safeguard has been applied at level j.

While the application of a cybersecurity safeguard strengthens the defense of the healthcare organization, it is associated with two types of cost namely; *indirect* and *direct*. Examples of indirect cost are System Performance and Usability. We express the indirect cost of an application level j by the random variable $C\colon \mathcal{C} \times \mathcal{L} \times \mathcal{U} \rightarrow \mathbb{Z}^{+}$. Note that $C(j,i)$ adheres to the defined property for any safeguard against a user group i. Further, the indirect cost increases with an increase in the level of application of the safeguard i.e.,

$$j > j' \Leftrightarrow C(j,i) \geq C(j',i), \quad \forall j \neq j'. \tag{3}$$

From the above, we derive the *overall expected loss* of the organization when application level j is applied on user group i as

$$\sum_{i=1}^{|\mathcal{U}|} L(j,i) + C(j,i). \tag{4}$$

Each level has also a direct cost expressed by the random variable $F\colon \mathcal{L} \rightarrow \mathbb{Z}^{+}$ that maps the safeguards and application levels to the monetary cost of the plan. In this paper, we refer the direct cost to be the available investment budget of the organization. For reference purposes, the symbols used throughout this paper are described in Table 1.

3.2 Game-Theoretic Model for Selection of Safeguards Levels

This section presents a formal model for the selection of safeguard implementation levels for each of the available safeguards. The Defender chooses to implement (or apply as in this paper we use these two terms interchangeably) a cyber hygiene safeguard from $\mathcal{S}$, while the Attacker chooses to attack a user group from $\mathcal{U}$. The Defender must decide to apply this safeguard at a specific level (pure strategy) or combination of different levels (mixed strategy) both from $\mathcal{L}$. The higher the level, the greater is the applied degree of a cyber hygiene safeguard. We refer to the application of a safeguard s at a certain level j as *cybersecurity safeguard plan.* This strategic interaction is modeled as a game where the Defender chooses the level of a safeguard to implement rather than the safeguards from $\mathcal{S}$.

We define the Cyber Safeguard Game (CSG) between Defender and Attacker, as an *one-shot, bimatrix* game of *complete information* played for any of the

Table 1. List of symbols

Symbol	Description
$\mathcal{S}$	Set of safeguards
$\mathcal{U}$	Set of users
$\mathcal{L}$	Set of safeguard implementation levels
R_i	Probability of group i to be attacked
S_i	Security level of group i
A_i	Asset value that group i has access to
λ	Maximum application level
U_d	Utility of the Defender
U_a	Utility of the Attacker
$\boldsymbol{\delta}_{\sigma,j}$	Randomized Safeguard Strategy for safeguard σ at application level j
$\boldsymbol{\alpha}$	Randomized Attacking Strategy
$\boldsymbol{\alpha}(i)$	Probability of attacking group i
L_i	Expected loss from group i
$L(j,i)$	Expected loss from group i when choosing application level j
$L(\boldsymbol{\delta}_{\sigma,j},i)$	Expected loss from group i when choosing Safeguards Plan $\boldsymbol{\delta}_{\sigma,j}$
$C(j,i)$	Indirect cost of level j when applied to group i
$E(j,i)$	Efficacy of application level j on group i
$E(\boldsymbol{\delta}_{\sigma,j},i)$	Efficacy of safeguards plan $\boldsymbol{\delta}_{\sigma,j}$ on group i
$\Gamma_{\sigma,\lambda}$	Cyber Safeguard Game for safeguard σ and maximum application level λ
$\boldsymbol{\delta}_{\sigma,\lambda}^{NE}$	Nash Safeguards Plan
$F(\boldsymbol{\delta}_{\sigma,j})$	Financial cost of Safeguards Plan $\boldsymbol{\delta}_{\sigma,j}$
$F(\sigma,j)$	Financial cost of safeguard σ when applied at level j
B	Available financial budget to invest in Nash Safeguards Plans

safeguards leading to a total number of $\mathcal{S}$ independent games. For simplicity, we have assumed no inter-dependencies between the safeguards, i.e., each safeguard mitigates a portion of the overall risk inflicted by the Attacker [25].

The set of pure strategies of the Defender consists of all possible application levels, $j \in \mathcal{L}$, while the Attacker's pure strategies are the different user groups $i \in \mathcal{U}$ which could be targeted using attacks such as social engineering. Thus, in CSG a pure strategy profile is a pair of Defender and Attacker actions, $(j,i) \in \mathcal{L} \times \mathcal{U}$ giving a pure strategy space of size $|\mathcal{L}| \times |\mathcal{U}|$. For the rest of the paper, we adopt the convention where the Defender is the row player and the Attacker is the column player.

Each player's preferences are specified by a *payoff function* defined as $U_d : (j,i) \rightarrow \mathbb{R}_-$ and $U_a : (j,i) \rightarrow \mathbb{R}_+$ for the Defender and Attacker, respectively, for the pure strategy profile (j,i). According to [26], we define a *preference relation* $\succsim$, when i is chosen by the Attacker, defined by $j \succsim j'$, if and only if $U_d(j,i) \geq U_d(j',i)$. In general, given the set $\mathcal{L}$ of all available application levels of a safeguard, a rational Defender can choose a level (i.e., pure strategy) j^* that is *feasible*, that is $j^* \in \mathcal{L}$, and *optimal* in the sense that $j^* \succsim j, \forall j \in \mathcal{L}, j \neq j^*$; alternatively she solves the problem $\max_{j \in \mathcal{L}} U_d(l,i)$, for a user group $i \in \mathcal{U}$. Likewise, we define the preference relation for the Attacker, where $i \succsim i' \iff U_a(j,i) \geq U_a(j,i')$, for an application level $j \in \mathcal{L}$. CSG is a game defined for each cyber hygiene safeguard and it is realistic to assume that all levels may be available for selection by the Defender. Their availability depends on the investment budget of the Defender and the overall financial cost of the game solution.

To derive optimal strategies for the Defender, we deploy the notion of *mixed strategies*. Since players act independently, we can enlarge their strategy spaces to allow them to base their decisions on the outcome of random events that create uncertainty to the opponent about individual strategic choices maximizing their payoffs. Hence, both Defender and Attacker deploy randomized (i.e., mixed) strategies. The mixed strategy $\boldsymbol{\delta}$ of the Defender is a probability distribution over the different application levels (i.e. pure strategies) where $\boldsymbol{\delta}(j)$ is the probability of applying level j under mixed strategy $\boldsymbol{\delta}$. We refer to a mixed strategy of the Defender as a *Randomized Safeguard Strategy* (RSS). For the finite nonempty set $\mathcal{L}$, let $\Pi_{\mathcal{L}}$ represent the set of all probability distributions over it, i.e.,

$$\Pi_{\mathcal{L}} := \{\boldsymbol{\delta} \in \mathbb{R}^{+R} | \sum_{j \in \mathcal{L}} \boldsymbol{\delta}(j) = 1\}. \tag{5}$$

Therefore a member of $\Pi_{\mathcal{L}}$ is a mixed strategy of the Defender. Likewise, the Attacker's mixed strategy is a probability distribution over the different available user groups. This is denoted by $\boldsymbol{\alpha}$, where $\boldsymbol{\alpha}(i)$ is the probability of attacking the i-th user group under mixed strategy $\boldsymbol{\alpha}$. We refer to a mixed strategy of the Attacker as the *Randomized Attacking Strategy* (RAS). Alike (5), we express $\Pi_{\mathcal{U}}$ as the set of all probability distributions over the set of all Attacker's pure strategies (i.e., given by $\mathcal{U}$). Therefore, a member of $\Pi_{\mathcal{U}}$ is as a mixed strategy of the Attacker. From the above, the set of mixed strategy profiles of CSG is the Cartesian product of the individual mixed strategy sets, $\Pi_{\mathcal{L}} \times \Pi_{\mathcal{U}}$.

Definition 1. *(Support of RSS) The support of $\boldsymbol{\delta}$ is the set of application levels $\{j|\boldsymbol{\delta}(j) > 0\}$, and it is denoted by supp($\boldsymbol{\delta}$).*

Definition 2. *(Support of RAS) The support of $\boldsymbol{\alpha}$ is the set of healthcare user groups $\{i|\boldsymbol{\alpha}(i) > 0\}$, and it is denoted by supp($\boldsymbol{\alpha}$).*

The above definitions state that the subset of applications levels (resp. user groups) that are assigned positive probability by the mixed strategy $\boldsymbol{\delta}$ (resp. $\boldsymbol{\alpha}$) is called the *support* of $\boldsymbol{\delta}$ (resp. $\boldsymbol{\alpha}$)). Note that a pure strategy is a special case of a mixed strategy, in which the support is a single action.

Now that we have defined the mixed strategies of the players, we define CSG as the finite strategic game

$$\Gamma := \langle (\text{Defender, Attacker}), \Pi_{\mathcal{L}} \times \Pi_{\mathcal{U}}, (U_d, U_a) \rangle. \tag{6}$$

For a given mixed strategy profile $(\boldsymbol{\delta}, \boldsymbol{\alpha}) \in \Pi_{\mathcal{L}} \times \Pi_{\mathcal{U}}$, we denote by $U_d(\boldsymbol{\delta}, \boldsymbol{\alpha})$, and $U_a(\boldsymbol{\delta}, \boldsymbol{\alpha})$ the expected payoff values of the Defender and Attacker, where the expectation is due to the independent randomization according to mixed strategies $\boldsymbol{\delta}$, and $\boldsymbol{\alpha}$. This can be formally represented as

$$U_d(\boldsymbol{\delta}, \boldsymbol{\alpha}) := \sum_{j \in \mathcal{L}} \sum_{i \in \mathcal{U}} U_d(j, i)\, \boldsymbol{\delta}(j)\, \boldsymbol{\alpha}(i), \tag{7}$$

and similarly

$$U_a(\boldsymbol{\delta}, \boldsymbol{\alpha}) := \sum_{j \in \mathcal{L}} \sum_{i \in \mathcal{U}} U_a(j,i)\, \boldsymbol{\delta}(j)\, \boldsymbol{\alpha}(i). \tag{8}$$

By using the preference relation we can say that, for an Attacker's mixed strategy $\boldsymbol{\alpha}$, the Defender prefers to follow the RSS $\boldsymbol{\delta}$ as opposed to $\boldsymbol{\delta}'$ (i.e., $\boldsymbol{\delta} \succsim \boldsymbol{\delta}'$), if and only if $U_d(\boldsymbol{\delta}, \boldsymbol{\alpha}) \geq U_d(\boldsymbol{\delta}', \boldsymbol{\alpha})$.

Definition 3. *The Defender's (resp. Attacker's) best response to the mixed strategy $\boldsymbol{\alpha}$ (resp. $\boldsymbol{\delta}$) of the Attacker (resp. Defender) is an RSS (resp. RAS) $\boldsymbol{\delta}^{\mathrm{BR}} \in \Pi_{\mathcal{L}}$ (resp. $\boldsymbol{\alpha}^{\mathrm{BR}} \in \Pi_{\mathcal{U}}$) such that $U_d(\boldsymbol{\delta}^{\mathrm{BR}}, \boldsymbol{\alpha}) \geq U_d(\boldsymbol{\delta}, \boldsymbol{\alpha}), \forall \boldsymbol{\delta} \in \Pi_{\mathcal{L}}$ (resp. $U_a(\boldsymbol{\delta}, \boldsymbol{\alpha}^{\mathrm{BR}}) \geq U_d(\boldsymbol{\delta}, \boldsymbol{\alpha}), \forall \boldsymbol{\alpha} \in \Pi_{\mathcal{U}}$).*

Remark 2. The game-theoretic solutions that we propose in the next section involve *randomization*. For instance, in a mixed equilibrium, each player's randomization leaves the other *indifferent* across her randomization support. These choices can be deliberately randomized, however these are not the only equilibria interpretations. For instance, the probabilities over the pure actions (i.e., application level or user group pure selections) can represent (i) time averages of an "adaptive" player, (ii) a vector of fractions of a "population", where each player type adopts pure strategies and, (iii) a "belief" vector that each player has about the other regarding their behavior.

3.3 CSG Solutions

Given the definition of CSG and its components, we derive optimal strategies for the Defender. First, we investigate the problem of determining best RSSs and RASs (i.e., mixed strategies), for the Defender and the Attacker respectively, when both players are strategic and play simultaneously.

As we have not explicitly defined the *strategic type* of Attacker, we consider different types of solutions based on various Attacker behaviors. This analysis will allow us to draw robust conclusions regarding the *overall optimal* Defender strategy, which will minimize expected damages *regardless of the Attacker type.*

The most commonly used solution concept in game theory is that of *Nash Equilibrium* (NE) [26]. This concept captures a steady state of the play of the CSG in which both Defender and Attacker hold the correct expectation about the other players' behavior and they act rationally. A NE dictates optimal responses to each other's actions, keeping the others' strategies fixed, i.e., strategy profiles that are resistant against unilateral deviations of players.

Definition 4. *In any Cyber Safeguard Game, a mixed strategy profile $(\boldsymbol{\delta}^{\mathrm{NE}}, \boldsymbol{\alpha}^{\mathrm{NE}})$ of Γ is a mixed NE if and only if*

1. $\boldsymbol{\delta}^{\mathrm{NE}} \succsim \boldsymbol{\delta}, \forall \boldsymbol{\delta} \in \Pi_{\mathcal{L}}$, when the Attacker chooses $\boldsymbol{\alpha}^{\mathrm{NE}}$, i.e.

$$U_d(\boldsymbol{\delta}^{\mathrm{NE}}, \boldsymbol{\alpha}^{\mathrm{NE}}) \geq_{\forall \boldsymbol{\delta} \in \Pi_{\mathcal{L}}} U_d(\boldsymbol{\delta}, \boldsymbol{\alpha}^{\mathrm{NE}}); \tag{9}$$

2. $\boldsymbol{\alpha}^{\text{NE}} \succsim \boldsymbol{\alpha}, \forall \boldsymbol{\alpha} \in \Pi_{\mathcal{U}}$*, when the Defender chooses* $\boldsymbol{\delta}^{\text{NE}}$*, i.e.*

$$U_a(\boldsymbol{\delta}^{\text{NE}}, \boldsymbol{\alpha}^{\text{NE}}) \geq_{\forall \boldsymbol{\alpha} \in \Pi_{\mathcal{U}}} U_a(\boldsymbol{\delta}^{\text{NE}}, \boldsymbol{\alpha}). \tag{10}$$

Definition 5. *The Nash Safeguards Plan (NSP), denoted by* $\boldsymbol{\delta}^{\text{NE}}$*, is a probability distribution over the different levels, as determined by the NE of the CSG.*

Example 1. For a safeguard with 3 application levels including level 0, which corresponds to not applying the safeguard at all, an NSP $(0, 0.2, 0.8)$ dictates that 20% of the users will be strengthened (e.g., trained) at $j = 1$ (e.g., once when they join the organization), while 80% of the users will be applied a higher level of the safeguard $j = 2$ (e.g., attending training once per year).

3.4 Optimality Analysis

We model *complete information* Nash CSGs, according to which both players know the game matrix, which contains the utilities of both players for each pure strategy profile. The utility function of the Defender is determined by the probability of failing to protect a user group and the indirect costs of the chosen application levels. We consider a *zero-sum* CSG, where the Attacker's utility is the opposite of the Defender's utility. The rationale behind the zero-sum CSG is that when the Defender is uncertain about the Attacker type, she considers the *worst case scenario*, which can be formulated by a zero-sum game where the Attacker can cause her *maximum damage*. The idea behind a zero-sum game like this is that the Attacker focuses on causing maximum corruption to cyberspace, while the Defender aims at minimizing the damage. Due to the Attacker's goal being conflicting to the Defender's objective, the application of game theory to study the selection of safeguards application levels is convenient. While in most security situations the interests of the players are neither in strong conflict nor in complete identity, the zero-sum game provides important insights into the notion of "optimal play", which is closely related to the *minimax theorem* [27].

In the zero-sum CSG,

$$\Gamma_0 = \langle \{d, a\}, \mathcal{L} \times \mathcal{U}, \{U_d, -U_d\} \rangle, \tag{11}$$

the Attacker's gain is equal to the Defender's security loss, and vice versa. We define the utility of the Defender in Γ_0 as

$$U_d^{\Gamma_0}(j, i) := -w_L \, L(j, i) - w_C \, C(j, i). \tag{12}$$

The first term of (12) is the expected loss of the Defender inflicted by the Attacker when attempting to compromise user group i, while the second term expresses the aggregated indirect cost of the safeguard application irrespective of the attacking strategy. Let $w_L, w_C \in [0, 1]$ are importance weights, which can facilitate the Defender with setting her preferences in terms of security loss, and indirect cost, accordingly.

For a mixed profile $(\boldsymbol{\delta}, \boldsymbol{\alpha})$, the utility of the Defender equals

$$\begin{aligned} U_d^{\Gamma_0}(\boldsymbol{\delta}, \boldsymbol{\alpha}) &\overset{(7)}{=} \sum_{j \in \mathcal{L}} \sum_{i \in \mathcal{U}} U_d^{\Gamma_0}(j,i) \boldsymbol{\delta}(j)\, \boldsymbol{\alpha}(i) \\ &\overset{(12)}{=} \sum_{j \in \mathcal{L}} \sum_{i \in \mathcal{U}} [-w_L\, L(j,i) - w_c\, C(j)]\, \boldsymbol{\delta}(j)\, \boldsymbol{\alpha}(i) \\ &= -w_L \sum_{j \in \mathcal{L}} \sum_{i \in \mathcal{U}} L(j,i)\, \boldsymbol{\delta}(j)\, \boldsymbol{\alpha}(i) - w_C \sum_{j \in \mathcal{L}} C(j,i)\, \boldsymbol{\delta}(j). \end{aligned} \tag{13}$$

As Γ_0 is a zero-sum game, the Attacker's utility is given by $U_a^{\Gamma_0}(\boldsymbol{\delta}, \boldsymbol{\alpha}) = -U_d^{\Gamma_0}(\boldsymbol{\delta}, \boldsymbol{\alpha})$. Since the Defender's equilibrium strategies maximize her utility, given that the Attacker maximizes her own utility, we will refer to them as *optimal strategies*.

As Γ_0 is a two-person zero-sum game with a finite number of actions for both players, according to Nash [28], it admits at least a NE in mixed strategies and saddle-points correspond to Nash equilibria as discussed in [29] (p. 42). The following result from [30], establishes the existence of a saddle (equilibrium) solution in the games, we examine and summarizes their properties.

Definition 6 (Saddle point of the CSG). *The Γ_0 Cyber Safeguard Game (CSG) admits a saddle point in mixed strategies, $(\boldsymbol{\delta}_{\Gamma_0}^{\text{NE}}, \boldsymbol{\alpha}_{\Gamma_0}^{\text{NE}})$, with the property that*

- $\boldsymbol{\delta}_{\Gamma_0}^{\text{NE}} = \arg\max_{\boldsymbol{\delta} \in \Delta_{\mathcal{L}}} \min_{\boldsymbol{\alpha} \in \Delta_{\mathcal{U}}} U_d^{\Gamma_0}(\boldsymbol{\delta}, \boldsymbol{\alpha})$, $\forall \boldsymbol{\alpha}$, *and*
- $\boldsymbol{\alpha}_{\Gamma_0}^{\text{NE}} = \arg\max_{\boldsymbol{\alpha} \in \Delta_{\mathcal{U}}} \min_{\boldsymbol{\delta} \in \Delta_{\mathcal{L}}} U_a^{\Gamma_0}(\boldsymbol{\delta}, \boldsymbol{\alpha})$, $\forall \boldsymbol{\delta}$.

Then, due to the zero-sum nature of the game, the minimax theorem [27] holds, i.e. $\max_{\boldsymbol{\delta} \in \Delta_{\mathcal{L}}} \min_{\boldsymbol{\alpha} \in \Delta_{\mathcal{U}}} U_d^{\Gamma_0}(\boldsymbol{\delta}, \boldsymbol{\alpha}) = \min_{\boldsymbol{\alpha} \in \Delta_{\mathcal{U}}} \max_{\boldsymbol{\delta} \in \Delta_{\mathcal{L}}} U_d^{\Gamma_0}(\boldsymbol{\delta}, \boldsymbol{\alpha})$.

The pair of saddle point strategies $(\boldsymbol{\delta}_{\Gamma_0}^{\text{NE}}, \boldsymbol{\alpha}_{\Gamma_0}^{\text{NE}})$ are at the same time security strategies for the players, i.e. they ensure a minimum performance regardless of the actions of the other. Furthermore, if the game admits multiple saddle points (and strategies), they have the ordered interchangeability property, i.e. the player achieves the same performance level independent from the other player's choice of saddle point strategy.

The minimax theorem [27] states that for zero-sum games, NE and minimax solutions coincide. Therefore, $\boldsymbol{\delta}_{\Gamma_0}^{\text{NE}} = \arg\min_{\boldsymbol{\delta} \in \Delta_{\mathcal{L}}} \max_{\boldsymbol{\alpha} \in \Delta_{\mathcal{U}}} U_a^{\Gamma_0}(\boldsymbol{\delta}, \boldsymbol{\alpha})$. This means that regardless of the strategy the Attacker chooses, NSP is the Defender's security strategy that guarantees a minimum performance.

Formally, the Defender seeks to solve the following LP:

$$\begin{aligned} &\max_{\boldsymbol{\delta} \in \Delta_{\mathcal{L}}} \min_{\boldsymbol{\alpha} \in \Delta_{\mathcal{U}}} U_d^{\Gamma_0}(\boldsymbol{\delta}, \hat{\imath}) \\ &\text{subject to} \begin{cases} U_d^{\Gamma_0}(\boldsymbol{\delta}, 1) - \min_{\boldsymbol{\alpha} \in \Delta_{\mathcal{U}}} U_d^{\Gamma_0}(\boldsymbol{\delta}, \hat{\imath}) e \geq 0 \\ \vdots \\ U_d^{\Gamma_0}(\boldsymbol{\delta}, |\mathcal{U}|) - \min_{\boldsymbol{\alpha} \in \Delta_{\mathcal{U}}} U_d^{\Gamma_0}(\boldsymbol{\delta}, \hat{\imath}) e \geq 0 \\ \boldsymbol{\delta} e = 1 \\ \boldsymbol{\delta} \geq 0. \end{cases} \end{aligned} \tag{14}$$

In this problem, e is a vector of ones of size $|\mathcal{U}|$.

3.5 Multiple Games per Safeguard

Given that we have to allocate a budget in applying different safeguards, we may come across the challenge of not having enough monetary resources to select some of the equilibria of the CSG. Therefore, one has to derive the financial cost of equilibrium and assess its feasibility by comparing its financial cost to the available remaining budget. We refer to "remaining" budget as we expect that the Defender will have to select among a number of equilibria, one per safeguard, as we show later in this section.

To provide to the Defender a wider variety, in terms of financial cost, of equilibria per safeguard, we define a number of CSGs per safeguard. Each of these games has a different number of application levels available to the Defender. Aligned with [3], for each safeguard σ, we study $|\mathcal{L}|$ CSGs.

Definition 7. *To differentiate among different safeguards and implementations levels, we denote the CSG by $\Gamma_{\sigma,\lambda}$, where the safeguard σ can be applied up to $\lambda \in [0, |\mathcal{L}|]$.*

Note that we allow $\lambda = 0$ so that the Defender has the option to avoid selecting a safeguard should this violate some budget constraints. A Knapsack optimisation is used in the second phase of the model to select the equilibria, at most one per safeguard. In this way, we manage to have $|\mathcal{L}|$ NSPs per safeguard, each of a different financial cost. Each $\Gamma_{\sigma,\lambda}$ is a game where (i) Defender's pure strategies correspond to consecutive application levels of safeguard σ starting always from 0 and including all levels up to λ and, (ii) Attacker's pure strategies are the different targets akin to user groups. Figure 1 illustrates the different Cybersecurity Safeguards Games along with the utilities of the Defender.

Let $\boldsymbol{\delta}^{NE}_{\sigma,\lambda}$ be the equilibrium of $\Gamma_{\sigma,\lambda}$ then

$$\boldsymbol{\delta}^{NE}_{\sigma,\lambda} = [\delta^{NE}_{\sigma,0}, \delta^{NE}_{\sigma,1}, \ldots, \delta^{NE}_{\sigma,\lambda}]. \quad (15)$$

Let $F(\boldsymbol{\delta}_{\sigma,\lambda})$ be the financial cost of the safeguards plan $\boldsymbol{\delta}_{\sigma,\lambda}$ which can be derived by summing the financial costs of all application levels $j \in \{1, 2, \ldots, \lambda\}$ for safeguard σ contributed proportionally by using the corresponding probability from $\boldsymbol{\delta}_{\sigma,\lambda}$, i.e., $\delta_{\sigma,j}$. Let $F(\sigma, j)$ denote the financial cost of safeguard σ then

$$F(\boldsymbol{\delta}_{\sigma,\lambda}) = \sum_{j \in \{1,2,\ldots,\lambda\}} \delta_{\sigma,j}\, F(\sigma, j). \quad (16)$$

3.6 Investment in Nash Safeguards Plans

Let $\mathcal{S}$ be the set of all available safeguards to the Defender. We can solve all $|\mathcal{S}| \times |\mathcal{L}|$ CSGs and derive a set of equilibria per safeguard σ represented as follows

$$\{\boldsymbol{\delta}^{NE}_{\sigma,1}, \boldsymbol{\delta}^{NE}_{\sigma,2}, \ldots, \boldsymbol{\delta}^{NE}_{\sigma,|\mathcal{L}|}\}. \quad (17)$$

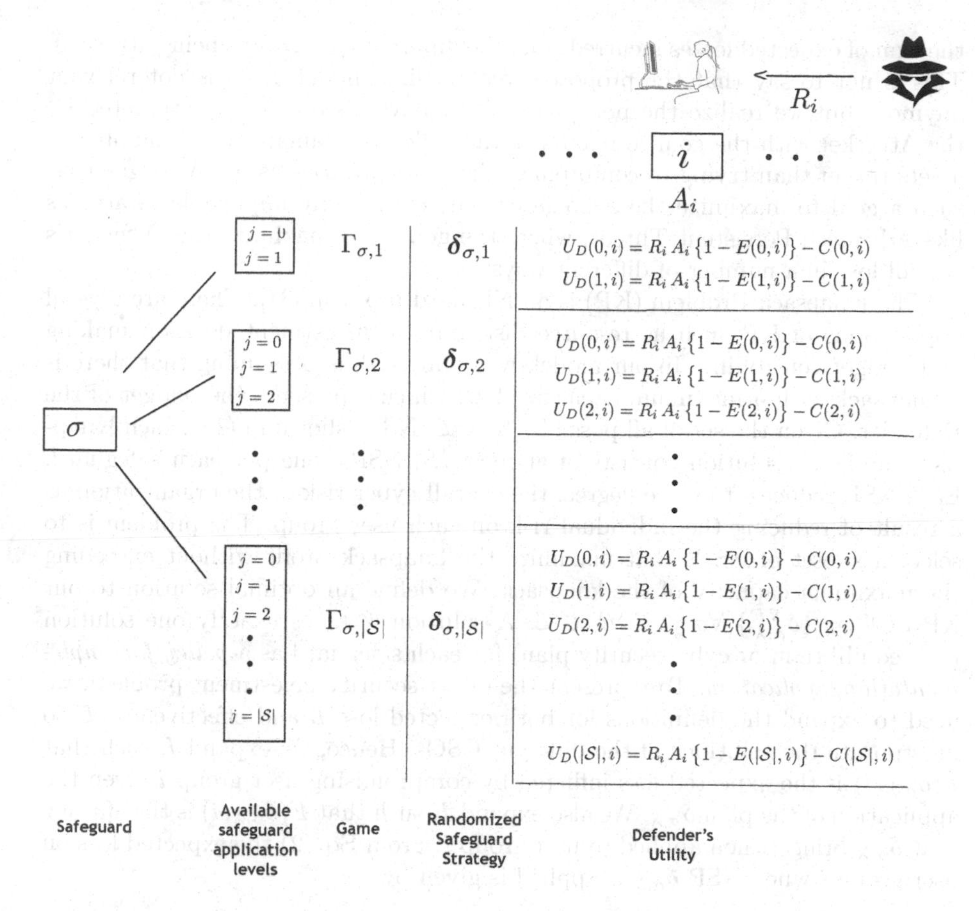

Fig. 1. Illustration of the safeguard-centered model of OST used to devise game-theoretic strategies for the Defender.

For all safeguards $\{1, 2, \ldots, |S|\}$ the following set of sets of equilibria, i.e., NSPs, is available

$$\left\{\{\delta_{1,0}^{NE}, \delta_{1,1}^{NE}, \ldots, \delta_{1,|\mathcal{L}|}^{NE}\}, \{\delta_{2,0}^{NE}, \delta_{2,1}^{NE}, \ldots, \delta_{2,|\mathcal{L}|}^{NE}\}, \ldots, \{\delta_{|S|,0}^{NE}, \delta_{|S|,1}^{NE}, \ldots, \delta_{|S|,|\mathcal{L}|}^{NE}\}\right\}. \tag{18}$$

Optimal budget allocation in cybersecurity can be tackled by combinatorial optimization as previously investigated by Smeraldi and Malacaria [25]. We are concerned with the challenge of protecting multiple targets, in our case user groups, with the use of a number of NSPs that interact between them in different ways. In the following, we model the challenge of investing in these different NSPs in a way that at most one NSP per safeguard is chosen and the sum of financial costs of these NSPs fits an available cybersecurity budget. We have used 0-1 Knapsack Optimization to solve this problem. As opposed to the solution provided in [3], we have chosen the objective function of the Defender to consider the sum of expected losses incurred from the different user groups being attacked. This is not to say that the proposed weakest-link model in [3] is not relevant anymore but we realize the potential risk to have all user groups targeted by the Attacker with the goal to maximize the collective damage over a number of assets rather than trying to compromise the most precious asset. We argue that such a goal to maximize the aggregated damage is more applicable in attacks like Advanced Persistent Threat, where the goal is to maximize the Defender's overall loss in a number of different ways.

The Knapsack Problem (KP) is an NP-hard problem [31]. There are several applications of KP such as resource distribution, investment decision making and budget controlling. In our model, we define KP as: Assuming that there is a knapsack with a maximum capacity of B, which represents the budget of the Defender. Given the set of all possible $|\mathcal{S}| \times |\mathcal{L}|$ NSPs shown in (18), each Knapsack candidate solution consists of at most $|\mathcal{S}|$ NSPs, one per each safeguard. Each NSP reduces, to some degree, the overall cyber risk of the organization as a result of reducing the individual risk on each user group. The problem is to select a subset of NSPs that maximize the knapsack profit without exceeding the maximum capacity of the knapsack. We define an optimal solution to our KP as $\Psi = \{\boldsymbol{\delta}^{NE}_{\sigma,\lambda}\}, \forall \sigma \in \mathcal{S}, \forall \lambda \in \mathcal{L}$. A solution Ψ takes exactly one solution (i.e., equilibrium or cybersecurity plan) for each safeguard as a *policy for implementation/application*. To represent the cyber security investment problem, we need to expand the definitions for both expected loss L and effectiveness E to incorporate the solutions of the different CSGs. Hence, we expand L such that $L(\boldsymbol{\delta}_{\sigma,\lambda}, i)$ is the expected loss inflicted by compromising user group i given the application of the plan $\boldsymbol{\delta}_{\sigma,\lambda}$. We also expand E such that $E(\boldsymbol{\delta}_{\sigma,\lambda}, i)$ is the efficacy that $\boldsymbol{\delta}_{\sigma,\lambda}$ brings when applied to user group i. From Eq. (2) the expected loss on user group i when NSP $\boldsymbol{\delta}_{\sigma,\lambda}$ is applied is given by

$$L(\boldsymbol{\delta}_{\sigma,\lambda}, i) = R_i\, A_i\, [1 - E(\boldsymbol{\delta}_{\sigma,\lambda}, i)] \tag{19}$$

A natural approach is the KP to seek a set of NSPs that minimize the aggregated expected risks across all user groups. We assume that each NSP may protect more than one user groups. We then seek optimal safeguards allocation for a series of user groups each of which can be protected by a different set of NSPs. The latter may not necessarily have an additive efficacy. The following illustrated example considers two NSPs and explains how we have decided to combine their efficacy in a single formula that we then use in KP formulation.

Example 2. By slightly abusing notation, assume two NSPs $\boldsymbol{\delta}, \boldsymbol{\delta}'$ that mitigate 20% and 30% of the same user group risk, respectively. If the NSPs had additive efficacy the total expected loss on user group i when applying both $\boldsymbol{\delta}, \boldsymbol{\delta}'$ equals $R_i\, A_i \left\{1 - \{E(\boldsymbol{\delta}, i) + E(\boldsymbol{\delta}', i)\}\right\} = R_i\, A_i\, (1 - 0.2 - 0.3) = 0.5\, R_i\, A_i$. In this paper, we assume a more conservative expected loss mitigation function when combining two or more NSPs as follows $R_i\, A_i \left\{\{1 - E(\boldsymbol{\delta}, i)\}\,\{1 - E(\boldsymbol{\delta}', i)\}\right\} = R_i\, A_i \cdot (1 - 0.2)\,(1 - 0.3) = R_i\, A_i \cdot 0.8 \cdot 0.7 = 0.56\, R_i\, A_i$.

Given the above, if we represent the solution Ψ by the bitvector $\boldsymbol{z}$, we can then represent the 0-1 KP as

$$\max_{\boldsymbol{z}} \sum_{i=0}^{|\mathcal{U}|} A_i\, R_i \left\{ \prod_{\sigma=1}^{|\mathcal{S}|} \left\{ 1 - \sum_{j=0}^{\lambda} E(\boldsymbol{\delta}_{\sigma,\lambda}^{NE}, i)\, z_{\sigma,\lambda} \right\} \right\}$$

$$\text{s.t.} \sum_{\sigma=1}^{|\mathcal{U}|} \sum_{\lambda=0}^{|\mathcal{L}|} F(\boldsymbol{\delta}_{\sigma,\lambda})\, z_{\sigma,\lambda} \leq B,$$

$$\sum_{\lambda=0}^{|\mathcal{L}|} z_{\sigma,\lambda} = 1,\, z_{\sigma,\lambda} \in \{0,1\}, \forall \sigma = 1, 2, \ldots, |\mathcal{S}|. \tag{20}$$

where B is the available budget of the Defender to be spent in cyber safeguards and $z_{\sigma,\lambda} = 1$ holds when $\boldsymbol{\delta}_{\sigma,\lambda}^{NE} \in \Psi$. Among KP solutions that all maximize the overall expected loss, we choose the solution with the lowest financial cost as this will be, in overall, the best advice to the defender producing same benefit for lower price.

4 Model Evaluation

We have developed the proposed models as part of the Optimal Safeguards Tool (OST) proposed in [1]. OST computes Nash Safeguards Plans as well as the Knapsack solutions. OST aims at offering realistic actionable advice to healthcare organizations. The following represents a case study based on Critical Internet Security (CIS) 17 Control "Implement a Security Awareness and Training Program".

4.1 Use Case

User Groups. Here, we assume a representative (non-exhaustive) set of three user groups, denoted by i, in decreasing order of *access privileges*:

- $i = 1$; **ICT:** The information and communication technology professionals responsible for the systems, networks and software. They set up digital systems, support staff who use them, diagnose and address faults, as well as set up and maintain security provisions. In addition to the ICT infrastructure, they may also interact with medical devices and electronic healthcare record systems. We consider the value of corresponding assets that can be affected by an attack on this group to be the highest possible, $A_1 = 100$ (e.g., \$100k). At the same time, due to limited interaction with the public, this is the group with the lowest visibility to attacks targeting the human, and as such we can consider it as lower risk, $R_1 = 0.2$.
- $i = 2$; **Clinical:** Nurses, doctors and other clinical staff have access to medical devices and electronic healthcare records. We consider the value of corresponding assets that can be affected by an attack on this group to be $A_2 = 50$ (e.g., \$50k). As a result of visibility due to interaction with the patients and presence on the hospital's website, this group has a moderate risk, $R_2 = 0.5$.

- $i = 3$; **Administration:** Receptionists, medical secretaries and other administration roles involve access to electronic healthcare records. We consider the value of corresponding assets that can be affected by an attack on this group to be $A_3 = 25$ (e.g., \$25k). This group of users may have high interaction with the public and volume of email traffic (e.g., appointment requests) and as such high risk, $R_3 = 0.8$.

Table 2. Evaluation parameters for control CIS-17.4.

Control level\Role		ICT	Clinical	Administration
Low (once per year)	E	0.35	0.3	0.3
	C	1	30	10
Medium (twice per year)	E	0.6	0.5	0.5
	C	2	60	20
High (once per month)	E	0.8	0.7	0.7
	C	12	360	120

We have assumed a user group ratio of size 1:30:10 that loosely follows the corresponding breakdown of hospital workforce in the United States[9]: 81,790 computer, information system and security managers and analysts; 2,437,540 healthcare practitioners; 737,750 receptionists, healthcare record information clerks and other office and administrative support staff.

Table 3. Evaluation parameters for control CIS-17.6.

Control level\Role		ICT	Clinical	Administration
Low (Tests)	E	0.25	0.2	0.2
	C	1	30	10
Medium (Videos)	E	0.7	0.6	0.6
	C	2	60	20
High (Games)	E	0.6	0.5	0.5
	C	4	120	40

Safeguards. As safeguards, we have considered a representative pair from the SANS institute's CIS-17 group of critical security controls[10]: CIS-17.4 "Update

[9] https://www.bls.gov/oes/current/naics3_622000.htm.
[10] https://www.cisecurity.org/controls/implement-a-security-awareness-and-training-program/.

Awareness Content Frequently" and CIS-17.6 "Train Workforce on Identifying Social Engineering Attacks". All values used in this case study, for these two safeguards, are presented in Tables 2 and 3.

For CIS-17.4, we set the frequency of completion of the updated training (once per year, twice per year, or once per month - i.e., 12 times per year) as the level of control. As indirect cost $C(j, i)$, we consider the total time spent in training by the employees in group i at application level j (in this case is *frequency*), which is proportionate to the size of the group and the frequency of the training. This time can be translated to some financial cost (in $) resulting from loss of productive working hours. In this way, the indirect cost can be subtracted from the expected loss comprising the final utility value of the Defender in each cell of the game utility matrix.

For CIS-17.6, we set the nature of the work-based training (tests, videos, games) as the levels of control. Further, we set the corresponding efficacy values for each type roughly equivalent to their importance in helping predict user susceptibility to semantic social engineering attacks. Specifically, [15] has identified work-based security training with videos as the best predictor out of the three. In terms of efficacy values, we have differentiated slightly between groups based on our perceived rate of adoption of controls in each one. Specifically, we assume that adoption is greater for ICT than for clinical and administration employees. This is only for illustration purposes, so that the model can also take into account the group at each level of control. We also assume, further, that the primary indirect cost is employee time required, with a ratio of 1:2:4 for the three control levels.

4.2 Comparison with Alternative Defense Strategies

In the following, we analyze the proposed model in two phases; (i) the game-theoretic; and (ii) the 0-1 Knapsack optimization. The *first phase* evaluates different cybersecurity safeguard selection strategies using the utility table of the investigated Cybersecurity Safeguards Games (CSG) based on the use case discussed in the previous section. To evaluate our approach, we have created a simulated environment in Python which performs the attack sampling. For all comparisons performed, a sample size of 1,000 attacks was used. Such a sample is referred to as a *run* in the results. In the following, we present the results, where 25 runs have been performed in each case and the average Defender Utility (in $) seen across the runs have been plotted.

More specifically, we have simulated $\Gamma_{\sigma,2}$ and $\Gamma_{\sigma,3}$ (please see Table 1 for the notation) for the two different safeguards presented in the use case, i.e., CIS 17.4 (denoted as $\sigma = 1$) and 17.6 (denoted as $\sigma = 2$). The games $\Gamma_{1,2}$, $\Gamma_{2,2}$ exhibit maximum safeguard application level of 2 (Medium), while the games $\Gamma_{1,3}$, $\Gamma_{2,3}$ are investigated up to application level 3 (High). Each CSG generates a utility table that we use to derive three different Defender application level selection strategies:

- *Nash* Safeguard Strategy (NSS), as described in Sect. 3 and computed using the the open source *Nashpy* Python library[11].
- the *Weighted* Safeguard Strategy (WSS), which distributes the choice of a safeguard level over the weighted expected utility of the CSG by computing probability $\delta_{\sigma,j}$ of choosing application level j of safeguard σ as follows:

$$\delta_{\sigma,j} := \frac{\sum_{i=1}^{|\mathcal{U}|} U_d(j,i)}{\sum_{j=1}^{|\mathcal{L}|} \sum_{i=1}^{|\mathcal{U}|} U_d(j,i)}$$

- the *Cautious* Safeguard Strategy (CSS), which always prefers the *highest* application level of a safeguard.

Regarding adversarial strategies, we consider three profiles:

- the *Nash* Attacker who plays the Nash Attacking Strategy (NAS), presented in Sect. 3 and computed using the *Nashpy* Python library.
- a *Weighted* Attacker who plays the Weighted Attacking Strategy (WAS) by attacking a user group i with probability $\frac{A_i}{\sum_{i \in \mathcal{U}} A_i}$, i.e. the Attacker attacks the different user groups proportionally based on the asset values they have access to.
- the *Opportunistic* Attacker who uniformly chooses the different user groups to attack.

Figure 2 illustrates the performance of NSS against WSS and CSS in terms of average Defender's utility over the 1,000 attacks for 25 runs. In all cases, we contrast between Attackers who follow NAS and WAS.

Nash Attacker. The results, in Fig. 2(a), show that NSS outperforms both WSS and CSS when the Attacker chooses NAS. More specifically, the percentage improvement values, seen when choosing NSS, in comparison to WSS for the different games $[\Gamma_{1,2}, \Gamma_{1,3}, \Gamma_{2,2}, \Gamma_{2,3}]$ are [20.2%, 79.78%, 16%, 52.12%], respectively. Likewise, when choosing NSS over CSS, we observe improvement values of [34.48%, 87.07%, 28.57%, 62.26%] for the different games $[\Gamma_{1,2}, \Gamma_{1,3}, \Gamma_{2,2}, \Gamma_{2,3}]$, respectively.

Remark 3. These results demonstrate an average improvement of approximately 42% of NSS over WSS and 53% over CSS.

Comparably, the smallest average improvement for NSS over WSS is around 16% when playing Control 17.6 at the maximum application level of 2 ($\lambda = 2$). Likewise, the minimum improvement of NSS over CSS, approximately equal to 28%, is for the same control and $\lambda = 2$. On the other hand, the maximum improvement seen in NSS over CSS is approximately 87%, where the maximum improvement over CSS does not exceed 80%, for Control 17.4. and $\lambda = 3$.

[11] https://nashpy.readthedocs.io/en/stable/index.html.

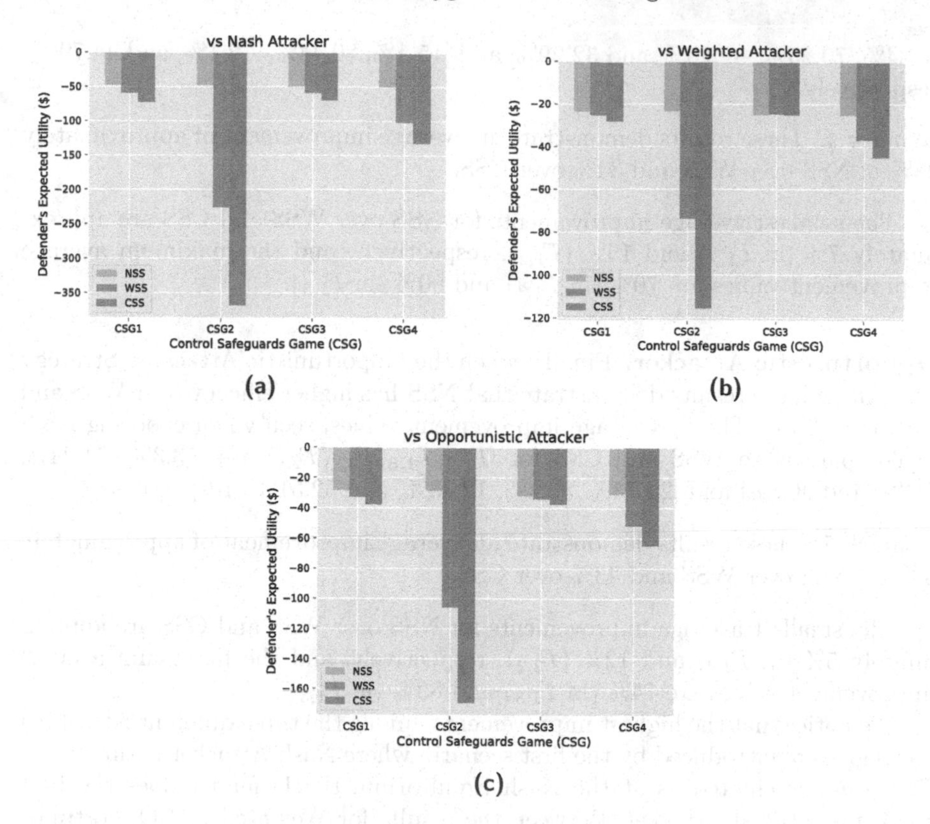

Fig. 2. Game-theoretic optimization results: Average Utility of the Defender over 1,000 attacks for 25 runs. for various CSGs.

One of the primary reasons why naive-deterministic safeguard selection approaches perform poorly against the Nash Defending strategy is that they fail to incorporate the opponent's strategies. At the same time, we have considered CSG as a zero-sum game. The class of zero-sum games offers a degree of freedom as it can be shown that assuming that the adversary's intentions are exactly opposite to the defender's assets, i.e., the Attacker seeks to cause maximum damage, any other incentive of the Attacker can only improve the Defender's situation [32].

Weighted Attacker. When the Weighted Attacking Strategy is simulated, the results demonstrate that NSS has higher efficacy over WSS and CSS apart from one game $\Gamma_{2,2}$ in which both WSS and CSS perform approximately 2% better than NSS (Fig. 2(b)). This difference is negligible making NSS being at least as good as the rest of the Defending strategies in all investigated games. Despite the performance of NSS in $\Gamma_{2,2}$, for the rest of the games, NSS performs significantly better than WSS and CSS. The percentage improvement values, seen when choosing NSS, in comparison to WSS and CSS for $[\Gamma_{1,2}, \Gamma_{1,3}, \Gamma_{2,2}, \Gamma_{2,3}]$ are

[7.34%, 70.25%, −2.25%, and 32.29%] and [15.1%, 80.44%, −2.1%, and 44.79%], respectively.

Remark 4. These results demonstrate an average improvement of approximately 28% of NSS over WSS and 34% over CSS.

The smallest average improvements for NSS over WSS and CSS are approximately 7% (in $\Gamma_{1,2}$) and 15% ($\Gamma_{1,2}$), respectively, and the maximum average improvement values are 70% (in $\Gamma_{1,3}$) and 80% (in $\Gamma_{1,3}$).

Opportunistic Attacker. Finally, when the Opportunistic Attacking Strategy is simulated, the results demonstrate that NSS has higher efficacy over WSS and CSS (Fig. 2(c)). The percentage improvement values, seen when choosing NSS, in comparison to WSS and CSS for $[\Gamma_{1,2}, \Gamma_{1,3}, \Gamma_{2,2}, \Gamma_{2,3}]$ are [13.3%, 74.24%, 5.4%, and 40.8%] and [23.73%, 83.4%, 12.33%, and 52.51%], respectively.

Remark 5. These results demonstrate an average improvement of approximately 33% of NSS over WSS and 43% over CSS.

The smallest average improvements for NSS over WSS and CSS are approximately 5% (in $\Gamma_{2,2}$) and 12% ($\Gamma_{2,2}$), respectively, and the maximum average improvement values are 74% (in $\Gamma_{1,3}$) and 83% (in $\Gamma_{1,3}$).

We notice that the highest improvements among the three different Attacking strategies are introduced by the first scenario where Nash Attacker is simulated. This was anticipated as at the Nash Equilibrium the Defender does the best against a rational Attacker. Between the results for Weighted and Opportunistic Attacker, NSS is more efficient against an Opportunistic Attacker than a Weighted one.

4.3 Analysis of the Investment Problem

The Knapsack optimization phase investigates the optimal investment in Nash Safeguards Plans (NSPs) given a budget B (for details refer to Sect. 3.6). The Knapsack takes as input every NSP generated in the game-theoretic phase and recommends a single solution which minimizes the aggregated risk of all the user groups while satisfying the investment budget constraint. This is different to the weakest-link model investigated by [3].

Figure 3 presents the financial cost and aggregated risk overall users for each Knapsack candidate solution, i.e., a combination of NSPs for two different available budget values. We notice that there are multiple Knapsack optimal solutions, which are candidate solutions number 5, 6, 7 and 8. In the presence of multiple optimal solutions, the Knapsack solver, we have implemented, chooses the first option. For both budgets 40 and 100, the Knapsack optimization recommends investing in both CIS controls 17.4 and 17.6 at application level 1 i.e., Low (once per year) and Low (Tests), respectively. Note here that the small size of the use case effectively prohibits high variability of the parametric values, which led to the selection of only two types.

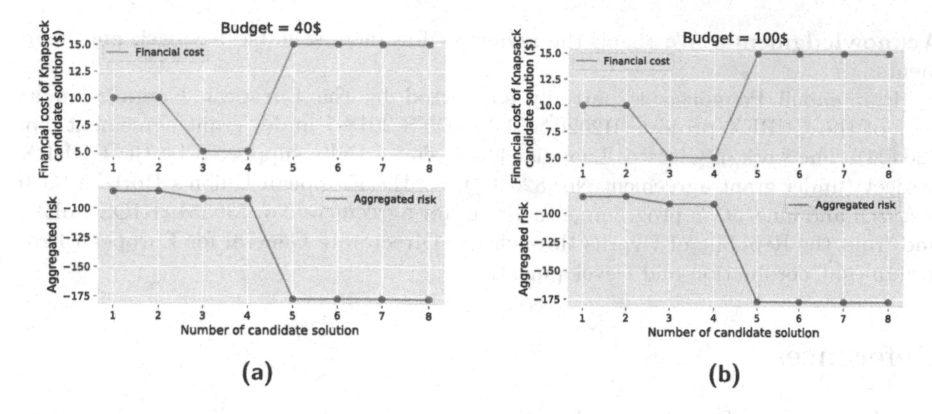

Fig. 3. Knapsack selection over available candidate solutions.

Note that the plots in Knapsack optimization only present the candidate solutions for the Nash Defender against Nash Attacker, in contrast to the plots in game-theoretic phase, (Fig. 2), which presents all three Defender strategies. This choice was made due to the Knapsack optimization not involving the notion of CSG. As a result of this, it does not optimize the overall indirect cost of safeguards when choosing NSPs which has been done in the previous phase. In addition, Knapsack does not consider the behavior of the Attacker characterizing all adversarial strategies as irrelevant to the Knapsack objective function.

5 Conclusions

In this paper, we have presented an approach, extending the previous work [3], which implements a cybersecurity safeguards selection model along with game-theoretic and Knapsack optimization tools. We have evaluated our model in a healthcare use case using the CIS group 17 controls which attend to implementation of security awareness and training programs for employees. The simulation results demonstrate that the Nash Safeguard Strategy comfortably outperforms common-sense selection strategies, such as the Weighted and Cautious, in terms of Defender's expected utility over a large number of attacks. This work is our step towards integrating the developed Optimal Safeguards Tool (OST) within cybersecurity risk management and investment environments.

An interesting extension to this work would be to capture the real-world uncertainty about an Attacker's type, for example considering a Bayesian game of application level selection. Furthermore, we plan to bring together several objective functions for Knapsack to compare the performance of the investment strategies. As the next steps, we aim at creating a use case with greater size of safeguards in collaboration with healthcare organizations. We also aim at using the well-known repository of cybersecurity safeguards like the 20 CIS controls or a list of Privacy Enhancing Technologies (PETs) to support our research.

Acknowledgments. We thank the reviewers for their valuable feedback and comments.

Emmanouil Panaousis is partially supported by the European Commission as part of the CUREX project (H2020-SC1-FA-DTS-2018-1 under grant agreement No. 826404). The work of Christos Laoudias has been partially supported by the CUREX project (under grant agreement No. 826404), by the European Union's Horizon 2020 research and innovation programme (under grant agreement No. 739551 (KIOS CoE)), and from the Republic of Cyprus through the Directorate General for European Programmes, Coordination and Development.

References

1. Mohammadi, F., Panou, A., Ntantogian, C., Karapistoli, E., Panaousis, E., Xenakis, C.: CUREX: seCUre and pRivate hEalth data eXchange. In: IEEE/WIC/ACM International Conference on Web Intelligence, vol. 24800, pp. 263–268 (2019)
2. Vishwanath, A., et al.: Cyber hygiene: the concept, its measure, and its initial tests. Decis. Supp. Syst. **128**, 113160 (2019)
3. Fielder, A., Panaousis, E., Malacaria, P., Hankin, C., Smeraldi, F.: Decision support approaches for cyber security investment. Decis. Supp. Syst. **86**, 13–23 (2016)
4. Kruse, C.S., Frederick, B., Jacobson, T., Monticone, D.K.: Cybersecurity in healthcare: a systematic review of modern threats and trends. Technol. Health Care **25**(1), 1–10 (2017)
5. Solans Fernández, O., et al.: Shared medical record, personal health folder and health and social integrated care in catalonia: ICT services for integrated care. In: Rinaldi, G. (ed.) New Perspectives in Medical Records. T, pp. 49–64. Springer, Cham (2017). https://doi.org/10.1007/978-3-319-28661-7_4
6. Coventry, L., Branley, D.: Cybersecurity in healthcare: a narrative review of trends, threats and ways forward. Maturitas **113**, 48–52 (2018)
7. Kotz, D., Gunter, C.A., Kumar, S., Weiner, J.P.: Privacy and security in mobile health: a research agenda. Computer **49**(6), 22–30 (2016)
8. Loukas, G.: Cyber-Physical Attacks: A Growing Invisible Threat. Butterworth-Heinemann, Oxford (2015)
9. Billingsley, L., McKee, S.A.: Cybersecurity in the clinical setting: Nurses' role in the expanding "internet of things". J. Contin. Educ. Nurs. **47**(8), 347–349 (2016)
10. Heartfield, R., Loukas, G.: Detecting semantic social engineering attacks with the weakest link: Implementation and empirical evaluation of a human-as-a-security-sensor framework. Comput. Secur. **76**, 101–127 (2018)
11. Such, J.M., Ciholas, P., Rashid, A., Vidler, J., Seabrook, T.: Basic cyber hygiene: does it work? Computer **52**(4), 21–31 (2019)
12. Zhou, L., Parmanto, B., Alfikri, Z., Bao, J.: A mobile app for assisting users to make informed selections in security settings for protecting personal health data: development and feasibility study. JMIR mHealth uHealth **6**(12), e11210 (2018)
13. Furnell, S., Sanders, P., Warren, M.: Addressing information security training and awareness within the european healthcare community. Stud. Health Technol. Inform. **43**, 707–711 (1997)
14. Heartfield, R., Loukas, G.: A taxonomy of attacks and a survey of defence mechanisms for semantic social engineering attacks. ACM Comput. Surv. **48**(3), 37 (2016)

15. Heartfield, R., Loukas, G., Gan, D.: You are probably not the weakest link: towards practical prediction of susceptibility to semantic social engineering attacks. IEEE Access **4**, 6910–6928 (2016)
16. Wash, R., Cooper, M.M.: Who provides phishing training?: facts, stories, and people like me. In Proceedings of the 2018 CHI Conference on Human Factors in Computing Systems, p. 492. ACM (2018)
17. Gordon, L.A., Loeb, M.P.: The economics of information security investment. ACM Trans. Inf. Syst. Secur. (TISSEC) **5**(4), 438–457 (2002)
18. Fielder, A., König, S., Panaousis, E., Schauer, S., Rass, S.: Risk assessment uncertainties in cybersecurity investments. Games **9**(2), 34 (2018)
19. Fielder, A., Panaousis, E., Malacaria, P., Hankin, C., Smeraldi, F.: Game Theory Meets Information Security Management. In: Cuppens-Boulahia, N., Cuppens, F., Jajodia, S., Abou El Kalam, A., Sans, T. (eds.) SEC 2014. IAICT, vol. 428, pp. 15–29. Springer, Heidelberg (2014). https://doi.org/10.1007/978-3-642-55415-5_2
20. Nagurney, A., Daniele, P., Shukla, S.: A supply chain network game theory model of cybersecurity investments with nonlinear budget constraints. Ann. Oper. Res. **248**(1–2), 405–427 (2017)
21. Wang, S.S.: Integrated framework for information security investment and cyber insurance. Pac.-Basin Finance J. **57**, 101173 (2019)
22. Chronopoulos, M., Panaousis, E., Grossklags, J.: An options approach to cybersecurity investment. IEEE Access **6**, 12175–12186 (2017)
23. Martinelli, F., Uuganbayar, G., Yautsiukhin, A.: Optimal security configuration for cyber insurance. In: Janczewski, L.J., Kutyłowski, M. (eds.) SEC 2018. IAICT, vol. 529, pp. 187–200. Springer, Cham (2018). https://doi.org/10.1007/978-3-319-99828-2_14
24. Whitman, M.E., Mattord, H.J.: Principles of Information Security. Cengage Learning, Boston (2011)
25. Smeraldi, F., Malacaria, P.: How to spend it: optimal investment for cyber security. In: Proceedings of the 1st International Workshop on Agents and CyberSecurity, p. 8. ACM (2014)
26. Osborne, M.J., Rubinstein, A.: A Course in Game Theory. MIT Press, Cambridge (1994)
27. Von Neumann, J., Morgenstern, O.: Theory of Games and Economic Behavior (60th Anniversary Commemorative Edition). Princeton University Press, Princeton (2007)
28. Nash, J.F.: Equilibrium points in n-person games. In: Proceedings of the National Academy of Sciences, pp. 48–49 (1950)
29. Alpcan, T., Basar, T.: Network Security: A Decision and Game-Theoretic Approach. Cambridge University Press, Cambridge (2010)
30. Basar, T., Olsder, G.J.: Dynamic Noncooperative Game Theory. Academic Press, London (1995)
31. Pisinger, D.: Where are the hard knapsack problems? Comput. Oper. Res. **32**(9), 2271–2284 (2005)
32. Rass, S., König, S.: Password security as a game of entropies. Entropy **20**(5), 312 (2018)

Author Index

Printed in the United States
By Bookmasters